Text Book Of

INDUSTRIAL HYDRAULICS AND PNEUMATICS

(22655)

For

Semester – VI

Third Year Diploma in Mechanical Engineering Group

As Per MSBTE's 'I' Scheme Syllabus

C. P. Murgudkar
B.E. (Mech.), M. Val. (RE) DBE, MIE, FIV
Industrial/Project Consultant
Kolhapur

NIRALI PRAKASHAN
ADVANCEMENT OF KNOWLEDGE

N4560

Industrial Hydraulics and Pneumatics ISBN 978-93-89686-82-1

First Edition : December 2019

Published By :

NIRALI PRAKASHAN

Abhyudaya Pragati, 1312, Shivaji Nagar

Off J.M. Road, PUNE – 411005

Tel - (020) 25512336/37/39, Fax - (020) 25511379

Email : niralipune@pragationline.com

➢ DISTRIBUTION CENTRES

PUNE

Nirali Prakashan (For orders within Pune) : 119, Budhwar Peth, Jogeshwari Mandir Lane, Pune 411002, Maharashtra
Tel : (020) 2445 2044; Mobile : 9657703145
Email : niralilocal@pragationline.com

Nirali Prakashan (For orders outside Pune) : S. No. 28/27, Dhayari, Near Asian College Pune 411041
Tel : (020) 24690204; Mobile : 9657703143
Email : bookorder@pragationline.com

MUMBAI

Nirali Prakashan : 385, S.V.P. Road, Rasdhara Co-op. Hsg. Society Ltd.,
Girgaum, Mumbai 400004, Maharashtra; Mobile : 9320129587
Tel : (022) 2385 6339 / 2386 9976, Fax : (022) 2386 9976
Email : niralimumbai@pragationline.com

➢ DISTRIBUTION BRANCHES

JALGAON

Nirali Prakashan : 34, V. V. Golani Market, Navi Peth, Jalgaon 425001, Maharashtra,
Tel : (0257) 222 0395, Mob : 94234 91860; Email : niralijalgaon@pragationline.com

KOLHAPUR

Nirali Prakashan : New Mahadvar Road, Kedar Plaza, 1st Floor Opp. IDBI Bank, Kolhapur 416 012
Maharashtra. Mob : 9850046155; Email : niralikolhapur@pragationline.com

NAGPUR

Nirali Prakashan : Above Maratha Mandir, Shop No. 3, First Floor,
Rani Jhanshi Square, Sitabuldi, Nagpur 440012, Maharashtra
Tel : (0712) 254 7129; Email : niralinagpur@pragationline.com

DELHI

Nirali Prakashan : 4593/15, Basement, Agarwal Lane, Ansari Road, Daryaganj
Near Times of India Building, New Delhi 110002 Mob : 08505972553
Email : niralidelhi@pragationline.com

BENGALURU

Nirali Prakashan : Maitri Ground Floor, Jaya Apartments, No. 99, 6th Cross, 6th Main,
Malleswaram, Bengaluru 560003, Karnataka; Mob : 9449043034
Email: niralibangalore@pragationline.com

Other Branches : Hyderabad, Chennai

Dedication ...

Dedication ...

Dedicated to

My family

Niharika, Uma, Hrishikesh, Devyani and Prachi

C. P. Murgudkar

Preface ...

This book is written as per New 'I' Scheme Syllabus Pattern prescribed by M.S.B.T.E. Mumbai for VIth Semester (3rd Year) Diploma in Mechanical Engineering Students.

First Topic gives idea about the basic Hydraulics and Pneumatics System. Second Topic tells about Pumps and Actuators. Third Topic illustrates the details of Control Valves used in Hydraulics and Pneumatics. The Fourth Topic is about Compressors and Fluid Power Accessories. The next two topics elaborate Hydraulics and Pneumatics Circuits.

As in the previous editions author have shown normal positions and actuated positions of all circuits and also given clear idea of how Symbols are constructed.

Great care has been taken to avoid errors in calculations and illustrated figures. However, if reader finds any error in this book, the suggestions/changes/corrections from that reader are always welcome.

While writing this book, the authors have referred good number of literature and titles written on Hydraulics and Pneumatics of renowned authors and publications. The author of this book takes this opportunity to acknowledge the direct and indirect contribution of all the authors in shaping this book.

I like to take this opportunity to thanks Mr. Dinesh K. Furia, Mr. Jignesh Furia, Mrs. Anagha Medhekar (Proof Reading), Mr. Akbar Shaikh (D.T.P.), Ms. Chaitali Takle (Fig. Drawing) and all the staff of Nirali Prakashan for their co-operation in making this book what it is.

I offer thanks from the bottom of our heart to our family members for always giving us support and the much needed time to write this book.

Finally, there was, is and always will be the love of my students whom I taught as well as our friends and our well-wishers.

While every care has been taken, to eliminate errors but they may creep in. Any valuable suggestions and further guidelines to improve this book will be highly appreciated.

Author

■■■

Syllabus ...

Unit I : Introduction to Hydraulic and Pneumatic Systems **(Teaching Hrs. 04, Marks 06)**

1.1 General Layout of Oil Hydraulics Maitain Pneumatic System
1.2 Applications, Merits, Limitation and Oil Hydraulic Systems and Pneumatics Systems
1.3 Properties of Fluids ISO and SAE Grades of Oil
1.4 ISO Symbols used in Hydraulic and Pneumatic Systems
1.5 Hazard and Safety in Industrial Hydraulics and Pneumatics

Unit II : Pumps and Actuators **(Teaching Hrs. 08, Marks 12)**

2.1 Classification of Pumps
2.2 Construction and Working of Gear, Vane, Screw, Piston Pumps (Axial and Redial)
2.3 Performance Characteristics and Selection of Pumps
2.4 Classification of Actuators and Pneumatic Actuators
2.5 Construction and Working of Linear and Rotary Actuators (Motors)

Unit III : Control Valves **(Teaching Hrs. 12, Marks 16)**

3.1 Classification of Control Valve
3.2 Pressure Control Valve Relief, Unloading, Sequence, Counter Balance, Pressure Reducing Valves
3.3 Direction Control Valve S-Check Valves 2/2, 3/2, 4/2, 4/3, 5/2, 5/3 D.C. Valves used in Hydraulics and Pneumatics
3.4 Standard Centre Positions Methods of Actuators
3.5 Flow Control Valves, Non-compensated Pressure and Temperature Compensated Valves

Unit IV : Compressor, Pneumatic Components and Accessories in Fluid System

(Teaching: Hrs. 08, Marks 12)

4.1 Types, Construction, Working Principle of Reciprocating and Rotary Compressor
4.2 Construction, Working Principle of FRL Unit, Duel (Twin) Pressure Valve, Shuttle Valve, Quick Exhaust Valve, Time Delay Valve
4.3 Accessories (Used in Pneumatics and Hydraulics) Pipes Hoses, Fittings, Oil Filters, Air Filters, Sealed and Gaskets, Intensifiers, Accumulators Heat Exchanger, Muffler

Unit V : Oil Hydraulic Circuits **(Teaching Hrs. 08, Marks 12)**

5.1 Simple Oil Hydraulic Circuits - Single Acting, Double Acting Hydraulic Cylinders, Motors
5.2 Speed Control Meter-in Meter-out Bleed Off Circuit
5.3 Regenerative, Synchronizing, Counter Balance, Sequencing, Two Pump Unloading Circuits
5.4 Hydraulic Circuit for Milling, Grinding and Shaping Machine Slotting Machine
5.5 Remedies and Fault Detection in Pneumatic Circuits

Unit VI : Pneumatic Circuits **(Teaching Hrs. 08, Marks 12)**

6.1 Direct/Indirect Control of Single and Double Acting Air Cylinders, Motors
6.2 Speed Control Circuit for Cylinders and Motors
6.3 Sequencing Circuit Logic AND/OR Circuit, Time Delay Circuit, Piston Continuous Back and Forth
6.4 Simple Hydro-pneumatic Applications
6.5 Simple Electro-Pneumatic Circuits
6.6 Remedies and Fault Detection in Pneumatic Circuits
6.7 Maintenance of Hydraulic and Pneumatic Systems

■■■

Contents ...

■■■

Chapter **1**

INTRODUCTION TO HYDRAULIC AND PNEUMATIC SYSTEMS

Weightage of Marks = 08, Teaching Hours = 06

Syllabus

1.1 General Layout of Oil Hydraulics and Pneumatic System
1.2 Applications, Merits, Limitation and Oil Hydraulic Systems and Pneumatic Systems
1.3 Properties of Fluids ISO and SAE Grades of Oil
1.4 ISO Symbols used in Hydraulic and Pneumatic Systems

About This Chapter

At the end of this chapter students will be able to:

- Sketch and Label General Layout of Oil Hydraulic System.
- Identify Components from their Symbols
- List the types of various Components in Simple Oil Hydraulic Circles
- Write the Basic Properties of Oils used as Hydraulic Fluid.

1.1 GENERAL LAYOUT OF HYDRAULIC SYSTEM

[S-09, W-10, S-11, W-13, S-15]

- General layout or standard layout of any hydraulic system is having following components/parts/equipments:
  (a) Oil storage tank or oil reservoir.
  (b) Oil conditioning elements such as filters, heat exchangers, heater and strainer.
  (c) Pump to pressurize oil.
  (d) Fluid controlling elements such as direction control valves, flow control valves, Pressure relief valves, Check valves etc.
  (e) Actuators – Reciprocating (linear actuators), Rotary (hydraulic motors).

Flow Chart of Hydraulic System [W-11]

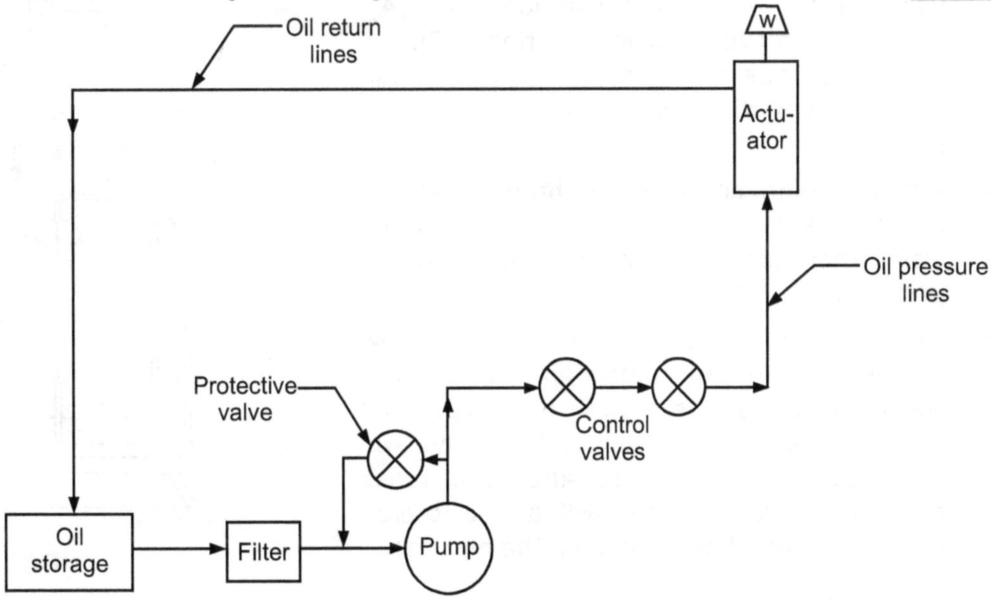

Fig. 1.1

(f) Fluid or oil conducting elements such as pipelines, connectors, hoses etc.
(g) Medium i.e. hydraulic oil.

1.2 FUNCTIONS OF EACH ELEMENT IN HYDRAULIC SYSTEM

(a) Oil Tank or Reservoir: This is an oil storage tank in which hydraulic oil is stored. The oil passes through various pipelines and after doing useful work in actuator, the oil returns back to oil tank. In the regions of low temperature, oil heaters are attached to air tank.

(b) Filter: This element filters the oil before going to next element i.e. pump.

(c) Pump: Hydraulic pump is a heart of any hydraulic system. Its main function is to create the flow of oil under pressure through entire hydraulic system and hence to assist transfer of power and motion (i.e. useful work).

(d) Direction Control Valves/Flow Control Valves/Pressure Relief Valves (Fluid Controlling Elements): These valves are fitted in hydraulic system at a particular locations. These valves control the flow of oil in the system. They also direct the flow of oil in system as also they control the speed of actuator.

(e) Actuators (Fluid Power Utilization Elements): These elements are known as actuators (either rotary or linear). The pressurized oil acts on actuator element. The oil gives or transfer its power to actuator to create useful work or mechanical advantage.

(f) Pipelines (Fluid Conducting Elements): It is the functional connection for oil flow in the hydraulic system. The efficiency of oil flow is greatly influenced by the physical characteristics of piping system.

There are two pipes:

(a) Pipe which carry pressurized oil (are called as pressure pipelines).
(b) Pipes which carry low pressurized oil or used oil (are called as return pipe lines).

Hoses, pipes, pipe fittings are the parts of fluid power pipeline.

1.3 ACTUAL HYDRAULIC SYSTEM STRUCTURE [W-16]

- To know more about hydraulic system we will see it with a closer view.
- We have already seen the functions of each element in hydraulic system. You can easily make out, the elements. Oil reservoir is oil storage tank. Pump sucks the oil from tank and pressurizes it. The oil is directed to port (A) or (B) of actuator by direction control valve. When oil enters through port (A), piston will move from left to right. The oil from other side of piston will return through port (B) via direction control valve to oil tank.
- When pressurized oil enters through port (B) piston will move from right to left and oil from other side of piston will return through port (A).
- Flow control valve will control the flow (lit./sec.) of oil to control or monitor the speed of piston (m/sec.). Pressure relief valve acts as a safety valve. This valve will not allow the system pressure to increase beyond set value. Filter will arrest foreign particles from oil and will keep the oil clean.

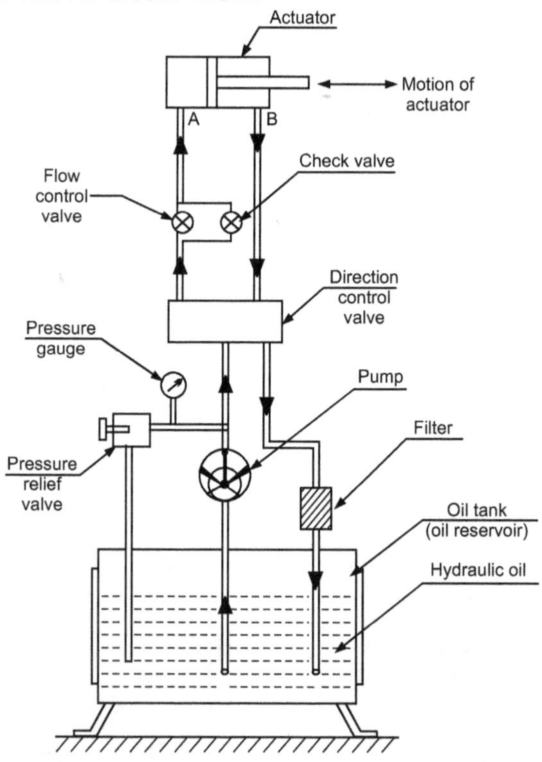

Fig. 1.2

1.4 ACTUAL HYDRAULIC SYSTEM STRUCTURE WITH THE HELP OF SYMBOLS

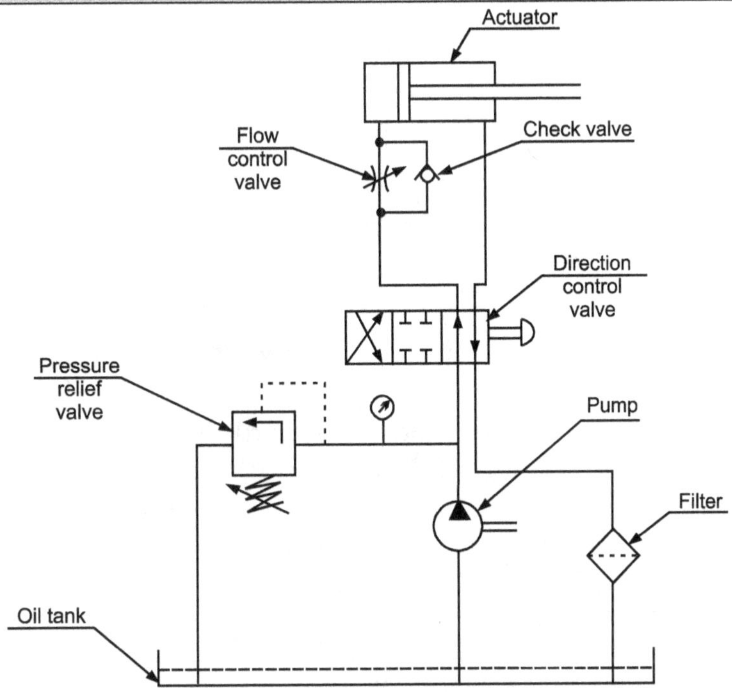

Fig. 1.3

- Fig. 1.3 shows symbolic representation of actual hydraulic system. Go through the symbols are depicted in above figure. We will have to study these symbols, in next topic.

1.5 MERITS AND LIMITATIONS OF HYDRAULIC SYSTEM

[W-09, 10, 16, S-14, 15]

(A) Merits (Advantages) of Hydraulic System

1. We can generate very high pressures in hydraulic system. Due to this nature of hydraulic system we can use this power to lift, hold, press very heavy loads. Rather we can say that, for very heavy duty earth moving equipments (excavators, levellers) we do not have alternative than to use fluid power. We can go upto 700 bar pressure with hydraulic system.
2. Weight to power ratio of a hydraulic system is comparatively less than that of an Electro-Mechanical System. Electric motor weighs appropriately 8.5 kg/kW whereas, same power hydraulic motor weighs 0.85 kg/kW only.
3. With hydraulic actuators we can achieve highly accurate and precise motions, so CNC/VMC machines use hydraulic controls for tool movements.
4. Hydraulic motor can be reversed (clockwise and anticlockwise) instantly. This peculiar property of hydraulic motors is very much useful in machine tools.
5. Limiting and Balancing of hydraulic forces can be easily performed.
6. The speed control of linear as well as rotary actuators can be achieved with ease. By merely adjusting small flow control valve, wide range of speed and feed can be obtained.
7. Hydraulic system uses oil. Hence it is a self-lubricating system. Special lubricating arrangements are thus not required at all.
8. Use of flexible connections in hydraulic systems permits generation of compound motion without gears and complicated mechanical drives.

9. The power source of hydraulic system need not be closer to the end user (machine). It can be kept away from machine. Due to this property, in limited space we can operate the machines comfortably.
10. Frictional resistance is very less in hydraulic systems.
11. The noise and vibration level produced by hydraulic pump is very low.
12. In any hydraulic systems division and distribution of power is simpler and easier.

(B) Limitations of Hydraulic System

1. Elements of hydraulic system have to be machined to high degree of precision which increases manufacturing cost of system.
2. Leakages of hydraulic oil during its flow in system causes heavy pressure drops.
3. The hydraulic system, due to oil leakages is 'dirty' and we cannot use this system in food and pharmaceutical industry.
4. Petroleum based hydraulic oils can create fire hazards if the temperature of the system goes beyond its 'flash point'.
5. In cold climate, hydraulic oil freezes. We have to use electrical heaters to liquefy the freezed oil. This increases the operating cost of the system.
6. The maintenance cost of hydraulic system is high.

1.6 PNEUMATICS: INTRODUCTION

- It is a technology which deals with power of air.
- In pneumatics, air is 'empowered' by compressing it and then its power is used to obtain mechanical advantage i.e. to do useful works.
- The use of air power was known to man from IInd World War.
- The industrial automation is done by using power of compressed air. The deburring tools, fettling tools are pneumatic tools.
- Pneumatic power is used widely in food processing industry. You might have visited a soft drink factory. In this factory, the movement of glass/plastic bottles are achieved in disciplined manner by using compressed air operated mechanisms called pneumatic circuits.
- In foundry, the casting needs cleaning. This cleaning operation is popularly known as fettling. In short, fettling is casting cleaning operation. To clean the surface of casting we need tools like straight grinders, angle grinders. These tools operate on pneumatic power.
- In road asphalting (making tar roads) process, we have to chip-off, the existing tar road. This chipping of the hard road is done by pneumatic chipper, which operates on compressed air power. The air is supplied to chipper by engine driven air compressor installed on truck or tractor.
- Similar example is as that of drilling the bore-well in ground. This drilling-rig operates on pneumatic power supplied by engine driven air compressor installed on truck.
- In printing press, especially on offset printing machine, you might have observed that, the single paper is lifted by plastic 'bellow' and is fed to machine for printing. This paper handling is a miracle of compressed air. In this mechanism, jet creates vacuum in 'bellow', to which single paper 'sticks' temporarily and when vacuum collapses, the paper is released.
- Siron which creates big sound to indicate 'lunch-recesses' timings to factory workers is another miracle of compressed air.
- This list is un-ending. In each industry, use of compressed air is becoming unavoidable. So we must understand, how pneumatic power is used.

- **Features of Air:** Power of Air has became so popular in industry automation, tools etc. because of some features of air (characteristics of air), which could not be found with any other medium. These important and notable features of air are as given below:
  - (a) Air is available free of cost.
  - (b) Wide availability of air.
  - (c) Air is absolutely fire proof (no danger of fire).
  - (d) Easy transportability i.e. through pipes.
  - (e) Easy storage i.e. in air receivers.
  - (f) The construction of pneumatic equipments is comparatively easy.
  - (g) The maintenance cost of equipments is low and maintenance is easy.
  - (h) The overall cost of compressed air and pneumatic equipments is low.

1.7 GENERAL LAYOUT OF PNEUMATIC SYSTEM

[S-09, 10, 11, 12, 14, 16; W-10, 12, 14]

- General layout or standard layout of any pneumatic system is having following components/ parts/equipments.
  - (a) Air compressor.
  - (b) Compressed air storage tank or air receiver.
  - (c) Moisture separators.
  - (d) Air filter – air pressure regulator – air lubricator unit (FRL) unit.
  - (e) Direction control valves/flow control valves/other values.
  - (f) Air actuators: reciprocating (linear actuators) and/or rotary actuators (air motors).
  - (g) Air conducting elements: pipes, valves, joints, bends etc.

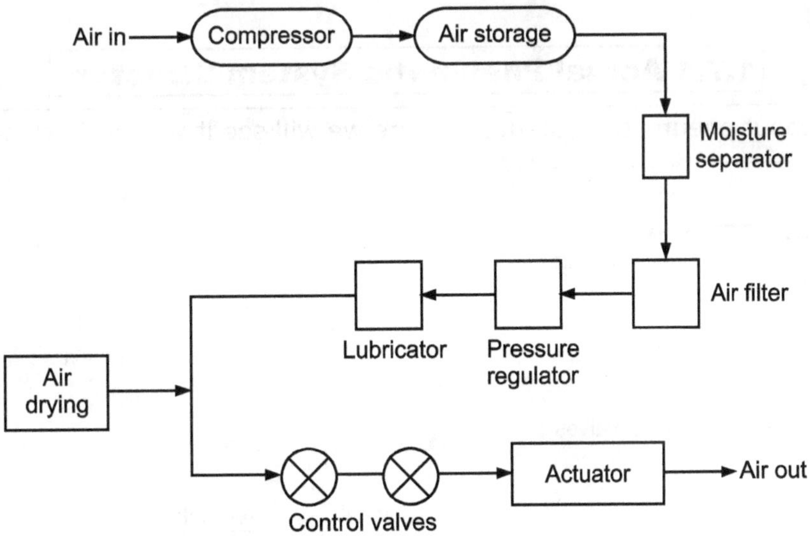

Fig. 1.4: Flow Chart of Pneumatic System

Functions of each element in Pneumatic System is discussed below:

(a) Compressor: This unit compresses the air which is at atmospheric pressure. Due to compression the pressure of air increases. This unit run either on electric motor or on diesel/petrol engine or gas/steam turbine.

Important:

- The compressors used in Engineering units/foundry are either reciprocating or screw compressor and runs on electric motors.

- The compressors used for 'bore well' machine are generally screw compressors and runs on diesel engine or directly on truck or tractor engine on which they are mounted.
- The compressors in sugar factory runs on steam turbine.

(b) Air tank/Air reservoir: This is a MS cylinder (popularly known as pressure vessel) similar to domestic gas cylinder. This tank stores compressed air which is output of air compressors.

(c) Moisture separator: Air contains particles of water called 'moisture'. These very fine particles of water must be removed from air otherwise, if they flow along with air, then they can rust the metal parts in system. 'Water trap' is a moisture separator. Other chemical or mechanical separators are also used. In these separators part moisture is separated.

(d) Air filter – Air pressure regulator – Air Lubricator (F.R.L): This is combined unit available readymade in the market. The air first gets filtered, then its pressure is regulated and then some fine particles of lubricating oil are mixed with air in lubricator. The particles help in proper lubrication of sliding parts/rotary parts of actuator and valves.

(e) Direction control valves/Flow control valves/Other valves (Control of air): These valves direct the flow of air, control the flow of air to control speed of actuators etc.

(f) Air actuators: These items are the heart of pneumatic system. Actuators are taking part in obtaining mechanical advantage or you may say, useful work from compressed air. These are either linear actuators or rotary actuators.

(g) Air conducting elements: The compressed air flows through pipes (GI pipes/Hoses etc.) There are on-off valves, pressure gages attached to pipe. All these elements comes under this head.

1.7.1 Actual Pneumatic System Structure [S-16] [W-11]

- To know more about pneumatic system structure we will see it with little closer view:

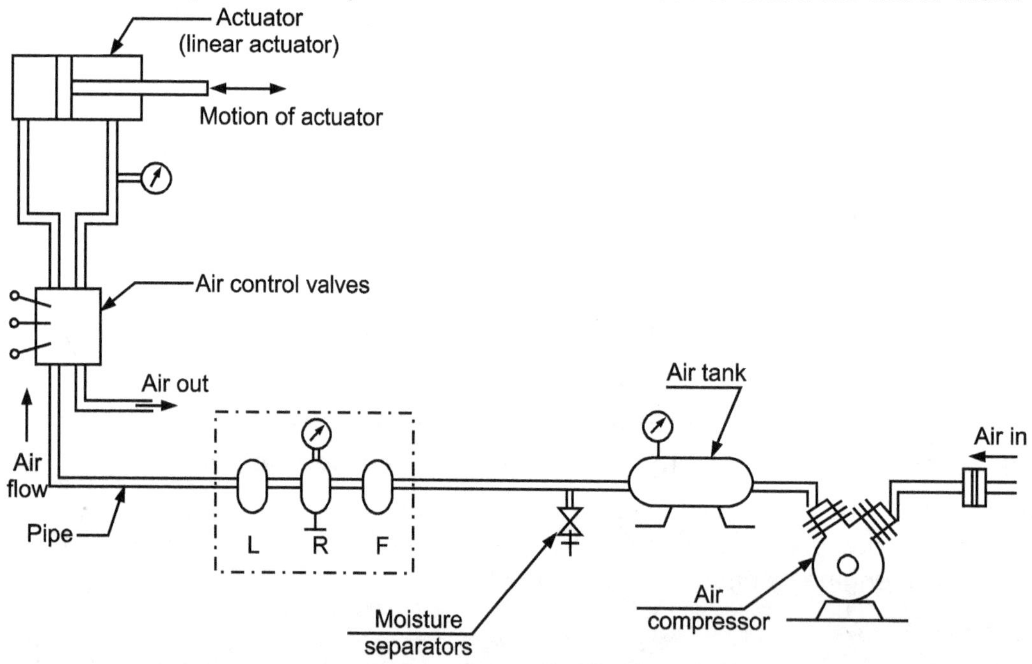

Fig. 1.5: Actual Pneumatic System Structure

1.7.2 Actual Pneumatic System Structure with the help of Symbols

[W-11, S-16]

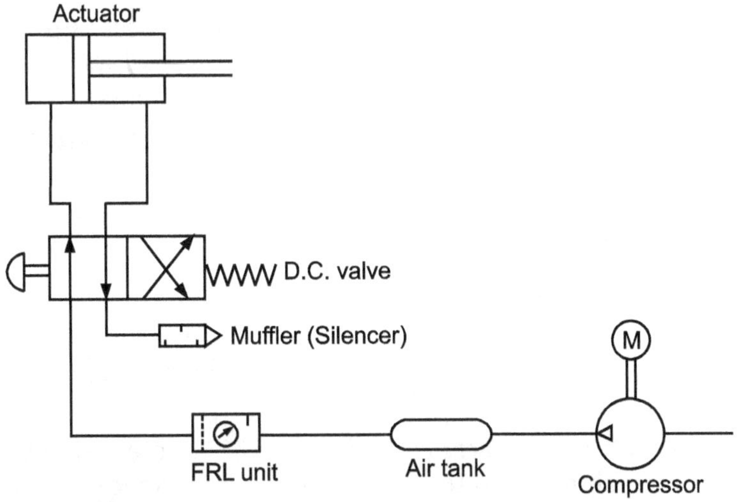

Fig. 1.6: Actual Pneumatic System with the Help of Symbol

1.8 MERITS AND LIMITATIONS OF PNEUMATIC SYSTEMS

[W-09, 12, S-14, 16]

(A) Merits of Pneumatic Systems:

1. The working medium of pneumatic system is air, which is easily and freely available.
2. Air is dry and hence any pneumatic system is clean. (In hydraulics systems the system is not clean due to presence of oil). Due to this property, pneumatic circuits are most widely used in the food industry automation and electronic industry automation.
3. Air cannot be exploded. (In hydraulic systems, if the temperature of system goes beyond the flash point of hydraulic oil then there is possibility of fire). Hence pneumatic circuits can be easily used in fire prone area.
4. The equipments used in pneumatic systems are easy in working and easy in manufacturing.
5. The pneumatic equipments such as chipper, grinders, rotary actuators, linear actuators, are having comparatively low cost and are light weight.
6. The maintenance cost of pneumatic equipments and circuit elements like compressors, FRL unit, DC valves, actuators etc. is low.
7. Pneumatic systems are not having return lines. Hence, the total cost of system is low.

   Return lines: In hydraulic systems, the medium i.e. oil is pressurized and then after doing useful work needs to be return to reservoir of soil. So we need special piping for flow of returning oil from actuators. These pipelines are known as return lines.

   In pneumatic systems, the compressed air after doing useful work in actuator, is exhausted to atmosphere. It need not to be returned.
8. Compressed air can be transported through light weight pipes.
9. The compressed air can be stored in Air tank under pressure. When air tank is full, the air can be used as and when required. During this period compressor is switched-off, which saves electric power (In hydraulic system, to pressurize oil we have to run the hydraulic pump continuously).

(B) Limitations of Pneumatic Systems: [S-10, 12, W-12, 16]

1. Pneumatic system can be used upto 10 bar pressure. If the circuit requires pressure below 10 bar (10 kg/cm^2) then we can use pneumatic circuit. But if the circuit requires pressure more than 10 bar we have to go with Hydraulic circuit. The Hydraulic circuit can operate upto 700 bar pressure. This is the main limitation of pneumatic systems.

2. If the pressure of stored air in air tank drops below required pressure, then pneumatic circuit stops working. It will start only after, required pressure is developed in air tank.
3. Due to compressing nature of air, accurate motions in actuator are not possible to obtain.

- In CNC/VMC machines, the tool movements needs to be very accurate. This accuracy can be in order of plus/minus 10 microns. This cannot be achievable in pneumatic circuits. So in VMC/CNC machines, hydraulic circuits are used for tool movements.

1.9 PRACTICAL APPLICATIONS OF HYDRAULIC SYSTEMS

[W-10, 13, 16, S-12, 16]

- Oil Hydraulic Systems to obtain useful work are having number of applications. Some of the important particle applications are listed below:

(a) Pressure Die Casting Machines are operated on oil hydraulic system. In this machine we are giving shape to hot aluminium ingots. These machine needs to create high pressure forces upto 400 Tons. Aluminium components like carburettors, links etc. are manufactured on this machine.

(b) Hydraulic Shearing Machines are used to shear off MS steel sheets upto 6 mm thickness. This machine operates on oil hydraulic system and has practically unlimited service life.

(c) Hydraulic Press Brake: Press Brake is a machine which bends the long sheets by using die-punch tool set. Hydraulic power is used to operate die-punch set. This machine can bend the sheets within the tolerance of ± 0.01 mm because hydraulic system gives the accuracy and consistency.

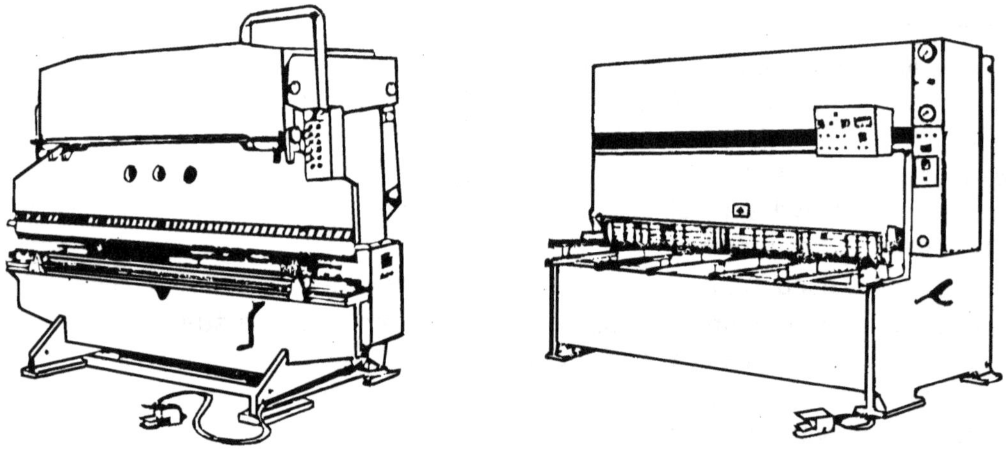

(a) (b)

(a) Hydraulic Shearing Machine **(b) Hydraulic Press Brake**

Fig. 1.7

(d) Power or Hydraulic Operated Bar Cutting Hack Saw: Bar cutting is a First Operation in many workshops. Reciprocating Hack Saw or Metal Band Saw is equipped with 'Automatic Bar Feeding Mechanism' which operates on Hydraulic Power. When piece of bar is cut, automatically bar advances for second piece to cut.

(e) Hydraulic Power Press: Fig. 1.8 shows Hydraulic Press which uses hydraulic energy created by way of pressurizing hydraulic oil. Pressurized oil will enter through port (A) and its force will act on piston. Piston will move down with force. The steel sheet kept on die on anvil will take a shape due to meeting of die and punch. Then pressurized oil will enter through port (B). Oil pressure will act on two pistons of lifting cylinder and main piston will move-up; so that we can take out pressed component.

Hydraulic presses are having wide applications. These machines are used in press shops to manufacture pressed components from CRCA sheets viz. crankcase, bottom cup of harmetically sealed compressor etc. These presses are used in press forging operation. For removal of axle or shaft bends, hydraulic presses are used.

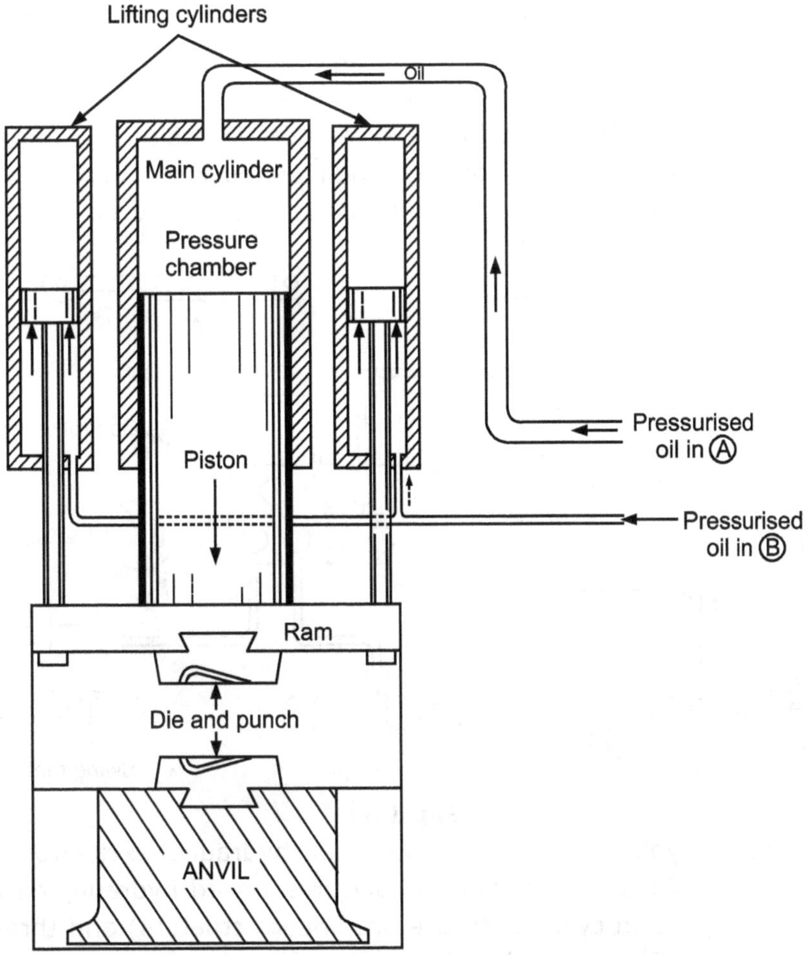

Fig. 1.8

(f) Broaching machine: Broaching is a method of removing metal by a tool (called Broach) having successively higher cutting edges in a fixed path. Generally, vertical Broaching machines are used in which there are two types – Pull type (Broach is pulled upwards) and Push type (Broach is pressed downwards). This push or pull action is done by oil Hydraulic Power because it requires very high pressure force to cut the metal.

(g) Hydraulic Thread Rolling Machine is used to cut threads on nuts/bolts/studs/screws. It is mass production machine. The main thread roll pressing action is done by hydraulic force. The machine operates on 40T pressure force created by hydraulic oil.

(h) Hydraulic Drilling/Reaming/Burnishing Machine: This machine is used specially for mass production. The movement of tools (up-down) is controlled by hydraulic power.

(i) CNC/VMC/HMC Machines: These are modern machines having state-of-art technology. These are turning centers and machining centers and are fast replacing traditional lathes, milling machines, shaping machines. The control of these machines is done by computer or microprocessor. For tool movements, automatic clamping or chucking, for pushing tool posts, turret movements, hydraulic power is used.

(j) Cylindrical Grinding Machine: Cylindrical grinding is a finishing operation. Bar grinding/Crankshaft pin grinding, Hole grinding are some of the final operations after heat treatment to control final close tolerance size of component. The table reciprocation movement of this machine is controlled by hydraulic power.

(k) Besides above application, hydraulic power is most commonly used in Special Purpose Machines (SPM) which are very useful in Mass Production.

(l) Plastic Injection Moulding: Plastic granuals (HDPE/LDPE/PP) are moulded in Plastic Injection Moulding Machines. Plastic chairs, stools, buckets, pots, tiffins are some of the products of plastic injection molding process. This machine uses hydraulic power.

(m) Earthmoving equipments like hydraulic excavators are using one and only one power i.e. hydraulic power! The excavator controls such as bucket lifting, arm swivelling, boom swivelling and extension all these actions are controlled by hydraulic circuits.

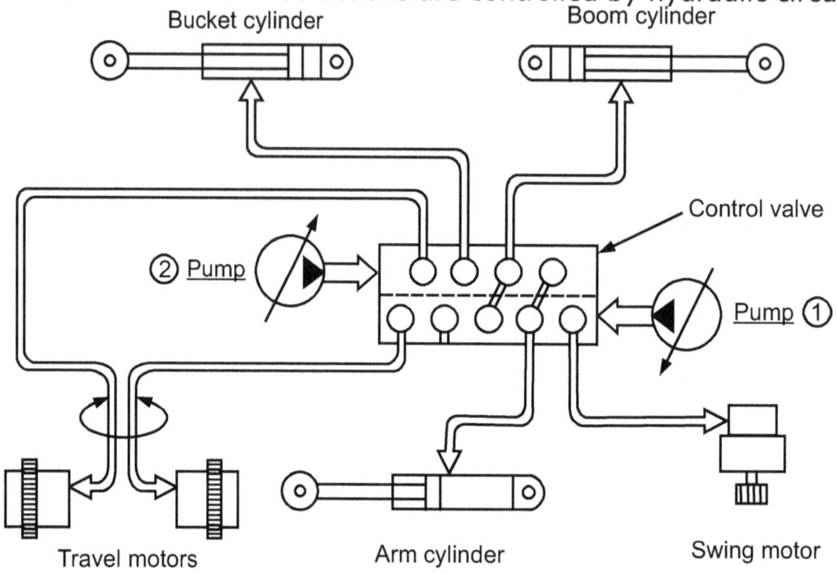

Fig. 1.9

- Fig. 1.9 shows typical hydraulic system of hydraulic excavator controlled on two hydraulic pumps. These two pumps supply pressurized hydraulic oil to bucket cylinder, boom cylinder and arm cylinder (these are linear actuators) and three hydraulic motors i.e. swing motor and Travel motors (these are rotary actuators).

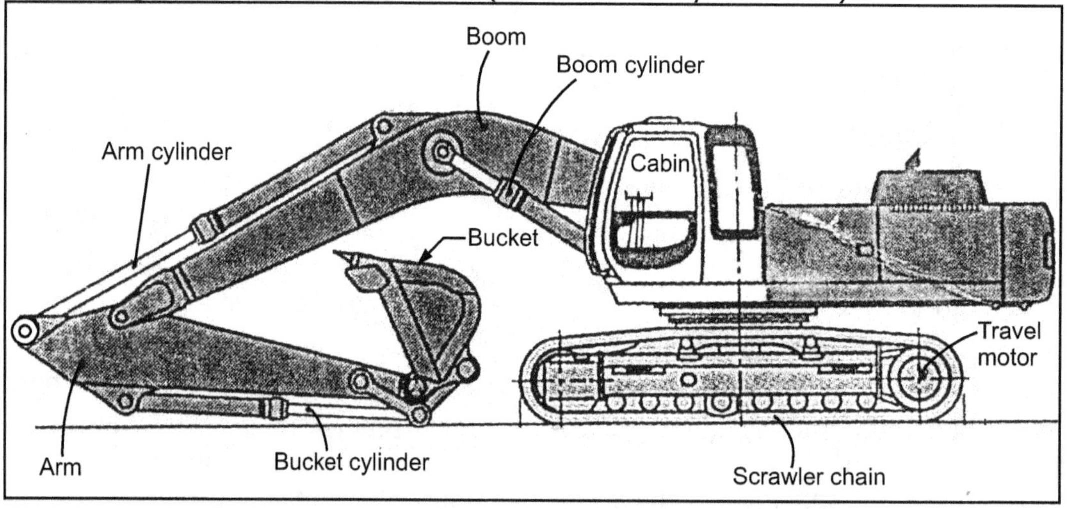

Fig. 1.10: Actual Line Diagram of Hydraulic Excavator

Hydraulic excavator cannot be transported on road. It is carried in heavy trucks to excavation/levelling sites.

(n) Hydraulic Torque Wrenches: Torque Wrench is a device similar to box spanner. It is used in tighting screws/nuts/bolts. But in case of Torque wrench, we can set the torque (Torque = Force × Distance. Unit of torque is N-mm) before tighting. Once the torque is set, the tighting will stop automatically on 'set' torque value. In case of hydraulic torque wrenches, the tighting force is created by hydraulic power.

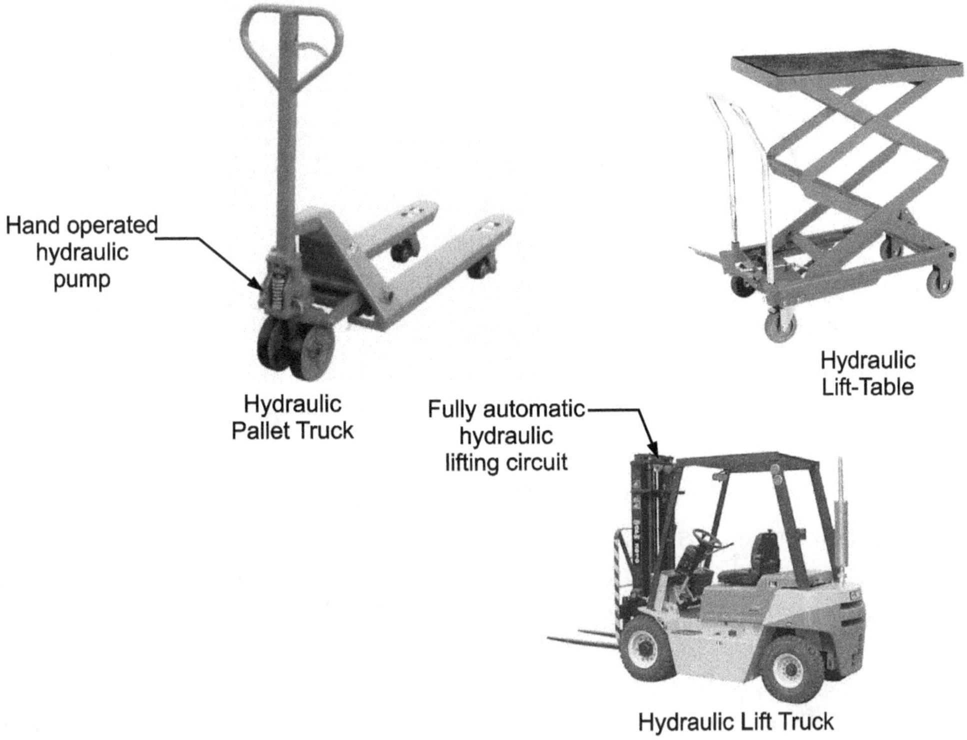

Fig. 1.11

(o) Backhole loader: This is similar to excavator but it has four rubber tyres and it can travel on road. So its transportation cost is low. It has bucket on one side and backhoe (big bucket) on other side. Entire control of two buckets is on hydraulic power.

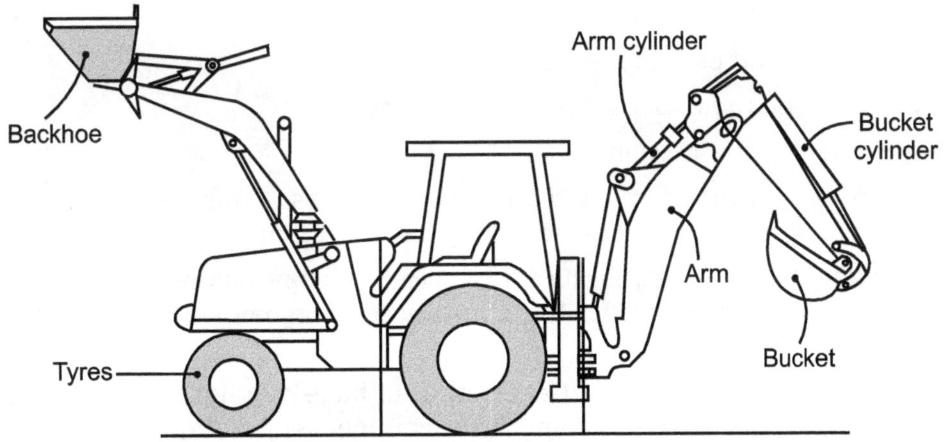

Fig. 1.12

(p) Vibrator Compactor: These are sophisticated modern road rollars. They create heavy pressures on road to compact hot tar metal on road to strengthen the road, while road asphalting. These machines have increased the speed of road making. Vibrator compactor uses hydraulic power to create high compacting pressures.

(q) Material Handling Equipments: In any mass production factory use of material handling equipments is unavoidable. hydraulic pallet truck (manual), hydraulic lift – table are some of the examples. All material handling lifters, pickers, stakers use hydraulic power.

(r) Hydraulic Cranes: These are generally truck mounted, telescopic hydraulic cranes. D-MAG is a German Brand of hydraulic cranes which can lift 80T of load. The lifting arm is telescopic (similar to radio antenna) and the extension of arm and lifting uses very heavy hydraulic power.

(s) Hydraulic Pullers: Pullers are the mechanical devices which are used for removing press fit parts on shaft such as bearings, pulley etc. Hydraulic pullers are having built-in hydraulic RAM assembly so that pulling force is easily created by hydraulic cylinders and pulling becomes smooth.

1.10 APPLICATIONS OF PNEUMATIC SYSTEMS

[S-09, 10, 15; W-10, 13]

- Following are some of the applications of Pneumatic Systems:

(a) Air brakes in heavy automobiles.

(b) Rubber tyres of automobiles are inflated by compressed air, then and then only they can run with speed on road.

(c) Blasting of rocks in mines: The blasting of rocks is done by drilling a hole in rock. This hole drilling in rock is achieved by compressed air.

(d) Pneumatic hammers used on construction site.

(e) Pneumatic screw drives, used in assembly lines of engines, machines etc.

(f) Pneumatic pistol drills used on construction site.

(g) Angle grinders, straight grinders used in fettling shop in foundry.

(h) Automation of food industry.

(i) Air jet looms: These are very high speed cloth weaving shuttleless looms, operating on pneumatic power.

(j) Pilfer proof (PP) caps manufacturing units (PP cap – cap on Medicine bottles made out of tin sheets).

(k) Injection and blow molding machines used for manufacturing plastic components.

(l) Paper handling in Offset Printing Press.

(m) Ground bore well and oil well development.

(n) Television, cathode ray oscilloscope assembly lines.

(o) Hydro-pneumatic press machines.

(p) Compressed air is used in manufacturing oxygen gas. (This gas is used in gas welding techniques and you all know it is life saving gas for terminally ill patients).

(q) It is interesting to note that, in 2008 Olympics, Abhinav Bindra has earned gold medal in 100 mm air rifal shooting event. The gun used by Abhinav was *'pre-loaded pneumatic air gun'*.

(r) Food Grain Industries (Floor Mills/Dal Mills) used huge overhead storage bins. The opening gate of these bins are at the bottom. There is huge weight of grains on this gate. To open and close these gates pneumatic power is used.

(s) Pneumatic files are used in tool rooms of engineering workshop. These files are used for deburring operations. These pneumatic files can generate 3500 strokes/min and works at 6 bar pressure and consumes 4 cfm if air.

(t) **Pneumatic nozzle cleaning stations:** In case of MIG/TIG welding, tourches needs frequent cleaning. This cleaning is done by using high pressure compressed air. (Air pressure 5 to 8 bar).

(u) Orbital rivetting machines are used for freely swivelling joints to clamp. These machines work on pneumatic power. The maximum air pressure is
7 bar.

(v) 'Through feed super finishing machines' are used for superfinishing operation on gudgeon pins, rollers, needles, shock absorbers, piston rods, computer printer rods. The components are continuously fed from one end of a pair of rollers and finished products are collected at other end. In this machine pneumatic power plays important role. The pneumatic oscillation system gives 1500 – 2000 strokes per min to superfinishing heads having rough to finish stones in progressive manner.

(w) Thermal spray coating: These are special machines to coat metal protective coatings on components like turbine rotors, marine engine pistons, ic engine valves, ball valves, paper mill rolls. In this machine electric arc is created and the coating material melts at tip of Arc. At that time compressed air is blown to create automized spray of coating metal.

- **A hydraulic fluid or hydraulic liquid** is the medium by which power is transferred in hydraulic machinery. Common hydraulic fluids are based on mineral oil or water. Examples of equipment that might use hydraulic fluids are excavators and backhoes, hydraulic brakes, power steering systems, transmissions, garbage trucks, aircraft flight control systems, lifts, and industrial machinery.
- Hydraulic systems like the ones mentioned above will work most efficiently if the hydraulic fluid used has zero compressibility. Hydraulic fluids can contain a wide range of chemical compounds, including: oils, butanol, esters (for example, phthalates, like DEHP and adipates, like bis (2-ethylhexyl adipate), polyalkylene glycols (PAG), organophosphate (e.g. tributyl-phosphate), silicones, alkylated aromatic hydrocarbons, polyalphaolefins (PAO) (for example, polyisobutenes), corrosion inhibitors (incl. acid scavengers), anti-erosion additives etc.)

Functions and Properties:

- The primary function of a hydraulic fluid is to convey power. In use, however, there are other important functions of hydraulic fluid such as protection of the hydraulic machine components. The table below lists the major functions of a hydraulic fluid and the properties of a fluid that affect its ability to perform that function:

Table 1.1

Function	Property
Medium for power transfer and control	• Non-compressible (high bulk modulus) • Fast air release • Low foaming tendency • Low volatility
Medium for heat transfer	• Good thermal capacity and conductivity
Sealing Medium	• Adequate viscosity and viscosity index • Shear stability
Lubricant	• Viscosity for film maintenance • Low temperature fluidity • Thermal and oxidative stability • Hydrolytic stability/water tolerance • Cleanliness and filterability • Demulsibility • Antiwear characteristics • Corrosion control
Pump efficiency	• Proper viscosity to minimize internal leakage • High viscosity index
Special function	• Fire resistance • Friction modifications • Radiation resistance
Environmental impact	• Low toxicity when new or decomposed • Biodegradability
Functioning life	• Material compatibility

1.11 PROPERTIES OF ISO HYDRAULIC OIL (MEDIUM) [W-12, S-16]

- In hydraulic system we are pressurising the fluid (i.e. oil) by pump. This oil convey (carry) energy and produces required force at actuators to obtain useful mechanical work.
- In the beginning of hydraulic system development era, the design engineers used water as a medium. (infact the word 'hydraulic' means 'water'). But water has following disadvantages.

  (a) It freezes at 0°C hence cannot be used in low temperature zones.

  (b) It is corrosive.

  (c) Its lubrication properties are poor.

  (d) Water cannot be used at temperature higher than 100°C because it starts boiling and converting into steam.
- In view of these disadvantages, modern fluids were specifically developed to suit hydraulic systems.
- **Demulsibility:** The ability of a fluid that is insoluble in water to separate from water with which it may be mixed in the form of emulsion.

  In other words it is an oils ability to release water.
- **Lubricity:** It is the measure of the reduction in friction of a lubricant. The study of lubrication and mechanism wear is called tribology.
- **Measurement of Lubricity:** Lubricity of a material cannot be directly measured, so tests are performed to quantify a lubricant's performance. This is done by determining how much wear is caused to a surface by a given friction-inducing object in a given amount of time. Other factors such as surface size, temperature and pressure are also specified. The greater the wear scar the worse the lubricity. For this reason lubricity is also termed as substance's anti-wear property.
- A satisfactory hydraulic oil must have following properties: [S-15]

  (i) **Good Lubrication Property:** In hydraulic system DC valve, actuators, pressure relief valves etc. are the components in which one surface moves/rotates with respect to other. The hydraulic fluid must separate and lubricate such surfaces, so that the life of components increases.

  (ii) **Good Chemical Stability:** The oil when pressurises, its temperature increases, when it moves in circuit its temperature further increases. Oil is also in constant contact with rubber seals, filter materials. In such instances, there is possibility of chemical reaction between oil and constant materials. This results in formation of sludges, carbon separation or precipitation. These contamination results into jamming of system and creates major problems. So a good hydraulic oil must be chemically stable and must be least responsive to chemical reactions under heat.

  (iii) **Non-corrosive:** A good hydraulic oil must be non-corrosive to metal surfaces in contact.

  (iv) **High Flash-Point:** Flash point is a temperature at which liquid catches fire automatically. The flash point of good hydraulic oil must be as high as possible so that fire possibility is nullified.

  (v) **System Compatibility:** We have already notified the concept of compatibility. Diesel engine is compatible to diesel as a fuel only. If we use petrol in diesel engine, the engine will not run. Similarly, if major changes/modifications in the hydraulic system are done, then also the original oil must perform its functions well. It must be compatible.

  (vi) **Minimum Toxicity:** A good hydraulic oil must be minimum Toxic to human beings working with them. Some fire resistant hydraulic oils are highly toxic which can cause occupational diseases.

  (vii) **Good heat Dissipation:** Heat dissipation means discharging the heat and becoming cool. When fluid is moving/pressurising, its temperature increases. The good oil must have property to fast dissipation of heat to atmosphere or cooler. That clearly means that oil must have high thermal conductivity.

(viii) Low Foaming Tendency: When oil returns to receiver, it comes in contact with air above the liquid surface. The oil has tendency to absorb air or gas and which results in foam formation. A good hydraulic oil must release the air/gas very quickly so that it does not form foam.

(ix) Fire Resistant: Hot oil can catch fire. A good oil must be fire resistant to avoid accidents.

(x) Low Coefficient of Expansion: A good hydraulic oil must have very low coefficient of expansion for minimum volume of oil required at operating temperature.

(xi) Low Specific Gravity: If the specific gravity of oil is low then overall weight of hydraulic system will be minimum.

(xii) Cost and Availability: The cost of oil must be affordable and at the same time it must be available in bulk.

1.12 HYDRAULIC OILS AVAILABLE

(A) Petroleum Based Hydraulic Oils: Most widely used (similar to Engine/Gear oil)

(B) Water Based Hydraulic Oils

(a) Water-in-oil emulsion.

(b) Oil-in-water emulsion.

(c) Water + Glycol mix.

(C) Synthetic Hydraulic Oils (Phosphate Esters)

1.12.1 Petroleum Based Hydraulic Oils

- These are thick oils (maroon coloured/chocolate coloured) similar to engine oils used in automobiles. These oils are most popular and used in major hydraulic circuits.
- Special additives are added to these oils so that its foaming tendency reduces, its lubrication properly improves and becomes user friendly.

Advantages of Petroleum based oils:

(i) Good lubricating property

(ii) Good rust preventive

(iii) Good heat dissipation tendency

(iv) Easy to keep clean

(v) Available with ease.

Disadvantages of Petroleum based oils:

(i) The oils are having tendency to ignite. So in hazardous places, these oils are having last choice.

(ii) These oils make the ground surface slippery which is due to leakages. The slipper ground can result into accidents.

1.12.2 Water Based Hydraulic Oils

- These fluids are water based i.e. water is base in which other liquids are mixed.

(i) Water-in-oil emulsion: This is a mixture of water (40%) and oil (60%).

(ii) Oil-in-water emulsion: This is a mixture of oil in water.

Advantages and Disadvantages of Water based Hydraulic Fluids:

These fluids are fire resistant. If water content reduces then the viscosity of these fluids reduce.

(iii) Water + Glycol: In this mixture water and polyglycols are mixed in 50%/50%.

1.12.3 Synthetic Hydraulic Oil (Phosphate Easters)

- These are synthetic fluids are also non-flammable and can be used at high temperature upto 150°C. In fact these fluids cannot work at lower temperatures. These are most expensive fluids and rarely used.

SAE Grades of Hydraulic Oils:

- The Society of Automotive Engineers (SAE) has established a numerical code system for grading motor oils according to their viscosity characteristics. This is why you would see something like this on the label: SAE 5W-30 with hydraulics, there are three grades for considerations - the viscosity grade and the hydraulic oil type grade 32-46-68. These specifications are typically determined by the weight and viscosity of the oil. Hydraulic Fluid Chart.
- The Society of Automotive Engineers (SAE) has established a numerical code system for grading motor oils according to their viscosity characteristics. Because the viscosity of oil changes with temperature, multigrade oils were developed to provide protection across a range of temperatures. This is why you would see something like this on the label: SAE 5W-30.
- In a 5W-30 for example, the number before the W describes the viscosity of the oil at low temperatures. The lower the number, the thinner the oil and the better the oil's cold temperature/cold start performance. The number after the W describes how thick the oil is at the engine's normal operating temperature.

ISO Grades of Oils used in Hydraulic Systems:

- ISO is known as the International Standards Organization Viscosity Grade, ISO VG for short. You do not have to listen very long in this field before someone says that viscosity is the most important physical property of a fluid when determining hydraulic oil requirements.

Some equivalent Grades of oil in SAE and ISO are given below:

ISO Grade	Equivalent SAE Grade	Kinematic Viscosity centiStokes
32	10 W	5.4
46	20	6.8
68	20 W	8.7

- **For example:** ISO 46 simply refers to the viscosity. AW stands for 'Anti Wear". Your AW Hydraulic Oil ISO 46 is going to be the same thing as AW46. It's just that the words are rearranged in a different order.

Table 1.2

In Viscosity Grade (ISO VG)	Kinematic Viscosity at 40°C (mm^2/s)		
	Minimum	Maximum	Mid-point
2	1.98	2.42	2.20
3	2.88	3.52	3.20
5	4.14	5.06	4.60
7	6.12	7.48	6.80
10	9.0	11.0	10.0
15	13.5	16.5	15.0
22	19.8	24.2	22.0
32	28.8	35.2	32.0
46	41.4	50.6	46.0
68	61.2	74.8	68.0
100	90.0	110	100
150	135	165	150
220	198	242	220
320	288	352	320
460	414	506	460
680	612	748	680
1000	900	1100	1000
1500	1350	1650	1500

1.13 SYMBOLS USED IN HYDRAULIC CIRCUITS

ISO SYMBOLISES USED IN HYDRAULIC CIRCUITS

Oil Reservoir/Oil Tank	[W-16] 
Filter	[W-09, 16] 
Heat Exchanger – Cooler	[S-10, W-16] 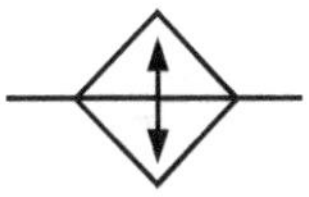
Heat Exchange – Heater	
Unidirectional Fixed Displacement Hydraulic Pump	[S-15, W-16] 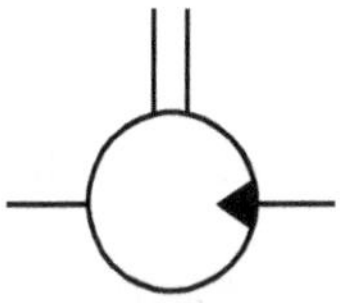
Bi-directional Fixed Displacement Hydraulic Pump	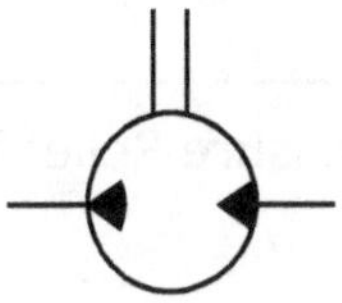
Unidirectional Variable Displacement Hydraulic Pump	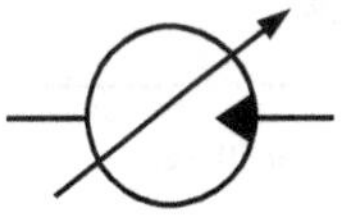
Bi-directional Variable Displacement Hydraulic Pump	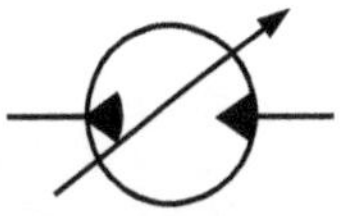
2 × 2 Direction Control Valve (2 × 2 DC Valve)	[W-16] 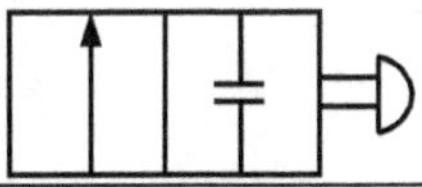
3 × 2 DC Valve	
4 × 2 DC Valve	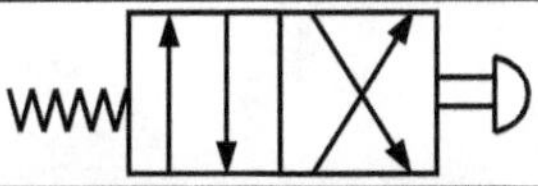
4 × 3 DC Valve (Centre Closed Position)	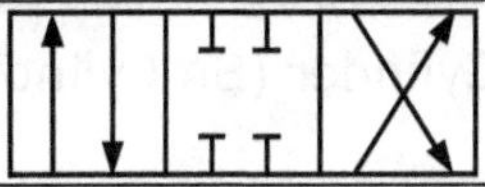

contd. ...

4 × 3 DC Valve (Open Centre Position)	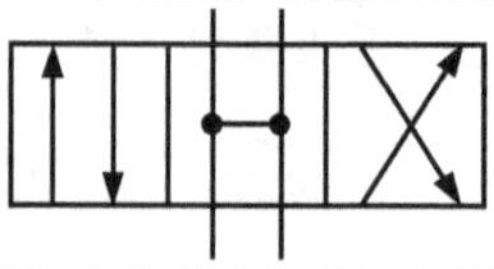
4 × 3 DC Valve (Tandem Centre Position)	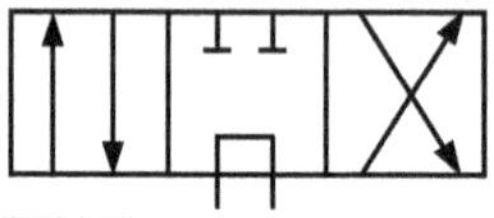
Flow Control Valve (Non-compensated)	[W-16] 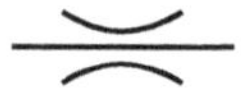
Variable Flow Control Valve (Non-compensated)	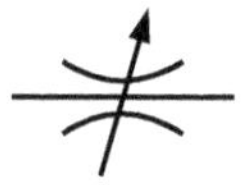
Pressure Compensated Flow Control Valve	[W-16] 
Simple Check Valve	[W - 09] 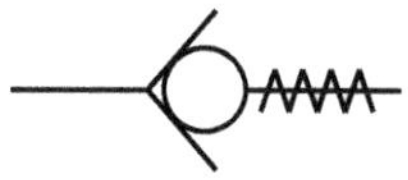
Pilot Operated Check Valve	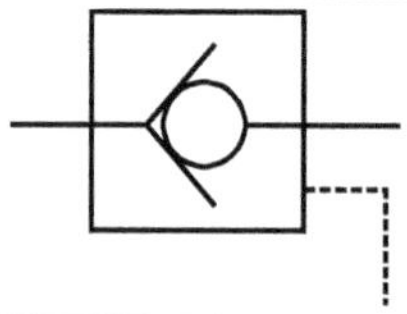
Directly Operated Pressure Relief Valve	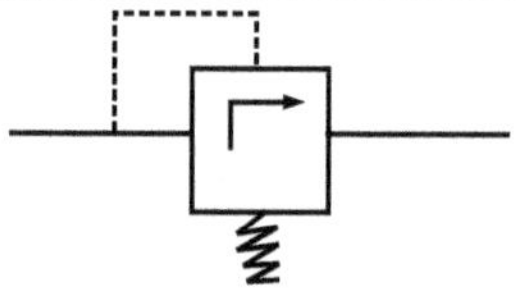
Pilot Operated Pressure Relief Valve	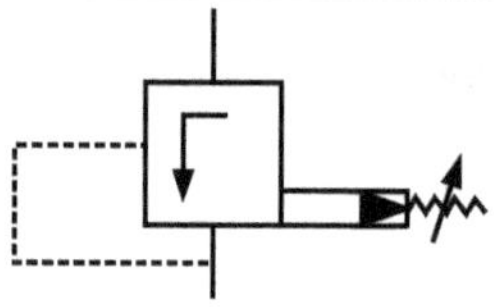
Pressure Reducing Valve	[W-16] 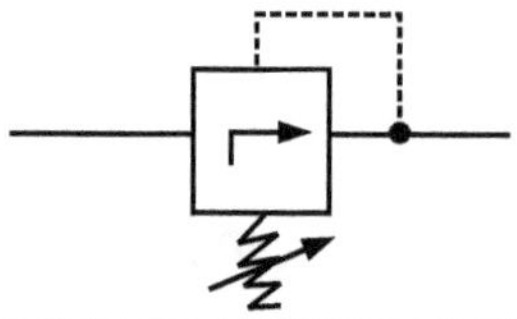
Unloading Valve	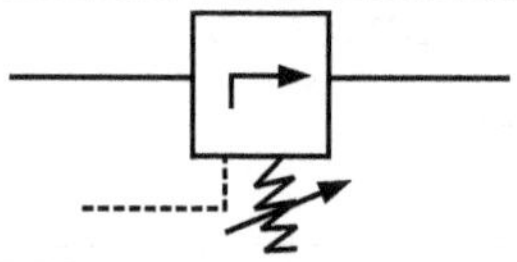
Sequence Valve	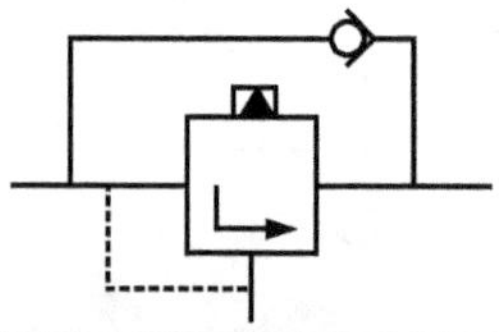
Single Acting Cylinder (SA Cylinder)	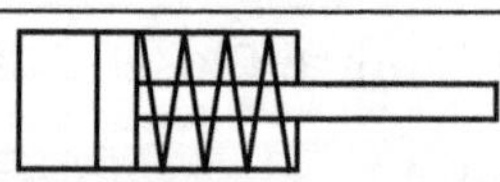

contd. ...

Double Acting Cylinder	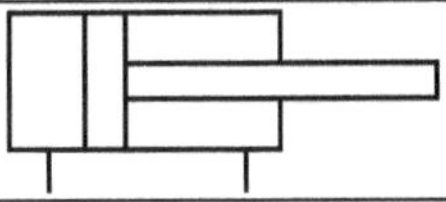
Hydraulic Ram Type SA Cylinder	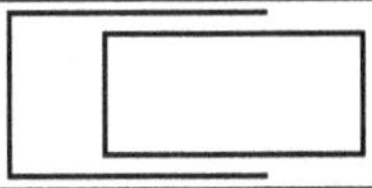
Double Acting Cylinder with Double Piston Rod	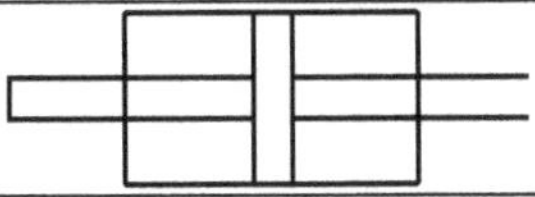
Tandem Cylinder	[S - 10] 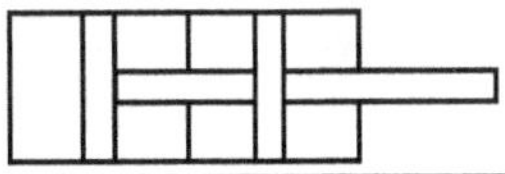
Telescopic Cylinder	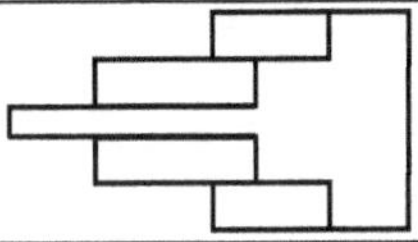
Uni-directional Air Motor	[W-10] 
Bi-directional Air Motor	
Limited Rotary Motion Motor	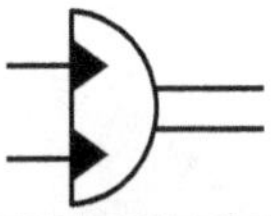
Dead Weight Accumulator	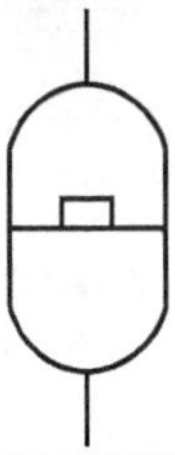
Spring Loaded Accumulator	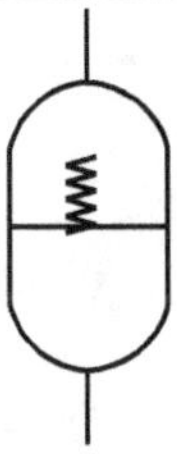
Gas Pressurized Accumulator	[W - 10] 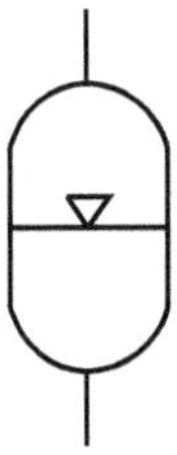

contd. ...

Working Pipe Line	
Drain Line/Pilot Line	
Flexible Line (Hose)	
Line Junction	
Pressure Gauge	
Push Button Control	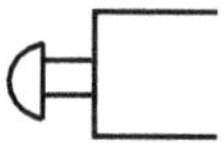
Pedal Control	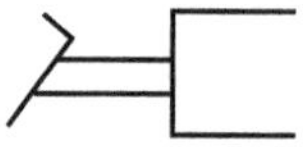
Roller Operated Control	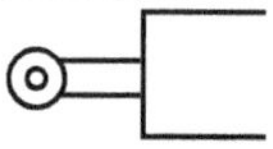
Plunger Control	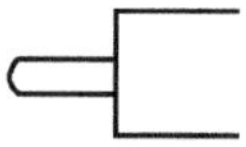
Spring Control	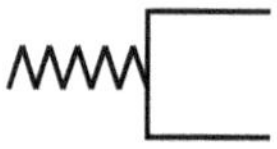
Hydraulic oil Signal Control (Pilot)	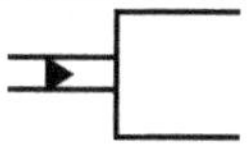
Solenoid Control	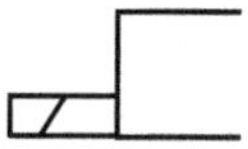

Views shown looking into Ports on the DC Valves:

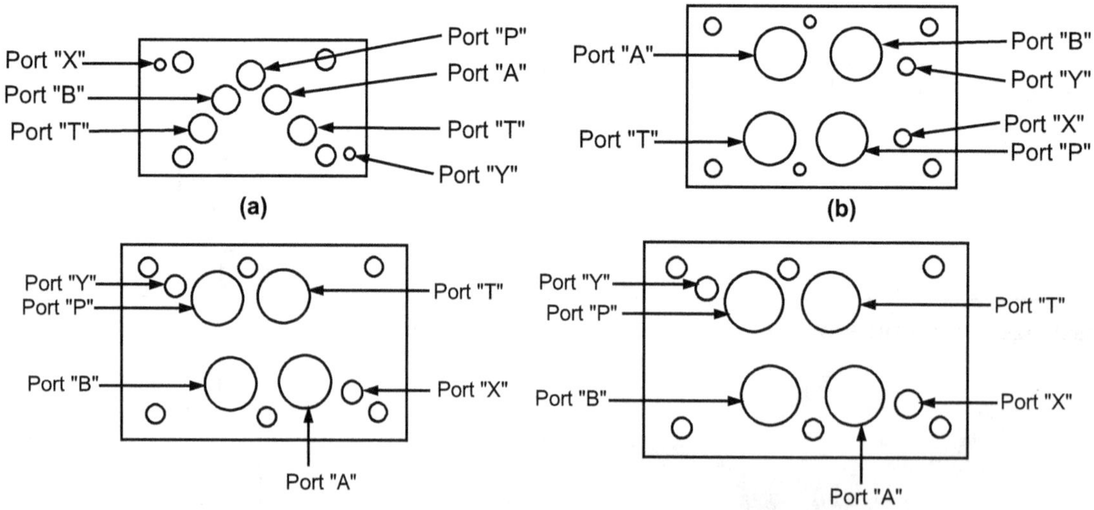

Fig. 1.13

1.14 ISO SYMBOLS USED IN PNEUMATIC CIRCUITS

(The order in which they appear in the topic)

Compressor (Irrespective of the type of compressor the symbol is same)	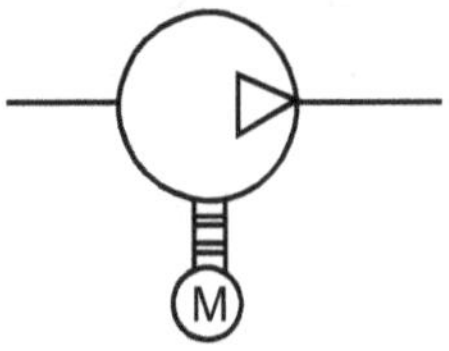
Air receiver or Air tank	
Water trap – Manual drain	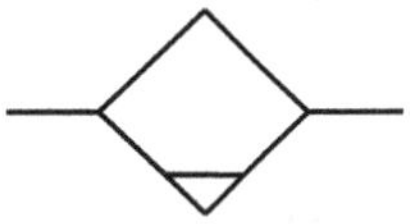
Water trap – Automatic drain	
Filter (F)	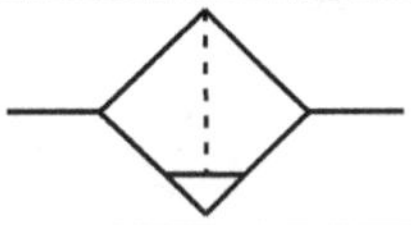
Regulator (R)	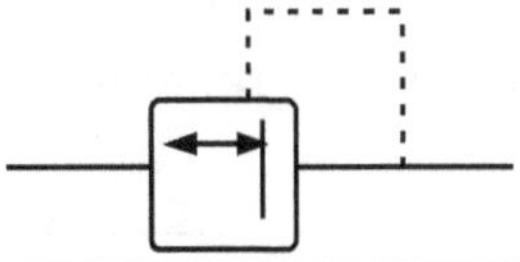
Lubricator (L)	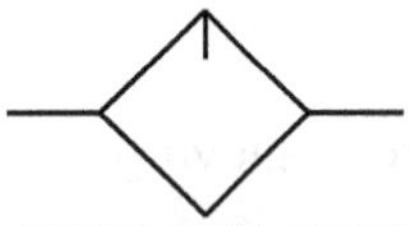
Combined F-R-L unit	
2 × 2 Direction control valve (2 × 2 DC valve)	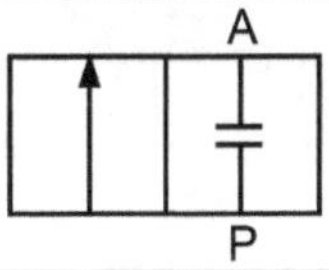
3 × 2 DC valve	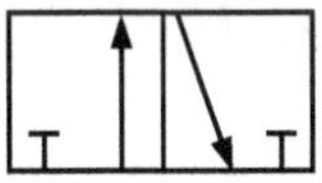
4 × 2 DC valve	[S - 10] 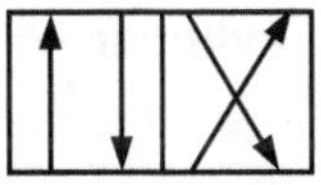
5 × 2 DC valve	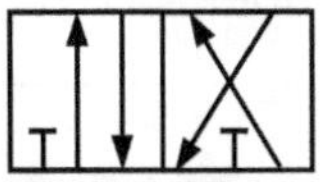
5 × 2 DC valve (Pilot operated)	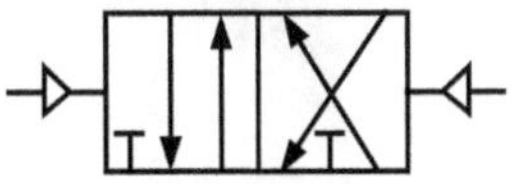
Non-return valve	

contd. ...

Fixed type flow control valve	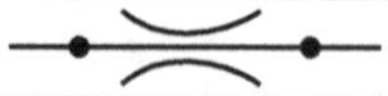
Variable flow control valve	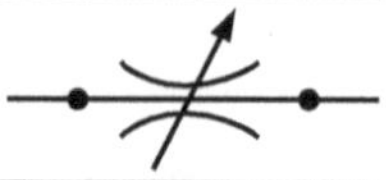
Variable flow control valve with built-in-check valve	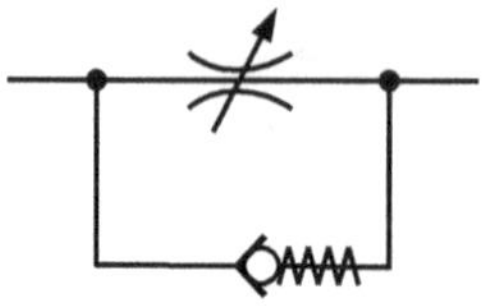
Pilot operated check valve	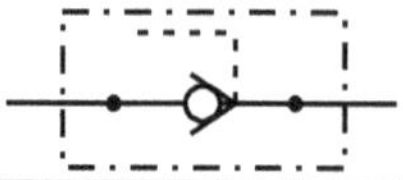
Shuttle Valve (double check valve)	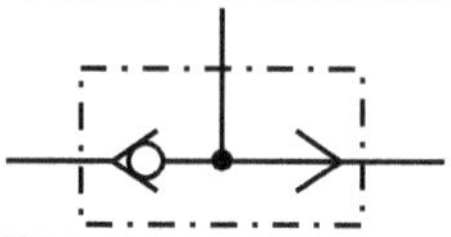
Pressure relief valve	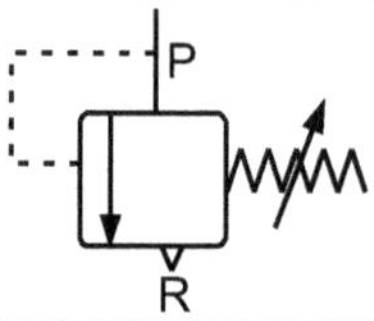
Pressure regulator non-relieving	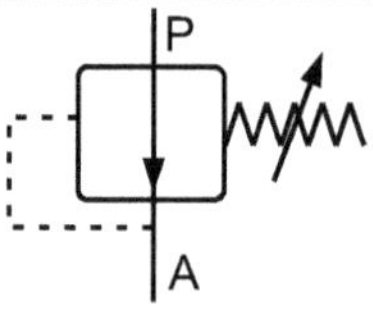
Pressure regulator relieving	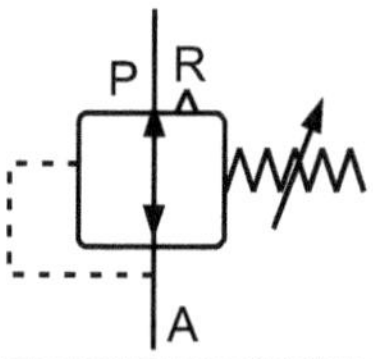
Single acting cylinder	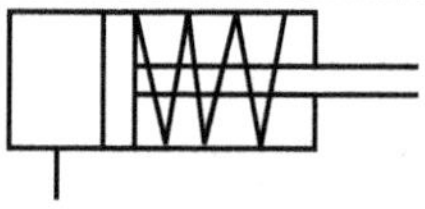
Diaphragm cylinder	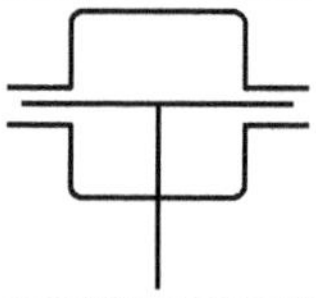
Rolling diaphragm cylinder	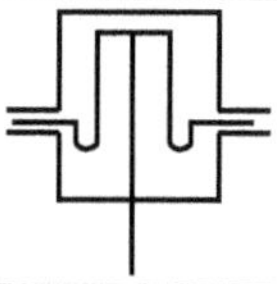
Double acting cylinder	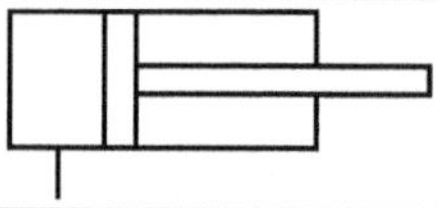
Turn cylinder	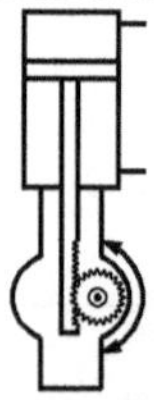

contd. ...

Tandem cylinder	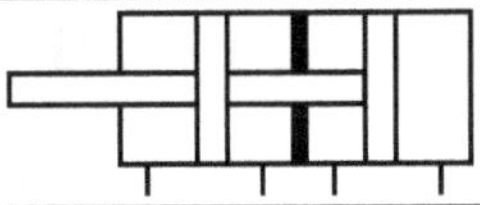
Bi-directional air motor	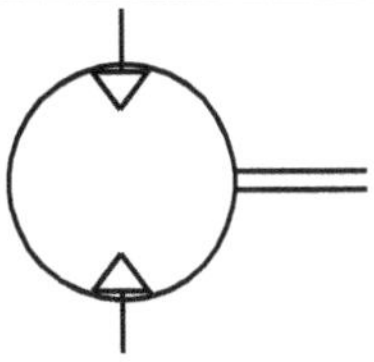
Uni-directional air motor	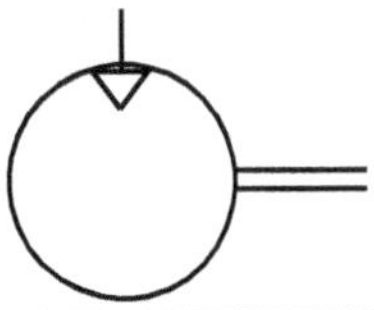
Oscillating air motor	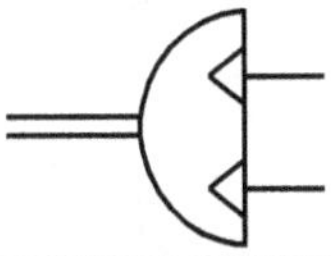
Shut-off valve	
Muffler (silencer)	[W-16] 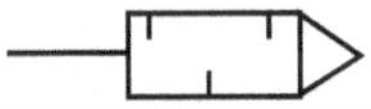
Working pipe line	
Drain line	
Flexible line (hose)	
Line junction	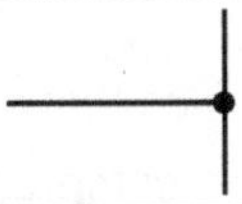
Pressure gauge	
Push button control (used with DC valves)	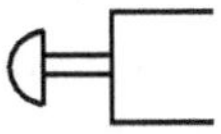
Pedal control	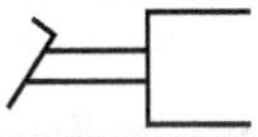
Roller operated control	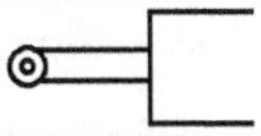
Plunger control	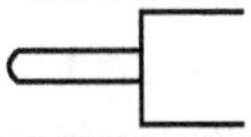
Spring control	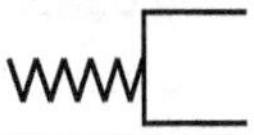
Air Signal control	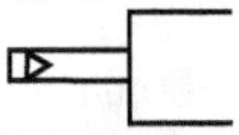
Solenoid control	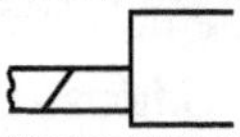

1.15 COMPARE HYDRAULIC AND PNEUMATIC SYSTEMS

Hydraulic System	Pneumatic System
(a) Energy carrying medium is oil.	(a) Energy carrying medium is air.
(b) Pump pressurises the fluid.	(b) Compressor compresses the air.
(c) Oil after pressurising needs to be used immediately. Storage of pressurised oil is not possible.	(c) Pressurised air can be stored in pressure vessel called air receiver.
(d) Can be used upto 700 bar pressures.	(d) Can be used upto 10 bar pressure.
(e) Return lines are essential to transport the used oil back to oil tank.	(e) Return lines are not required. Used air can be vented to atmosphere.
(f) These systems are having high costs.	(f) These systems are having much low costs.
(g) Motion accuracy is great. Hence used in CNC/VMC machine for tool move-ments.	(g) Due to uncontrolled expansion of air, motion accuracy is not good. Hence used in clamping like operations.
(h) Since hydraulic oil heats during the operations there is possibility of fire hazards.	(h) No fire hazards. Air is fire proof.
(i) Maintenance cost is high.	(i) Maintenance cost is low.
(j) Due to leakage of oil the overall system and space near by is not clean.	(j) Clean system.

Important Points

- Pneumatics is a branch that deals with power of air.
- In pneumatics air is compressed. The compressed air gains power (i.e. it will always try to expand). This power of compressed air is used to do mechanical work.
- Power of air has become popular in industrial automation.
- Air is available at free of cost.
- Air is fire proof and it is having easy transportability.
- Air can be stored easily in compressed state.
- Overall cost of compressed air and pneumatic equipments is lower.
- Some important applicants of pneumatic systems are:
  - (i) Air brakes in automobiles.
  - (ii) Fettling tools like pistol drill/angle grinder etc. are pneumatically operated.
  - (iii) Bore well drilling is pneumatic operation.
  - (iv) Automation of food industry.
  - (v) Paper handling in offset printing machine.
  - (vi) Hydro-pneumatic press machines.
  - (vii) Air guns.
  - (viii) Thermal spray coating.
  - (ix) Electronic assembly lines.
- Pneumatic system comprises of:
  - (i) Air compressor for compressing the air.

(ii) Air storage tank (Air receiver).
(iii) Moisture separator (Air dryers).
(iv) Filter regulator-lubricator unit (FRL).
(v) DC valves flow control valves.
(vi) Air actuators-for obtaining mechanical work.
(vii) Air conducting elements suction pipes etc.

- From 1920 oil hydraulics is in use by human being.
- Hydraulics – It means that science applied to energy transfer and control by using oil as a working medium.
- Practical applications of hydraulic systems are
  1. Pressure die casting.
  2. Hydraulic Shearing Machine.
  3. Hydraulic Press Break.
  4. Hydraulic Hack Saw
  5. Hydraulic Power Press.
  6. Broaching Machine.
  7. CNC/VMC/HMC Machines.
  8. Plastic Injection Moulding Machine.
  9. Hydraulic Excavator.
  10. Backhoe Loader.
  11. Hydraulic Cranes.
  12. Hydraulic Pullers.
  13. Material Handling Equipments.
- Oil Tank – Stores Oil.
- Filter – Filters the Oil.
- Pump – Heart of hydraulic system. It pressurizes the oil.
- Control valves – Controls the movements and pressure of oil in circuit.
- Actuators – the working elements of circuit.
- Pipe lines – carries the oil in the circuit.
- Up to 700 bar pressure can be created in the circuit by using hydraulic system.
- Weight to power ratio of hydraulic system is less.
- The limitation of hydraulic system are – In cold climate hydraulic oil freezes, the maintenance cost of hydraulic system is high, the cost of hydraulic system is high.
- Advantages of pneumatic systems are:
  (i) Air is working medium and which is freely and cheaply available.
  (ii) Air is dry and hence system is clean.
  (iii) No fire possibility (Fire proof system).
  (iv) Pneumatic equipments are easy in working, Low cost. Having very little maintenance cost.
  (v) No return lines. Hence circuit becomes less complicated.
  (vi) Piping network is also cheap because air can be transported through light weight pipes.
  (vii) Compressed air can be stored in air receiver under pressure. Hence when tank is full air can be used as an when required. Here we can save power.
- Limitations of pneumatic system are:
  (i) Pneumatic system can be used up to 10 bar pressure only. If we have to work on pressures higher than 10 bar then we have to go for hydraulic system.
  (ii) If pressure of air in air tank drops below certain limit the connected pneumatic equipments stops working.
  (iii) The accurate actuator movements are not possible to obtain.

Practice Questions

1. Define the term 'Hydraulics'.
2. Write any four applications of oil hydraulic systems.
3. Draw neat sketch of hydraulic power press and explain its working.
4. Draw flow chart of hydraulic system label the chart.
5. What is the function of: Oil Tank, (ii) Actuator, (iii) Pump, (iv) Filter?
6. Draw Actual Hydraulic System and explain its working.
7. Write any four merits of Oil Hydraulic System.
8. Write any four disadvantages of Oil Hydraulic System.
9. In cold climate why oil tank is equipped with oil heaters? Explain.
10. Why water cannot be used in hydraulic system as a medium?
11. State any four properties of hydraulic oil.
12. Write short note on available hydraulic fluids. State its classification.
13. Enlist any four applications of Hydraulic System.
14. Name any four basic elements used in Hydraulic circuit.
15. Define following term: (i) Viscosity, (ii) Lubricity.
16. Define Viscosity Index and Neutralization Number.
17. Enlist any four sources of heat generation in hydraulic system.
18. Draw symbol of: 2×2 DC Valve, (ii) Fixed type flow control valve, (iii) Pressure relief valve, (iv) Mufflur
19. Draw symbol of double check valve (OR Gate) and explain its working.
20. Draw symbol of: (i) Bi-directional motor, (ii) Uni-directional motor, (iii) Half rotation motor.
21. Draw symbol of: Pressure gauge, (ii) Air signal control, (iii) Water trap, (iv) Pilot operated check valve.
22. Draw the pneumatic symbol of: (i) Muffler (ii) 4×2 DC Valve.
23. Draw a symbol of 'Regulator'.
24. Draw detailed symbol of FRL unit and time delay valve.
25. What are the features of air? List any four features.
26. Draw the symbol for: (a) FRL unit (b) Check valve
27. Draw symbols: (a) Oil reservoir (b) Oil filter (c) Heat Exchanger (d) Unidirectional fixed displacement pump.
28. Draw symbols of: (a) Non-compensated flow control valve. (b) Non-compensated variable flow control valve. (c) Pressure compensated flow control valve. (d) Directly operated check valve.
29. Draw symbols of pilot operated pressure relief valve (Actuated position).
30. Draw symbol of: (i) Oil reservoir, (ii) Oil filter (iii) Heat exchanger, (iv) Unidirectional fixed displacement pump.
31. Draw the symbol of 'Single acting Cylinder' in Hydraulic circuit.
32. Draw the symbols of tandem cylinder.
33. Write any 4 properties of oils used in Hydraulic circuits.
34. What is ISO oil?
35. What is SAE oil?

■■■

Chapter **2**

PUMPS AND ACTUATORS

Weightage of Marks = 12, Teaching Hours = 08

Syllabus

2.1 Classification of Pumps

2.2 Construction and Working of Gear, Vane, Screw, Piston Pumps (Axial and Redial)

2.3 Performance Characteristics and Selection of Pumps

2.4 Classification of Actuators

2.5 Construction and Working of Linear and Rotary Actuators (Motors)

About This Chapter

At the end of this chapter students will be able to:

- Classify various types of pumps with justification
- Compare various types of pumps on the basis of given factors
- Select pump for the given application
- Classify various types of actuators with justification
- Sketch and explain construction of the given actuator(s)
- Select actuator for the given application with justification

2.1 PUMPS USED IN HYDRAULIC SYSTEM [W-10]

- Pump plays a specific and unique position in the system. It is a 'heart' of hydraulic system.
- The main function of pump is to take oil from tank and deliver it to rest of the hydraulic circuit. In doing so, it raises oil pressure to the required level.
- It is important to note that – fluid or oil is incompressible (cannot be compressed opposite to air in pneumatic systems). Then how pressure of oil increases ? The simple answer to this question is that, the load on fluid creates the pressure. If load is more, pressure is more. If load is less, pressure is less.

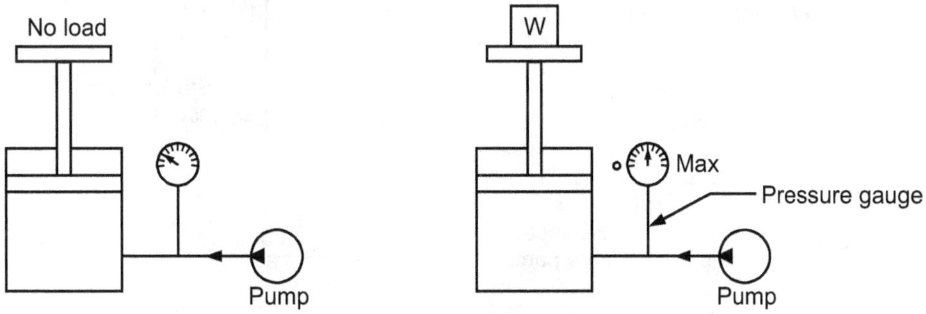

(a) Shows no Load on Fluid, hence Pressure is Very Low

(b) Shows Load 'W' Placed on, Then Pressure of Liquid Increases with same Pump

Fig. 2.1

2.2 SYMBOL OF PUMP [W-13]

1. Unidirectional Fixed displacement Hydraulic Pump [S-12, 15]

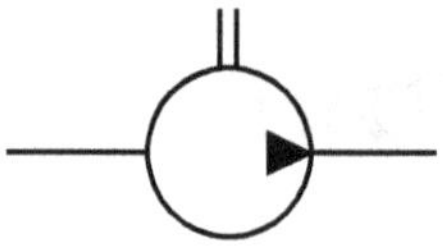

Fig. 2.2 (a)

2. Bi-directional Fixed displacement Hydraulic Pump

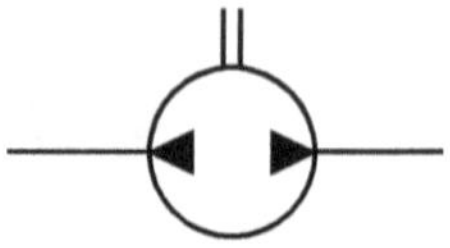

Fig. 2.2 (b)

3. Unidirectional and Bi-directional Variable Displacement Hydraulic Pump

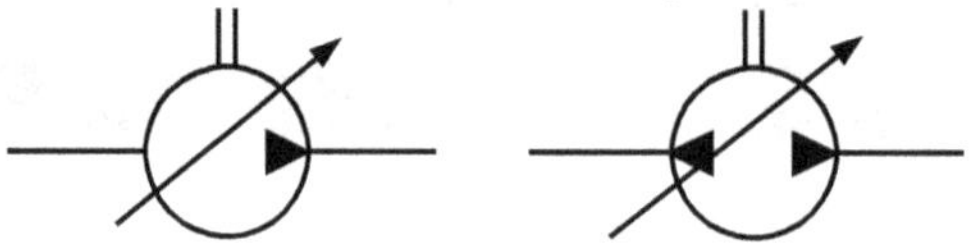

Fig. 2.2 (c)

2.3 CLASSIFICATION OF PUMPS [W-12, S-14]

- As we have seen in IVth semester subject 'Fluid Mechanics and Machinery' that pumps can be well classified into two major categories:

  (a) Positive Displacement Pumps.

  (b) Non-Positive Displacement Pumps (Rotodynamic).

- Pressure in major consideration in hydraulic system, hence majority pumps used in hydraulic systems are 'Positive Displacement Pumps'.
- The classification of positive displacement pumps is as under:

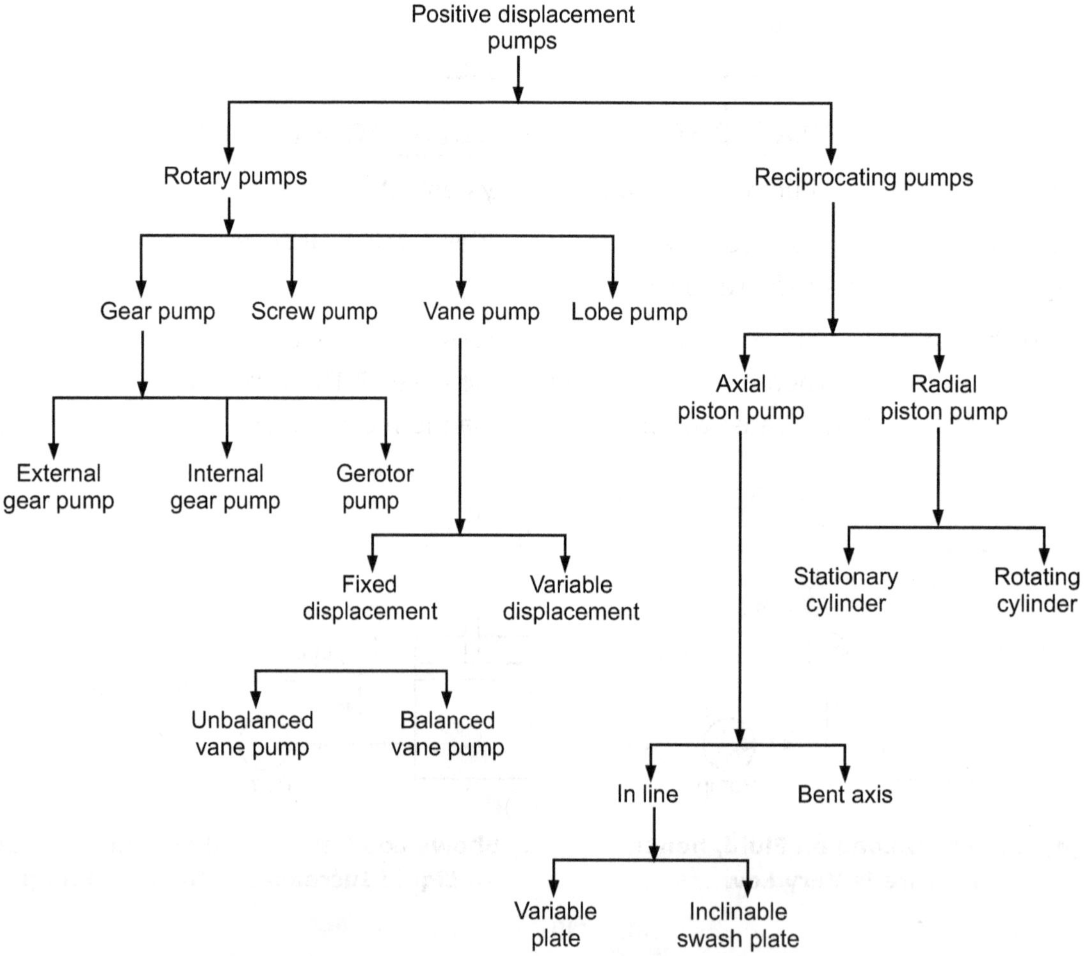

- The pressures that can be created by various positive displacement pumps and their efficiencies is given in the following table.

Pump Type	Pressure (Bar)	Efficiency
External Gear Pump	300	60%
Internal Gear Pump	350	85%
Fixed Vane Pump	175	75%
Variable Vane Pump	125	85%
Screw Pump	175	75%
Axial Piston Pumps	700	85%
Bent Axis Pumps	700	90%

Positive displacement pump	Rotodynamic pump
1. It delivers fluid in discrete volume/cycle.	1. The delivery is continuous.
2. After finishing one delivery stroke completely, only the next suction stroke can start.	2. Suction and delivery can keep going on continuously and simultaneously.
3. Discharge is independent of the resisting pressure at delivery.	3. As resistance increases, the discharge reduce.
4. Discharge depends only on speed.	4. Discharge depends on resisting pressure.
5. There is +(ve) sealing between the low pressure and high pressure regions (internal slippage).	5. The high and low pressure regions are not separate. They merge with each other.
6. Work done on the fluid is in the form of pressure energy.	6. Work done is in the form of kinetic energy.
7. Practically, there is no limit to the maximum pressure that can be built.	7. The maximum pressure that can be developed is limited.

- Non-positive displacement pumps are centrifugal pump, radial flow pump. These are used for very low pressure. In fact these pumps are very rearly used in Hydraulic System. These pumps cannot create pressure above 40 Bar (40 kg/cm^2).
- Now we will see important pumps one-by-one.

2.3.1 External Gear Pump

[W-10, 16, S-11, 16]

- **Construction:** It consists of two meshing gears, housing, two cover plates, bearings and shaft for gear mounting and a drive motor. The housing is having two ports (inlet and outlet). Gears are either spur or helical gears. If pressure angle is $14\frac{1}{2}^{\circ}$ then minimum number of teeth are 16.
- **Working:** Tooth space of one gear is filled by tooth of other gear. Motor drives one gear which in turn rotates other gear. As meshed gears start rotating, one tooth space after another is evacuated resulting in vacuum. Atmospheric pressure in the tank forces the oil into the tooth space. Now tooth space is full of oil. Due to rotation of gear, oil is carried around the periphery of the gear until the teeth again mesh and the oil is forced out through delivery port location on opposite side of suction port. This pump can create pressure upto 300 bar.

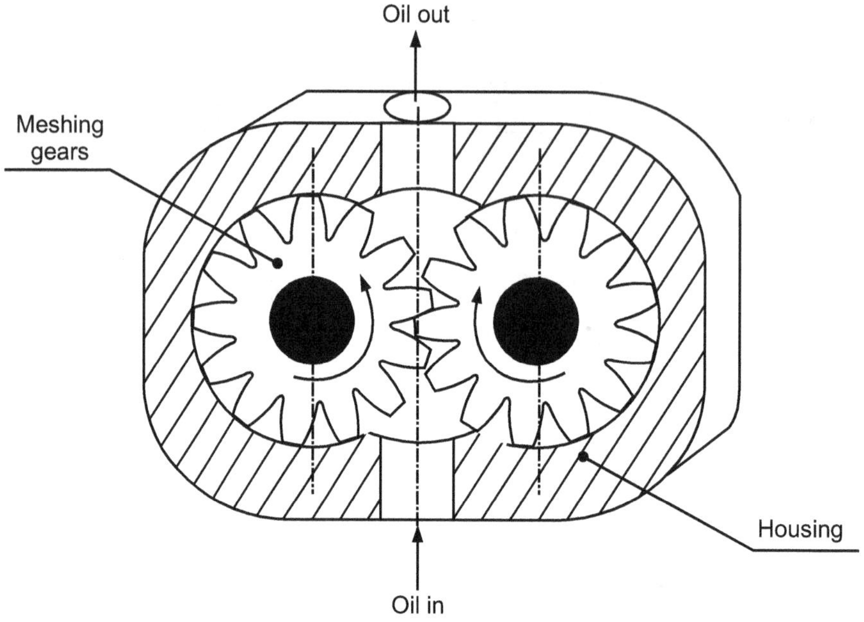

Fig. 2.3: External Gear Pump

2.3.2 Internal Gear Pump

[S-15, 16]

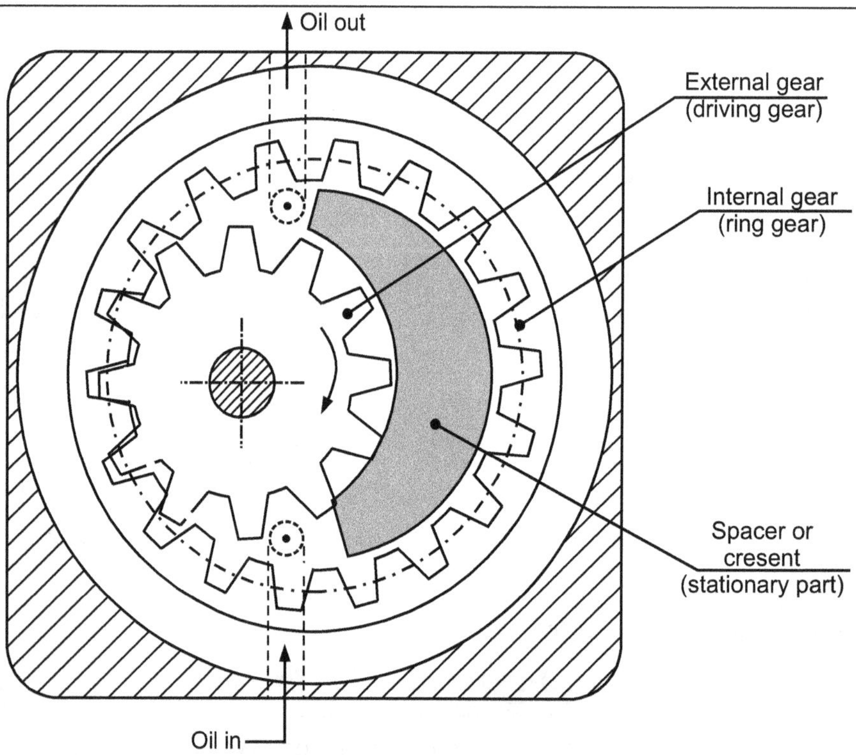

Fig. 2.4: Internal Gear Pump

- The basic principle of working is same as that of external gear.
- **Construction:** It consists of one external and one internal meshing gear pair. External gear is connected to electric motor and hence is driving gear. Internal gear or Ring gear is driven gear which rotates in same direction as that of external gear. Between two gears a spacer called 'cresent' is located which is a stationary piece connected to housing. Inlet and outlet ports are located in end plates.
- **Working:** External gear (driving gear) drives the Internal gear (Ring Gear). Portion where teeth starts meshing, a tight seal is created. Near inlet port the vacuum is created due to quick unmeshing and oil enters from oil tank through inlet port. Oil is trapped between the internal

and external gear teeth on both sides of cresent (spacer) and is then carried from inlet to outlet port. Meshing of gear near outlet port reduces the volume or gap and oil gets pressurised. These pumps can create pressure upto 350 bar and one other speciality of this pump is that it makes very less noise.

2.3.3 Gerotor Pump

[S-10, 11, 15, 16; W-12]

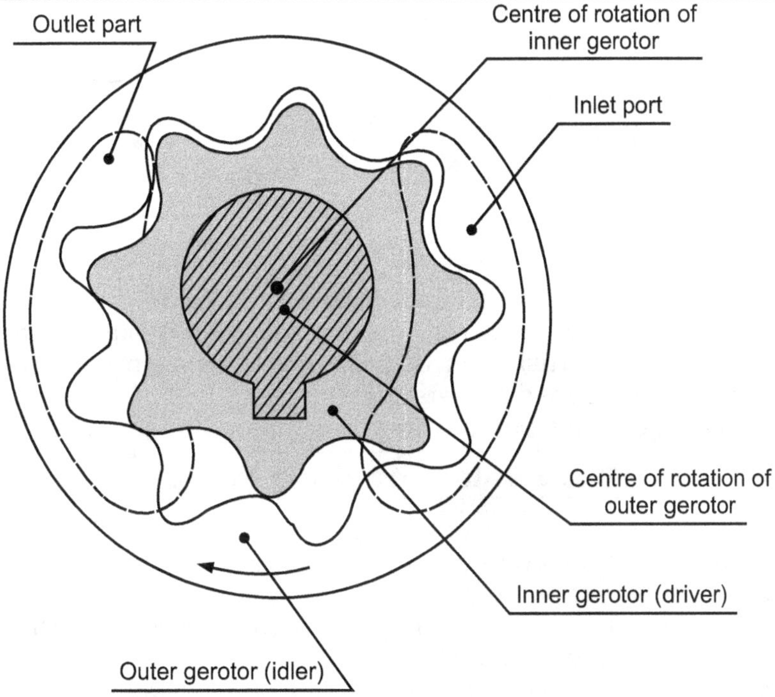

Fig. 2.5: Gerotor Pump

- **Construction:** It is similar to internal gear pump. But in this pump cresent is absent. It is dispenced with by using inner gerotor with one less tooth than outer gerotor. So also the centers around which both gerotors rotate are at some distance.
- **Working:** When inner gerotor rotates, the pockets go on increasing in size due to motion. These pockets create suction and oil comes in through inlet port. On outlet side these pocket area goes on decreasing and the trapped oil is forcefully discharge through the outlet port. So continuous motion causes continuous discharge. This pump can create pressure @120 Bar.

2.3.4 Unbalanced Vane Type Pump

- Vane pump operates on the principle of increasing and diminishing volume.
- **Construction:** The pump shown in Fig. 2.5 is unbalanced vane pump because its rotor is eccentric to housing. (There is distance between centre of rotor and centre of housing). Its main parts are:
  (a) Driving rotor with slots for vanes: The electric motor is coupled to the rotor.
  (b) Sliding vanes: Its steel flat piece shown in figure.
  (c) Cam ring to constrain the outward movement of vanes.
  (d) Housing with inlet and outlet ports.
- **Working:** The rotor is having number of vanes which are free to slide in or out in slots. When rotor starts rotating vanes flow out in slots due to centrifugal force. Now vanes touch the cam ring. As one can see number of oil trap chambers are formed between two consecutive vanes. Since rotor is offset with cam ring the area of these oil trap chambers diminishes and increases as rotor rotates.

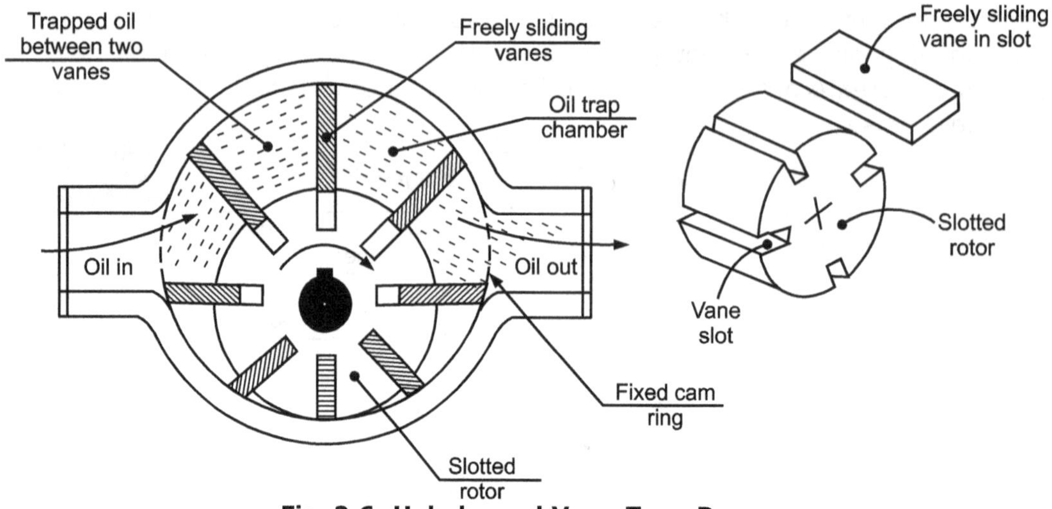

Fig. 2.6: Unbalanced Vane Type Pump

- In the beginning due to rotation of rotor, the air moves out quickly and partial vacuum is created near the inlet port. Oil rushes in and is trapped in each oil chamber. As the area of chamber reduces further the trapped oil gets pressurised.
- **Important:** The oil leakage in this pump is low as compared to gear pump.

2.3.5 Balanced Vane Pump

- A little consideration will show that in unbalanced vane pump, the pump is subjected to low pressure on inlet side and high pressure on outlet side. This pressure difference results in unbalanced pressure on vanes of pump and shaft.
- To avoid this problem unbalanced vane pump is modified to create balanced forces. Such a construction is called 'Balanced Vane Pump'.
- **Construction:** Similar to unbalanced vane pump it has following parts. The only difference is that instead of circular cam ring, an elliptical cam ring is used.
  (a) Driving rotor with slots for vanes,
  (b) Sliding vanes,
  (c) Elliptical cam ring,
  (d) Housing with crossed inlet and outlet ports (2 each).
- **Working:** In this type rotor and housing are concentric (having same centre, no eccentricity). When rotor rotates, vanes moves out due to centrifugal force and touch to cam ring. Oil trap chambers are formed between two consecutive vanes. In the beginning, air goes out and partial vacuum is created. The oil comes in through two inlet ports. Due to diminishing and increasing trap chamber volume, oil gets pressurised and goes out through two delivery ports. Due to pressure action created by two ports, the unbalanced force on vane and shaft is nullified.

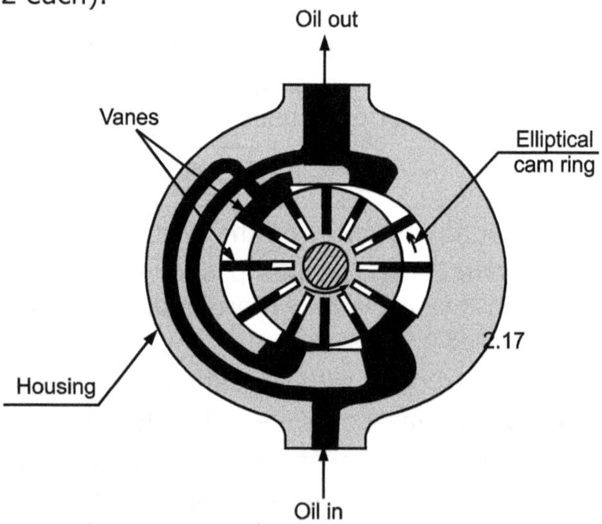

Fig. 2.7: Balanced Type Vane Pump

2.3.6 Variable Displacement Vane Pump

- In many hydraulic circuits, the flow rate of pump needs variable. This can be done by changing the speed (r.p.m.) of motor which drives the pump. But variable speed electric motor is a costly affair. An alternative method is to vary the displacement of pump. This is possible very easily with vane pump as depicted in Fig. 2.8 (a) above.

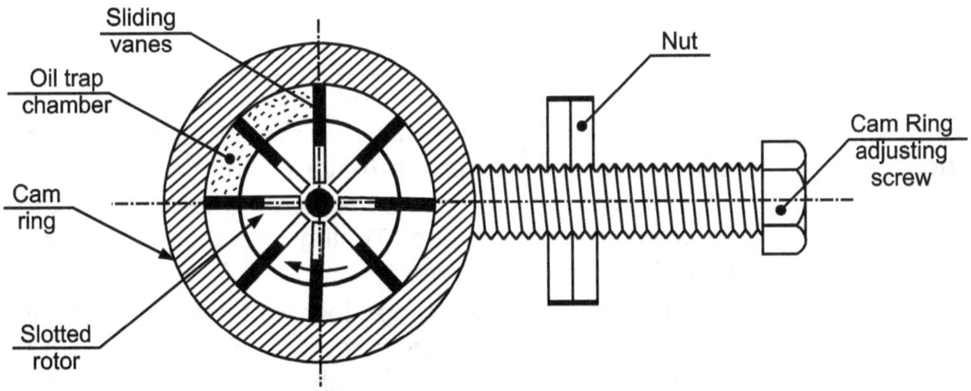

(a) Cam and Rotor Concentric

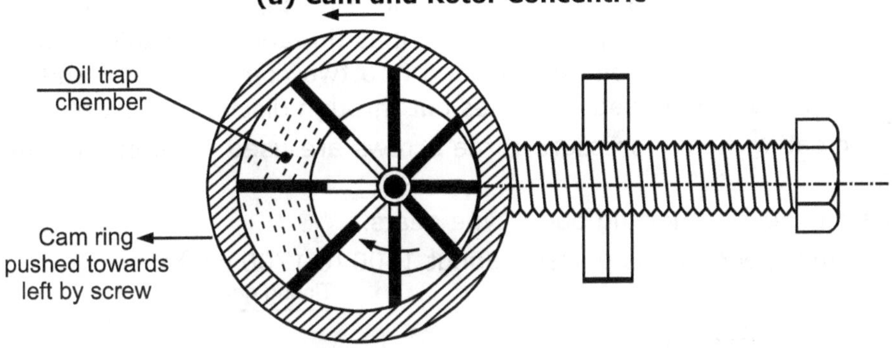

(b) Cam and Rotor Eccentric

Fig. 2.8

- If we change the geometric position of cam ring relative to slotted rotor, we can achieve delivery volume as per need of circuit. (See Fig. 2.8 (b))
- **Construction:** It is similar to unbalanced vane pump but the only difference is that we can adjust or shift the cam ring to left or right. Its main parts are:
  1. Driving motor with slots for vanes.
  2. Sliding vanes.
  3. Adjustable cam ring with adjusting screw.
  4. Housing with inlet and outlet ports.
- **Working:** Fig. 2.8 (a) shows cam ring and rotor are concentric. In this position, all oil trap chambers will be having same volume and oil will not be pressurised.
- Fig. 2.8 (b) shows, cam ring is shifted towards left by using screw and fixed nut pair. Now the pump will act exactly similar as unbalanced vane pump we have already seen. The volumes of 'trap oil chamber' will increase and diminish which will pressurise the oil.
- By adjusting screw we can increase or reduce the eccentricity between can ring and rotor. By doing so we can change or vary the flow output of pump.

Compare Gear Pump and Vane Pump used in Hydraulic Systems

Gear Pump	Vane Pump
(a) Types are: → External Gear Pump → Internal Gear Pump	(a) Types are: → Unbalanced Vane Pump → Balanced Vane Pump
(b) Gears are used for pressurising the oil.	(b) Sliding vanes are used in this pump and due to centrifugal force they fly away from centre and create space for pressuring the oil.
(c) Gear pumps are used when pressure requirement is between 350 to 300 bar.	(c) Vane pumps are adopted when pressure requirement is between 125 to 175 bar.
(d) Noisy operation	(d) Less noise
(e) Less expensive	(e) Cost is slightly higher

2.3.7 Screw Pump

- **Construction:** The special screws having 'right hand' and 'left hand' threads on same shank are used in this pump. Fig. 2.9 shows two such meshing screws. These are encased in closely machined housing. One screw is connected to prime move (motor). This is driving screw and since it meshes with other it is driven screw. Below the screw there is intake port; which is bottom of housing similar to crank case of engine. Timing gears on screw shaft provide the turning force between driving and the driven screws.
- **Working:** In this pump oil entering into the pump does not 'rotate' but moves 'linearly'. (similar to nut slides on rotating screw) as screw starts rotating, in the beginning air moves out and due to vacuum created oil draws in from suction chamber into pump housing. This oil moves on the helical grooves of screw. Due to close meshing of screw oil moves in tight gap (very small gap) between meshing threads as well as threads and inside surface of the housing. Due to special design of screw, oil gets divided into two compartments and advances towards the centre of pump and pressurised oil comes out through discharge port at centre.
- Sometimes for creating heavy pressure three screws are used. This pump can create pressure upto 400 bar.
- Timing gears are generally 'Herring bone' type gears.
- The usual operating speed of this pump is about 1500 to 1700 r.p.m.

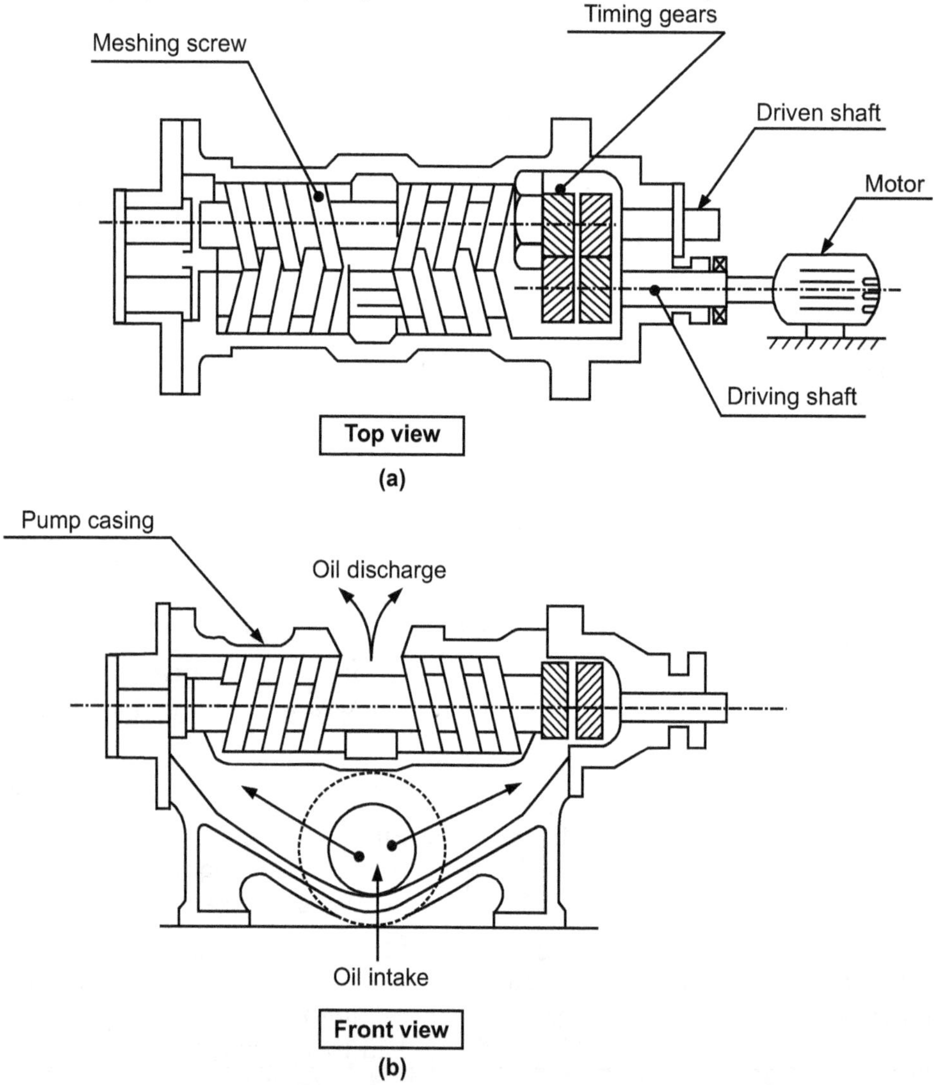

Fig. 2.9

2.3.8 Lobe Pump

- There are two lobes meshing with each other, driving lobe shaft is attached to motor. This pump is low pressure pump. When both lobes start rotating a vacuum created near inlet port draws the oil in. It is trapped between lobes and casing surface. Oil goes out with force due to high r.p.m. of lobes. This pump is used when more quantity of oil is required at low pressure.

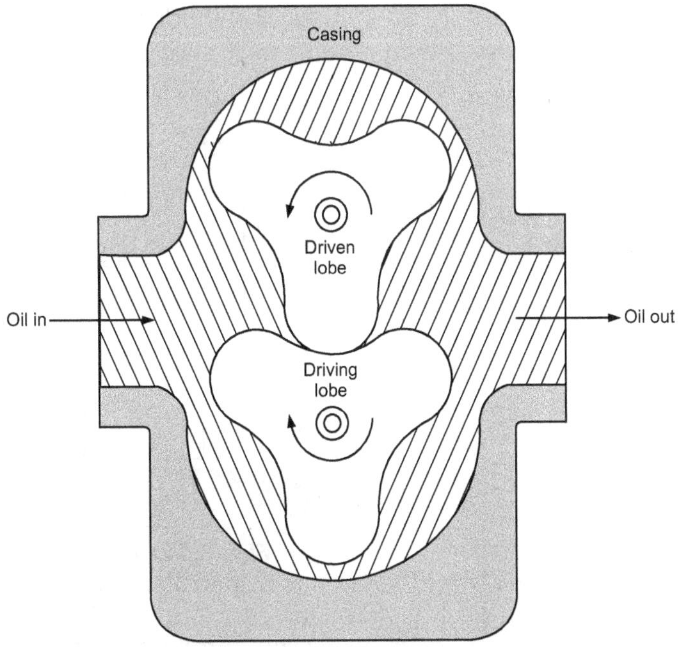

Fig. 2.10: Lobe Pump

2.3.9 Simple Piston Pump

[W-09, 11, S-14]

- This piston pump is exactly similar to reciprocating pump as well have studied in 4th semester.

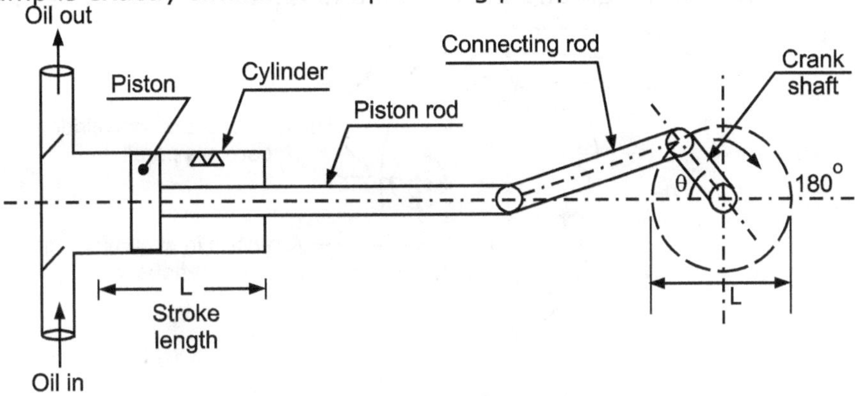

Fig. 2.11: Simple Piston Pump

- **Construction:** It consists of:
  - Finely finished cylinder barrel
  - Piston or plunger
  - 4 bar mechanism – piston rod, connecting rod, cranic shaft
  - Prime mover (generally electric motor) which drives crank shaft.
- **Working:** Motor drives crank shaft. Due to four bar mechanism, this rotary motion is converted into reciprocating motion of piston in the cylinder.
  - **When piston moves from Left to right:** Vacuum is created. Suction valve opens and oil comes in the cylinder cavity.
  - **When piston moves from Right to left (in second half):** Suction valve closes and delivery valve opens. The pressurised oil goes out.
  - As we know that in reciprocating pump, the piston travels distance equal to 'L' in half revolution of crank ($\theta = 0$ to $\theta = 180^{\circ}$). This is the stroke length. This distance is fixed and hence piston pump is fixed displacement pump.
  - We also know that in piston pump, piston moves in cylinder with acceleration and retardation. Hence delivery of oil is intermittent.
  - Piston pumps are adopted when pressure requirements are very high.

2.3.10 Multi Cylinder Piston Pump [S-11]

- In this type 2, 3, 4, 6 or 8 cylinders are either in line or in 'V' shape and their pistons are operated on the common crank shaft.

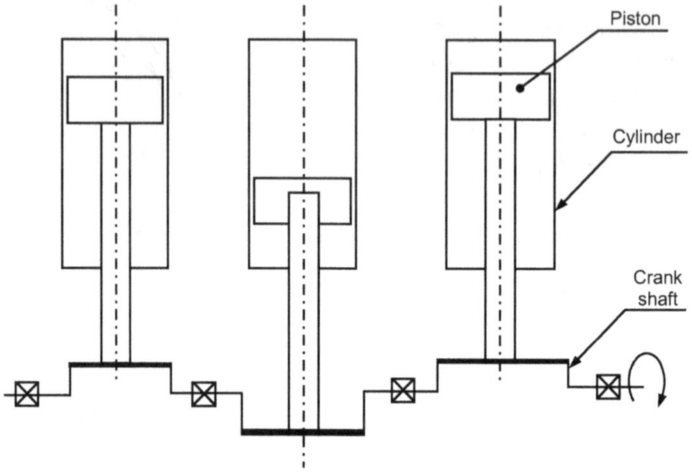

Fig. 2.12: Multi-cylinder Piston Pump

- The arrangement is similar to petrol or diesel engine block.
- These pumps can create pressure upto 450 bar.
- Generally two cylinder piston pumps are used in fork lifters, simple loaders etc.
- **Working:** Crank is rotating by motor. According to timer adjustment the piston reciprocates in definite order.
  - **Axial Piston Pumps:** In these type of pumps, pistons (or cylinder bores) are arranged axially parallel to each other around the circumferential periphery of the cylinder block.

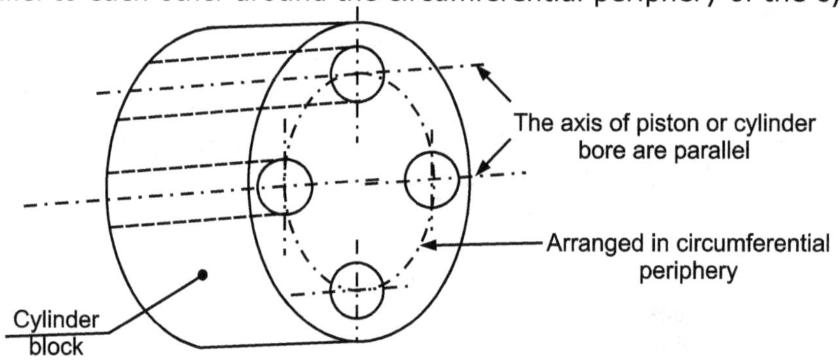

Fig. 2.13

Note: Number of cylinder bores are generally even numbers (6, 8 or 10)

- Axial piston pumps are subdivided into:
  (a) Swash plate axial piston pump.
  (b) Bent axis piston pump.
- The general Axial piston pump is depicted in the following Fig. 2.14:

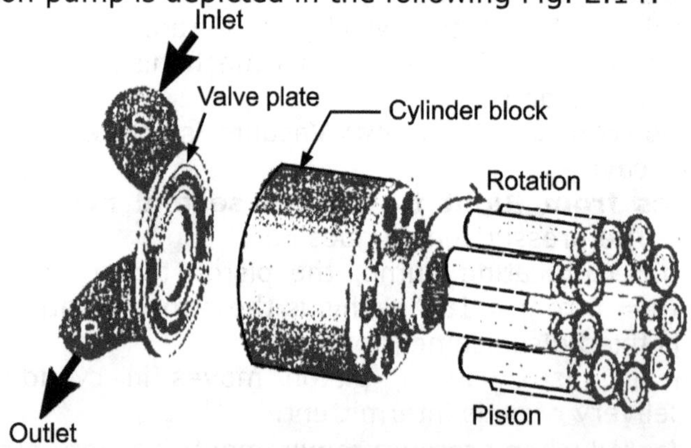

Fig. 2.14: Showing Valve Plate, Cylinder Block and Pistons

2.3.11 Swash Plate Axial Piston Pump

[W-10, S-11, 15]

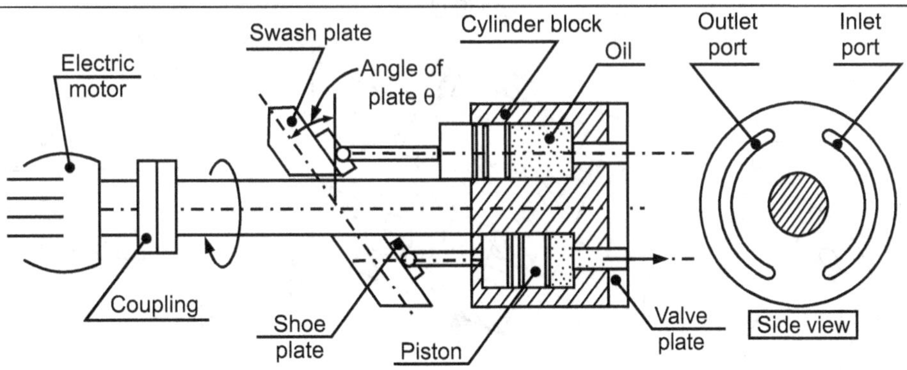

Fig. 2.15

- **Construction:** The pump consists of following parts:
  (a) Cylinder Barel having circumferencially arranged cylinder bores (6, 8 or 10).
  (b) Pistons (6, 8, or 10).
  (c) Swash plate with shoe plate for piston-rod connection. The angle of swash plate can be changed.
  (d) Prime mover (Electric motor).
- **Working:** Motor drives the shaft, which in turn rotates the entire cylinder block.
  - The pistons are connected to inclined swash plate through the piston rod.
  - Now one can easily understand that since swash plate is inclined and block is rotating, the pistons reciprocate inside the barrel.
  - The reciprocating motion of pistons causes suction and delivery of fluid through the inlet and outlet ports which come infront of outlet of piston.
  - We can change the angle of swash plate i.e. θ if
    (a) θ = 0 then no flow of oil, because pistons are at same level. When θ = 0 swash plate is vertical. No reciprocation of pistons, hence no flow (Refer Fig. 2.16 (a)).

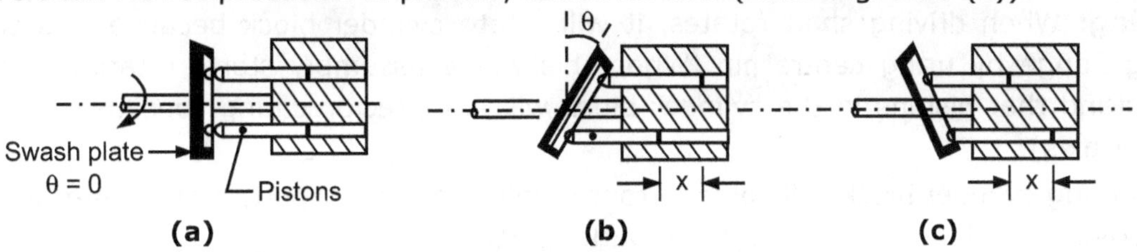

Fig. 2.16

    (b) θ = max or +ve, then x will be stroke length which is maximum and there will be maximum forward flow (Refer Fig. 2.16 (b)).
    (c) θ = – ve, then 'x' → i.e. stroke length will be maximum in reverse direction and hence there will be reverse flow. (Refer Fig. 2.16 (c)).
- By changing the swash plate angle we can vary the stroke length of the piston. It may be noted that maximum swash plate angle is limited to $17\frac{1}{2}^{\circ}$. By changing the angle of swash plate the pump output flow can be changed, hence we can use this pump as variable displacement pump also.

2.3.12 Bent Axis type Axial Piston Pump

- In swash plate axial pump, the axis of cylinder block is horizontal and angle is given to swash plate. Opposite to it is bent axis type axial piston pump, the cylinder block is at an angle and driving flange (similar to swash plate) is horizontal.

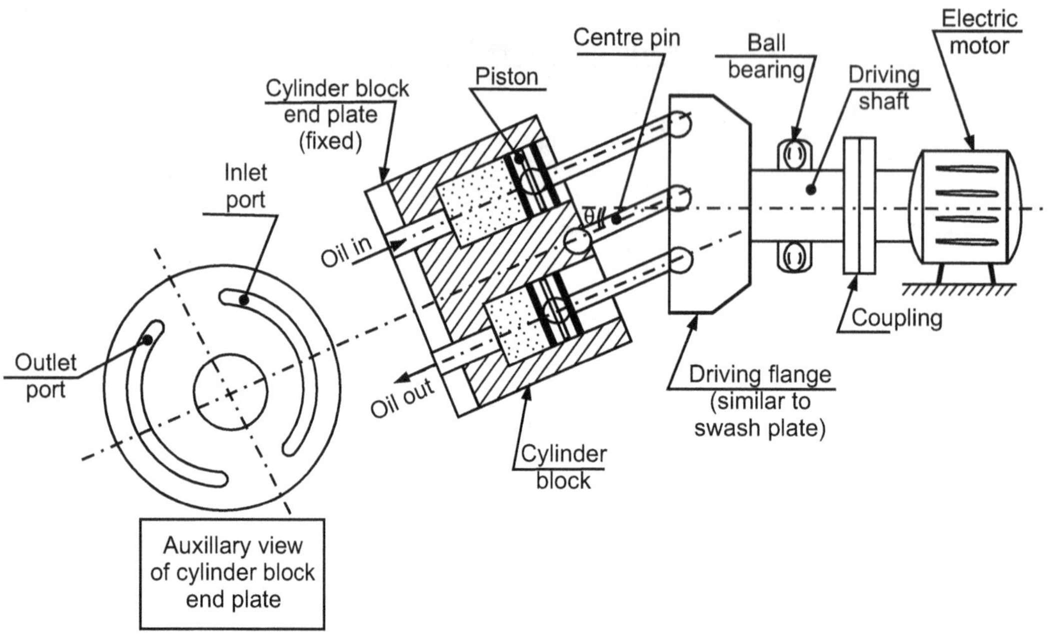

Fig. 2.17: Bent Axis Type Axial Piston Pump

- **Construction:** This pump consists of following parts/components:
  - (a) Cylinder block having circumferentially arranged cylinder bores (6, 8 or 10).
  - (b) Pistons (6, 8 or 10 Nos).
  - (c) Driving flange (similar to swash plate of axial piston pump).
  - (d) Cylinder block end plate in which inlet and outlet ports are arranged as shown in auxillary view.
  - (e) Prime mover (electric motor).
- In this pump axis of cylinder block is arranged in such a manner that it will form an angle θ with horizontal axis of driving staft and driving flange.
- **Working:** When driving shaft rotates, it will rotate cylinder block because it is connected to driving flange by using centre pin. When this whole assembly starts rotating one can easily understand that pistons in the cylinder block will start reciprocating (because of inclination of cylinder axis).
- The rotating cylinder block will come infront of inlet and outlet port of end plate (it is fixed) and pressurised oil will flow out through the outlet port.
  - The angle between driving shaft and cylinder block 'θ' can be fixed or we can change it.
  - The maximum value of θ is 30°.
  - Similar to swash plate axial pump of angle θ = 0 then axis of cylinder block and driving shaft coincide and there will be no flow.

2.3.13 Radial Piston Pump

- **Construction:** The radial piston pump consists of following components:
  - (a) Rotating cylinder block having number of cylinder bores radially created.
  - (b) Fixed or stationary cam which is eccentric with the cylinder block.
  - (c) Fixed or stationary housing in which suction oil and discharge oil moving paths are machined.
  - (d) Prime mover (electric motor).
  - (e) Each cylinder barrel is having hollow piston with spring for return as shown in Fig. 2.18.

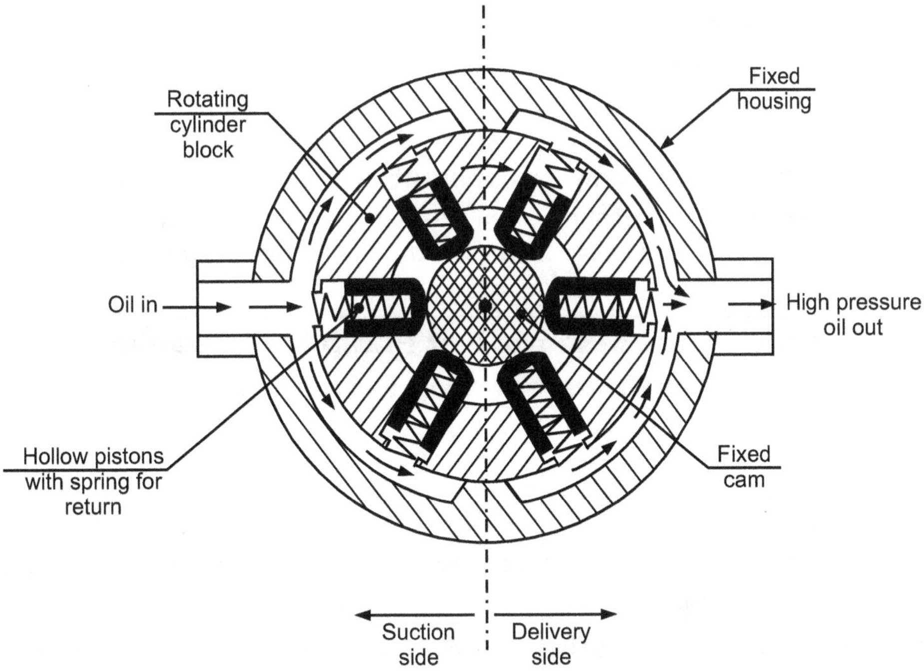

Fig. 2.18

- **Working:** When cylinder block starts rotating in clockwise direction, the eccentricity between fixed cam and rotating cylinder blocks causes the piston to reciprocate. Piston on suction side (left portion in figure) moves towards the centre of cam and sucks the oil in the cylinder. Pistons on delivery side (right portion in Fig. 2.18) moves away from centre of the cam thereby pressurizing the oil which goes out through outlet port.
  - When eccentricity between cam and cylinder block is fixed then flow is fixed or the pump is fixed displacement pump or constant delivery pump.
  - If we change the eccentricity by shifting the cam position then the flow of pump will change.
  - We can obtain pressures upto 500 bar with this pump.

2.3.14 Characteristic Curves for few Hydraulic Pumps

- Characteristic curve is a graphical representation of behaviour of pump under certain conditions on which pump operates.
- The graphs are obtained by testing the pumps in ideal condition as well as under actual conditions.

(a) Characteristic Curve of Vane Type Pump at Constant Speed:

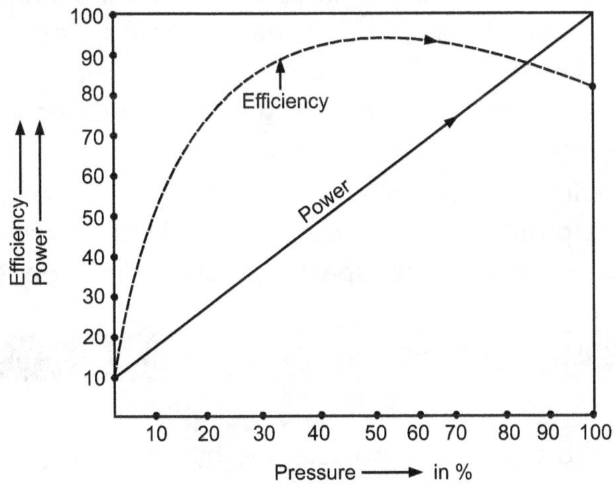

Fig. 2.19

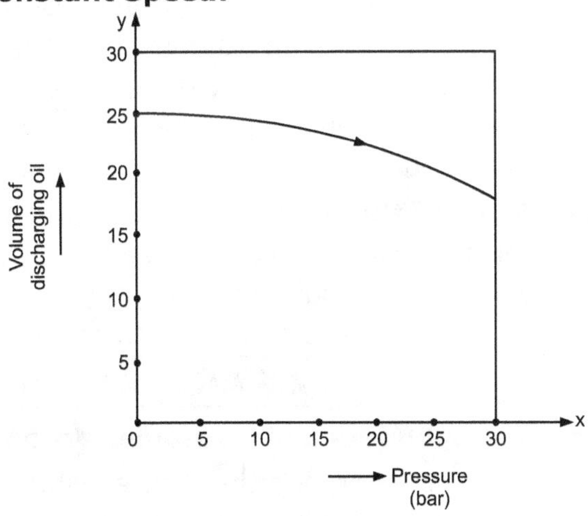

Fig. 2.20

- It is clear from above Fig. 2.19 that:
  - Power increases at higher pressure.
  - Efficiency increases at medium pressure but drops to some extent if pressure further increases.

(b) Characteristic Curve of Gear Pump at Constant Speed: (Refer Fig. 2.20).

- In this curve we have plotted pressure on x-axis and volume of discharging oil of y-axis. It is clear from the graph that when pressure is higher the discharge volume reduces to some extent.

2.3.15 Criteria for Selection of Pump in Hydraulic Circuit (or Basis of Pump Selection)

[S-10, 12, W-11]

- In hydraulic circuit we use actuators to do useful mechanical work. For every different application the actuators are different and so also load on the actuator is different. Hence selection of pump (it is heart of hydraulic system) is of high importance. Following are the criterias for selection of pump.
  - **(a) Pressure:** Pump is pressurizing the hydraulic oil to the level required by actuator. When pressures upto 300 bar are required, we can select gear pumps. If pressure requirement is still low then we can choose vane pump. But if pressure is very high (more than 500 bar) then we have to go for piston pump.
  - **(b) Flow of pressurised oil:** It is the volumetric output of pump measured in litres per min (LPM). The flow of oil is deciding the speed of actuator. Hence this criteria is also of importance.
  - **(c) Speed of pump:** If we increase the speed of the pump beyond its rated speed, we have to face the problems like cavitation or separation (as we have studied in reciprocating pump). These problems hamper the efficiency and overall the working of the hydraulic system. Hence proper speed pumps need to be selected.
  - **(d) Efficiency of the pump:** We should select the pump in such a manner that it should not work with low efficiency. Generally, three efficiencies are observed:

    (i) Volumetric efficiency $(\eta_v) - \frac{\text{Actual discharge volume}}{\text{Theoretical discharge volume}}$

    (ii) Mechanical efficiency $(\eta_m) - \frac{\text{Power output}}{\text{Power input}}$

    (iii) Overall efficiency $(\eta_o) - \eta_v \times \eta_m$
  - **(e) Oil compatibility:** The meaning of the word compatibility is 'acceptance'. If we use diesel in petrol engine as a fuel then the engine will not start. That means Petrol Engine is compatible only to petrol and not to diesel. On similar lines, each pump is compatible to specific hydraulic oil. If we use different oil then the pump will not give its rated and good performance. So oil compatibility is the most important criteria.
  - **(f) Price of the pump:** In modern age many varieties of pumps are available. Circuit designer must consider its cost. If we select high cost pump then the total cost of circuit increases and in working tenure of pump, the cost of the pump spares is also high. So this economical aspect must be carefully looked into.

2.4 ACTUATORS (LINEAR AND ROTARY) [W-12, S-14, 16]

(Hydraulic Cylinders and Hydraulic Motors)

- Hydraulic actuators are the devices which convert the power of pressurised hydraulic oil (pump does this task of pressurising the oil) into useful mechanical work. The useful work is obtained through force and motion of actuator.

- Hydraulic actuators are broadly classified into two categories:
  (i) Hydraulic cylinders also known as linear actuator (or linear motors).
  (ii) Hydraulic motors also known as rotary actuators.
  (iii)Limited rotary actuators.
- **Hydraulic Cylinders:** It is a device which converts fluid power (power of pressurised hydraulic oil) into linear mechanical force and motion. This motion is generally is linear direction and we can use this motion with force for carrying out operations like pushing, pulling etc.
- **Hydraulic Motor:** It is a device which converts fluid power into rotary mechanical energy. The rotations created by this energy can be used for winding/rewinding operations and many tasks in almost all engineering fields.
- **Limited Rotary Actuators:** These create half or $\frac{3}{4}$ rotations. Now we will see construction and working of all types of actuators.

2.4.1 Hydraulic Cylinders (Linear Actuators)

- Following are the types of hydraulic cylinders:
  (i) Single Acting Single Piston Rod Cylinder (SA Cylinder).
  (ii) Single Acting RAM.
  (iii) Double Acting Single Piston Rod Cylinder (DA Cylinder).
  (iv) Double Acting Double Piston Rod Cylinder.
  (v) Tandem Cylinder.
  (vi) Telescopic Cylinder.
- Above classification can also be plotted as under:

Linear Actuators (Hydraulic Cylinders)

(i) Single Acting Cylinder (ii) Single Acting RAM	(iii) Double Acting Cylinder (iv) Double Acting Cylinder with double piston rod	(v) Tandem Cylinder (vi) Telescopic Cylinder

- Now we will see construction and working of above linear actuators.

2.4.2 Single Acting Cylinder with Single Piston Rod

[W-13]

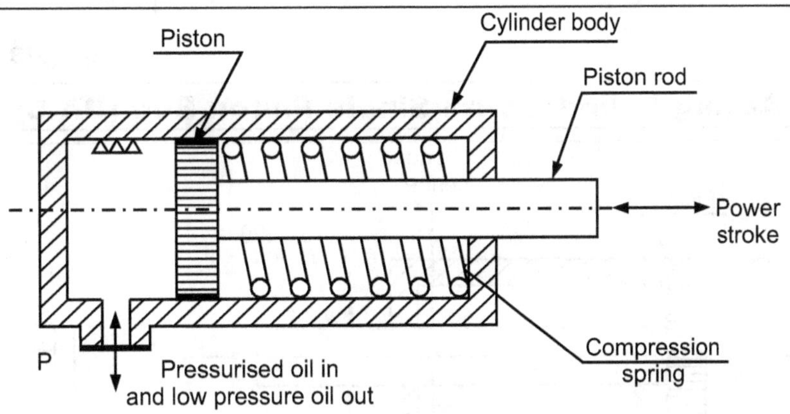

Fig. 2.21

- **Construction:** The single acting cylinder is shown in above Fig. 2.21. It consists of cylinder body in which cylinder bore is created. The inside surface is machined with close tolerances and is having high finish. Piston reciprocates in the bore. On backside of piston, compression spring is used around the piston rod.

- **Working:** Pressurized oil enters into the cylinder through port P. The pressure force (F = p × a) will push the piston towards right side thereby compressing the spring. This is the power stroke. Then the flow of oil stops. Then naturally compressed spring will expand and push the piston from right to left. During this movement of piston used oil is pushed out through port 'P' and the oil then returns to oil tank via filter.
- Now again the pressurised oil enters into cylinder the through port P and the cycle repeats.
  - During power stroke, useful work is done at the end of piston rod, viz. pushing the load.
  - This cylinder is having applications in sequencing circuit (Refer Chapter 3) or in clamping the work piece.

Symbol of SA cylinder:

- The symbol is similar to its construction.

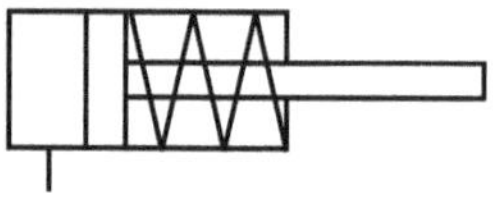

2.4.3 Single Acting RAM

[W-13]

- **Construction:** It consists of Big Plunger or RAM. It does not have separate piston and piston rod. RAM is a single body and its piston end is out of cylinder block. The load is kept on the piston end only. Piston itself is big and having uniform diameter to avoid buckling.
- **Working:** Pressurized oil comes into the cylinder bore through port 'A'. During this, port 'B' is closed. The pressurised oil starts occupying the space in the cylinder and due to pressure force, RAM starts moving upwards, thereby raising the load.

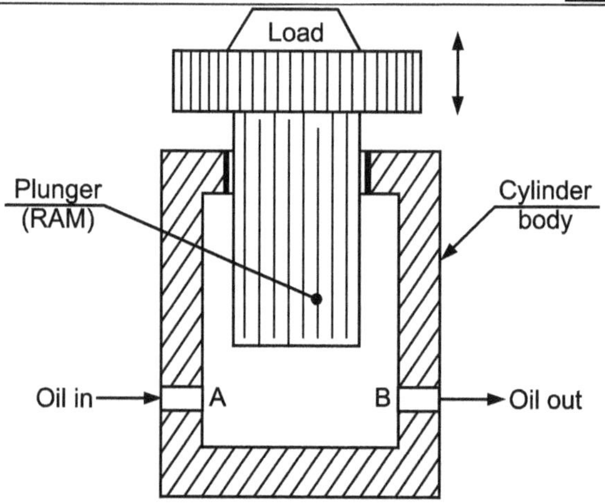

Fig. 2.22

- Now when RAM attains required height, the oil flow through port 'A' stops and this port is closed. Port 'B' opens, used oil goes out. The RAM comes down as oil goes out.
- This type of hydraulic RAM is used on shipyards to raise and lower the loads while loading it on cargo ships. This RAM is also used in material handling equipments used in stacking.

Symbol of Hydraulic RAM type SA cylinder

- This symbol also resembles with its construction.

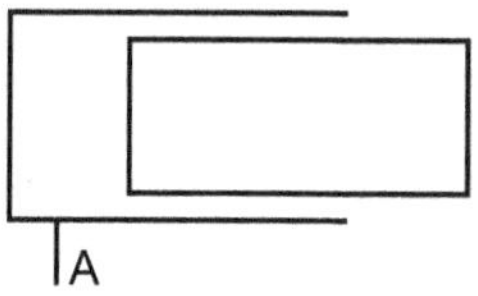

Fig. 2.23

2.4.4 Double Acting Cylinder with Single Piston Rod (DA Cylinder)

[S-10]

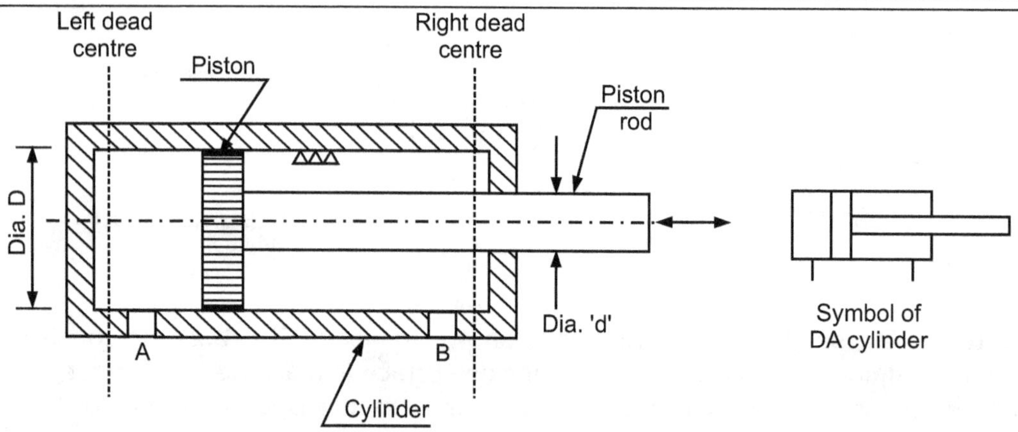

Fig. 2.24

- **Construction:** It consists of a cylinder body in which piston with single piston rod can reciprocate. Similar to single acting cylinder, the cylinder bore is highly finished surface having close tolerances. There are two ports 'A' and 'B' as shown.
- **Working:** The pressurised oil enters through port 'A'. Due to pressure force created by oil, piston will move towards right. When piston reaches 'right dead centre', flow through port 'A' stops and now pressurised oil starts entering into piston rod side of cylinder through port 'B'. Due to pressure force created by oil, piston will start moving from right and left. During this movement used oil on other side of piston will go out through port 'A' and will return to the oil reservoir. When piston reaches 'left dead centre', flow through port 'B' stops and flow through port 'A' starts. And cycle repeats which results into reciprocating motion of piston.
  - In DA cylinder spring is not required as in case we used in SA cylinder, because reciprocating motion is done by pressurised oil coming in through 2 ports.
  - In DA cylinder both strokes (right to left and left to right) are power strokes and we can use this power to do useful mechanical work.

2.4.5 Piston Force

- Consider DA cylinder having:
  - Piston Diameter = D
  - Piston Rod Diameter = d
  - Pressure of compressed air = P
  - Stroke length = L
- In a DA cylinder, when oil is admitted through piston side then piston will move from left to right. This stroke is called advance stroke and piston is said to be 'Advancing'.

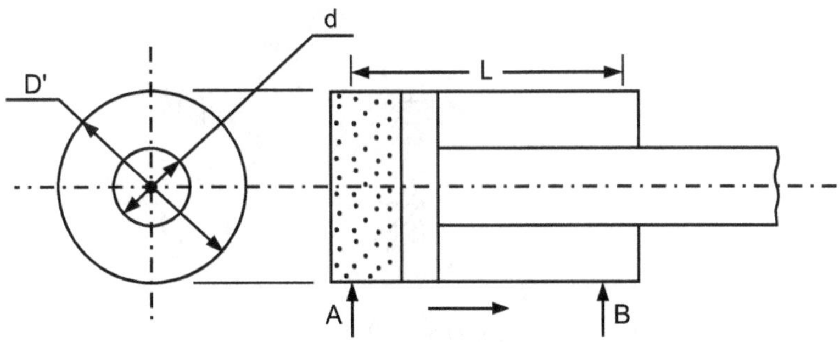

Fig. 2.25

- In a DA cylinder, when oil is admitted into DA cylinder from piston rod side then piston will move from right to left. Then this stroke is called retract stroke and piston is said to be 'retracting'.
- Now we will find out the force with which piston will advance (F_A) and force with which piston will retract (F_R).
- We know that:

$$\text{Pressure} = \frac{\text{Force}}{\text{Area}}$$

$$\therefore \quad \text{Force} = \text{Pressure} \times \text{Area}$$

- So from Fig. 2.75 given above force on piston is advance stroke which will be.

$$F_A = P \times \text{Area on which pressure is being applied}$$

$$\therefore \quad \boxed{F_A = P \times \frac{\pi}{4} D^2} \quad \ldots \text{D = diameter of piston}$$

- Force on piston is retract stroke which will be

$$F_R = P \times \text{Area on which pressure is being applied}$$

$$F_R = P \times \left(\frac{\pi}{4} D^2 - \frac{\pi}{4} d^2\right) \ldots \text{Deducting cross-sectional area of piston rod}$$

$$\therefore \quad \boxed{F_R = \frac{P\pi}{4}(D^2 - d^2)}$$

- In case of small pistons, the diameter of piston rod can be neglected.

2.4.6 Piston Speed

- In SA or DA cylinder (or any linear actuator), the piston reciprocates. That means it is having linear motion (motion in straight line) and hence it is having linear speed.

  Let, V = Speed of piston (m/sec)

  L = Stroke length in metre

  t = Time required by piston to cover length L in seconds

  Then $$V = \frac{L}{t} \text{ m/sec.}$$

- The piston speed in hydraulic system using SA or DA cylinder depends on:
  (a) Pressure of hydraulic oil created by the pump.
  (b) The stroke length of the actuator.
  (c) Flow rate of hydraulic oil (lit./sec.)

2.4.7 Actual Construction of Double Acting Cylinder

[S-11]

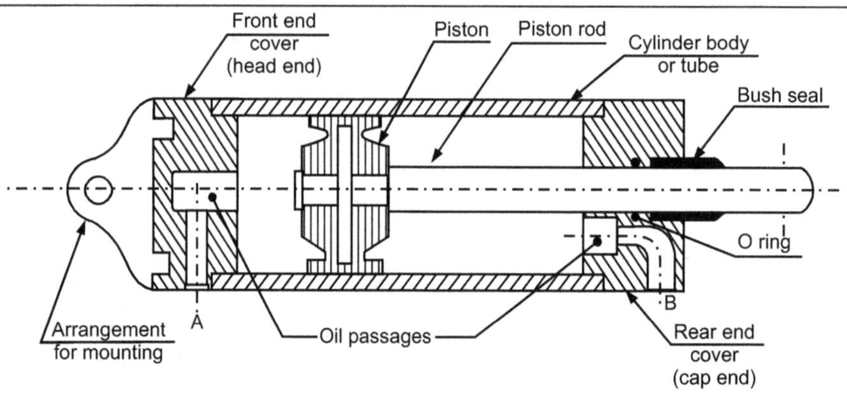

Fig. 2.26

- **Construction:** The actual double acting hydraulic cylinder (DA cylinder) is shown in the Fig. 2.76. Carefully observe the constructional details. It consists of:

  (a) Cylinder Body or Tube: It is a metal tube (small pressure vessel) in which piston reciprocates. Following materials are used for the tube. This tube is generally seamless (joint less) tube.
  (i) Aluminium alloys
  (ii) Cold drawn low carbon steel
  (iii) Stainless steel
  (iv) Cast iron
  (v) Titanium alloys.

  (b) Piston: Piston takes the pressure of oil coming into the cylinder. Hence it must be strong enough. It is made-up of low carbon steel.

  (c) Piston rod: The piston rod is fitted to piston and it also undergoes stresses created by pressure force of hydraulic oil. It is generally made-up of stainless steel bar material.

  (d) Front end cover (Head end): It is fitted in tube by press fit from front side. This cover inhouses oil passage (port A) as shown.

  (e) Rear end cover (Cap end): It is similar to head end. It is also fitted by press fit in tube and in houses oil passages (Port B) as shown. End covers are made-up of cast steel or S.G. iron.

  (f) Bush seal: Bush seal is fitted on cap end as shown to avoid leakage of oil. It is press fitted. It is made-up of brass, bronze, gun metal etc.

  (g) 'O' ring: It is also sealing for avoiding leakage. It is a rubber ring.

2.4.8 Designation of Hydraulic Cylinder

- Instead of writing all the dimensions of hydraulic cylinder on circuit drawing, one method of designation is adopted world wide.
- The designation is made in digits as under:

X/Y – Z

X ⇒ Diameter of piston in mm
Y ⇒ Diameter of piston rod in mm
Z ⇒ Stroke length in mm

For example, $\frac{70}{30}$ – 320 means of hydraulic cylinder having

Piston diameter ⇒ 70 mm
Piston rod diameter ⇒ 30 mm
Stroke length ⇒ 320 mm

2.5 DOUBLE ACTING CYLINDER WITH DOUBLE PISTON ROD [S-11]

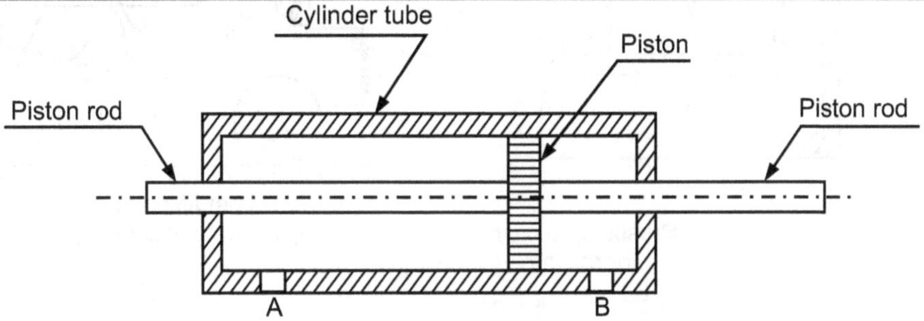

Fig. 2.27

- **Construction:** As stated in its name, this hydraulic cylinder is having 2 piston rods of same diameter attached to the piston. The piston is reciprocating in cylinder tube bore having high finish.
- **Working:** When pressurised oil enters through port A, piston will move towards right. When oil enters through port B, piston will move towards left and used oil on left-side of piston earlier stroke will return to oil tank through port A. The cycle repeats in each stroke.
- In this type of DA cylinder, we can obtain useful work at both ends of the cylinder.
- This DA cylinder is used in controlling table movement in cylindrical grinding machine and surface grinding machine.
- **Symbol:** Symbol is similar to the construction.

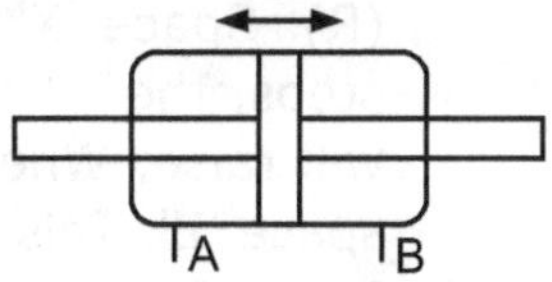

Fig. 2.28

2.6 TANDEM CYLINDER [W-13, S-15]

- This is a special DA cylinder.

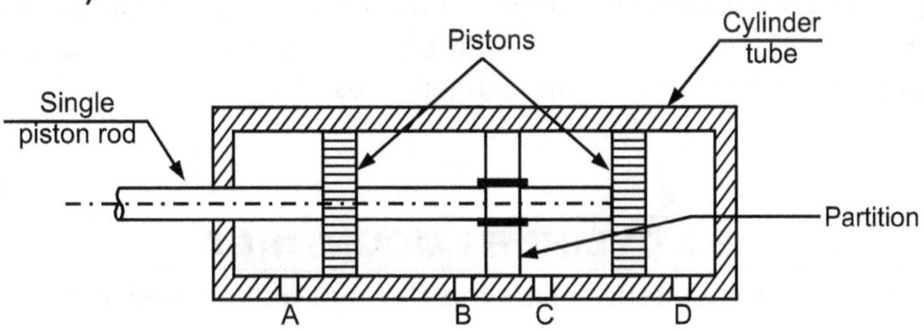

Fig. 2.29

- **Construction:** As shown in the Fig. 2.79 two cylinders are joined in 'series'. There is a sealed partition between two cylinders. Two pistons are fitted on one single piston rod. This arrangement is called 'tandem array'. There are four ports A, B, C and D as shown.

- **Working:** Pressurized oil is admitted at a time through ports (B) and (D). Piston will move towards left with **double force**.
- Similarly, pressurised oil is admitted at a time through ports (A) and (C), piston will move towards right with **double Force**.
- The main advantage of this DA cylinder is that we can obtain double force with same size of cylinder.

  Symbol: Its symbol is similar to its construction.

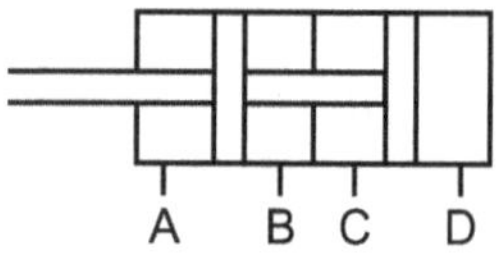

Fig. 2.30

2.7 TELESCOPE CYLINDER [W-11, S-10, 11, 12, 15]

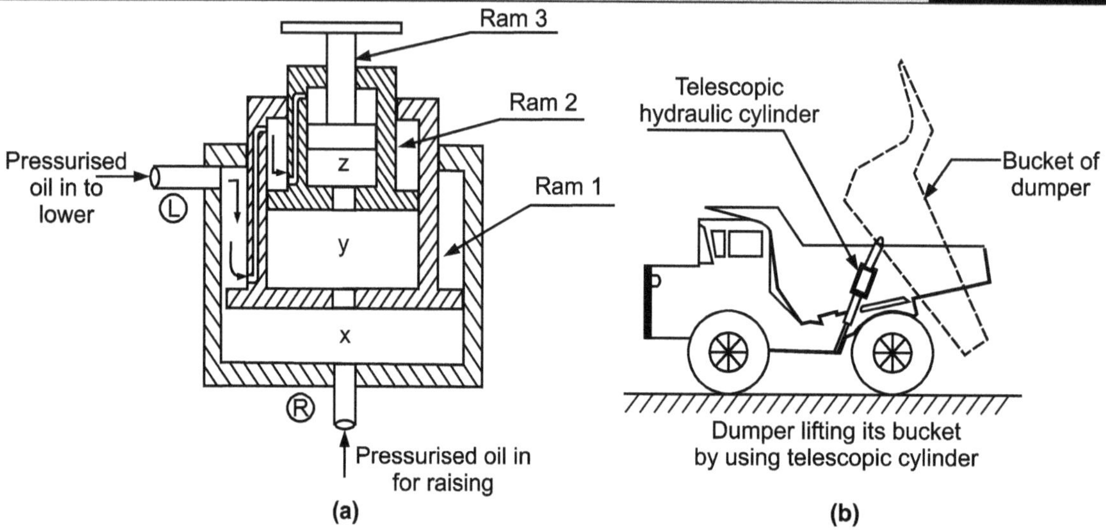

Fig. 2.31

- **Construction:** Fig. 2.31 shows three Rams assembled in each other like telescope (or Radio Antenna). This arrangement provides relatively long stroke with good mechanical strength. There are two inlet ports through which pressurised hydraulic oil enters. Port (R) is for raising the cylinder or extending the cylinders while port (L) is for cylinder lowering.
- **Working:**
  - **(i) Raising or Extending the cylinders:** Hydraulic oil under pressure will enter through port (R). Space 'X' will be filled by oil and Ram 1 will start raising upwards. When its raising stops, the oil now will start entering through and will occupy space 'Y'. Due to this Ram 2 will raise. When raising of Ram 2 stops, the oil will start entering through and will occupy space 'Z'. This will raise final Ram 3 upwards.
  - **(ii) Lowering the rams:** When pressurised oil will enter through port (L), then Ram 1 will come down. After its lowering Ram 2 will lower and then Ram 3 will lower.
- As shown in adjacent Fig. 2.32 the dumper which is having movable bucket carries heavy material like stone metal, stone boulders etc. The unloading of bucket is operated on telescopic cylinder.

  Symbol: Symbol is typical one and needs to be remembered.

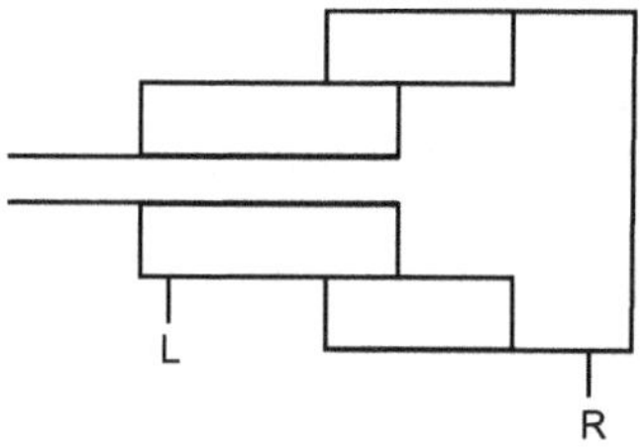

Fig. 2.32

2.8 CYLINDER MOUNTINGS [S-10]

- Word 'mounting' here means suitably installing the hydraulic cylinder in a hydraulic circuit for smooth running of circuit.
- There are several methods of cylinder mounting required for various applications.
- Mounting is an important parameter both from design and maintenance point of view.
- If we choose wrong mounting the entire hydraulic circuit starts giving troubles.
- Now we will see some important cylinder mountings.

2.8.1 Footing Mounting

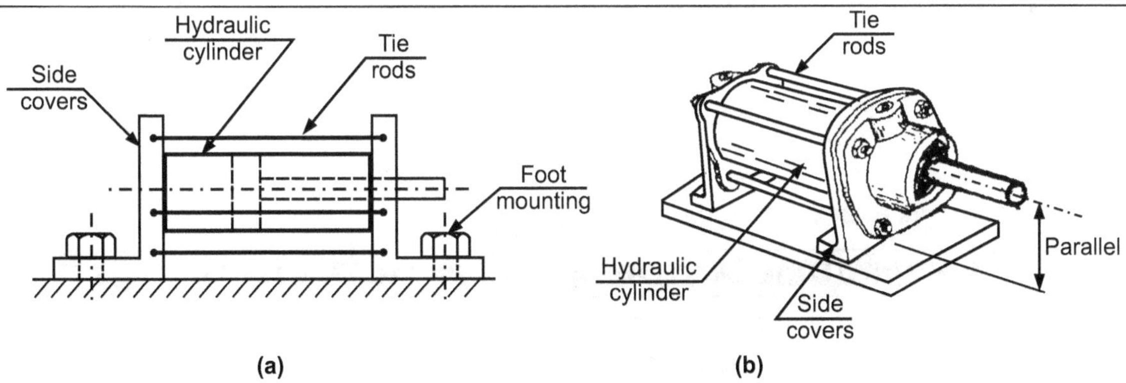

Fig. 2.33

- This mounting consists of flange type side cover to the cylinders. These covers can be fitted to parallel surface by means of nuts and bolts. This mounting is adopted when cylinders are to be mounted on the surface parallel to the axis of cylinder.
  Important: This mount tend to rotate or bend the cylinder about its mounting bolts.

2.8.2 Centreline Mounting

- It is the best method of mounting the cylinder. In this method, cylinder is mounted alongwith its centreline.
- In this mounting a special flange bolted at centre is used to mount the cylinder. When we use this mounting the bolts are under simple shear and tensile stress.

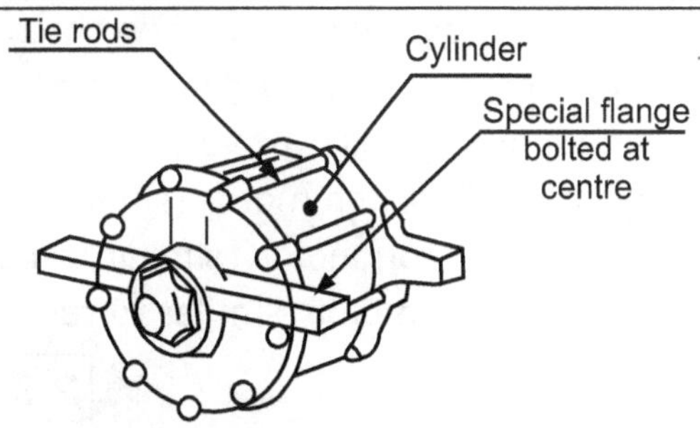

Fig. 2.34: Centreline Mounting

2.8.3 Pivot Mounting [W-10]

- In many industrial hydraulic circuits, a cylinder may have rotational freedom while it is reciprocating. To achieve this we can use two types of pivot mountings:
  (i) Clevis mounting
  (ii) Trunion mounting

(i) Clevis mounting:

- In this type of mounting, the end of cylinder is pivoted.

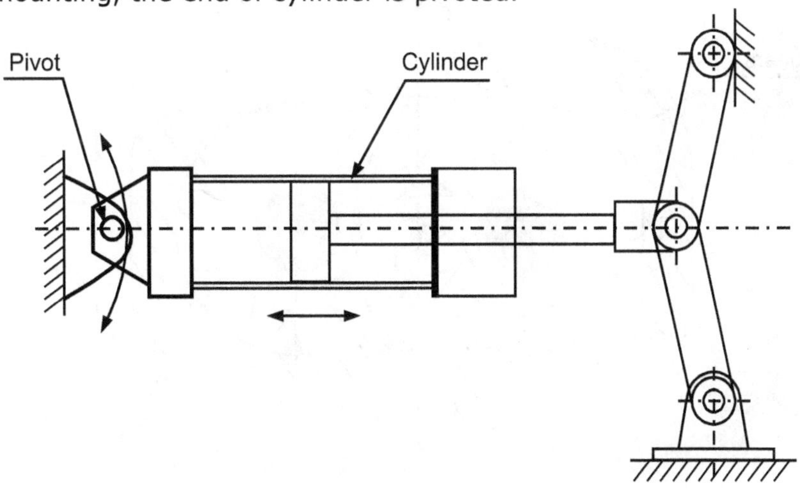

Fig. 2.35

- When piston reciprocates we can easily understand that, the cylinder can oscillate about pivot i.e. this freedom of oscillation is given to the cylinder.

(ii) Trunion mounting:

- Fig. 2.36 shows intermediate Trunion mounting. Trunion is a version of word 'Turning'. The cylinder is having pin type structure at the centre of its length. About axis A-A the cylinder can turn while reciprocating.

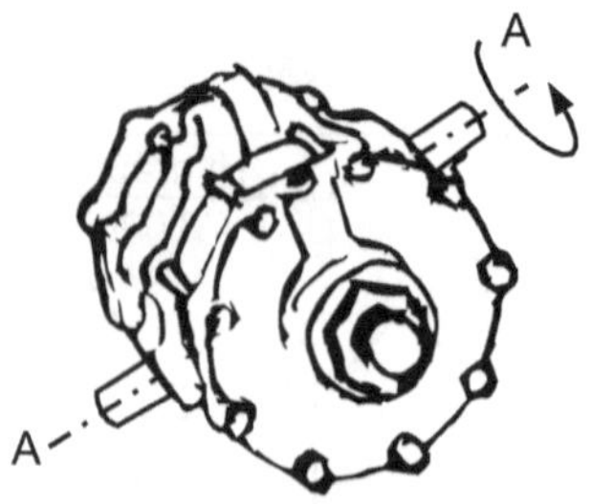

Fig. 2.36

2.9 HYDRAULIC MOTORS (ROTARY ACTUATORS) [W-09, S-12, W-12]

- Hydraulic motors provide rotational motion.
- Hydraulic motor converts hydraulic energy (processed by pressurised oil) into rotary mechanical energy used in many industrial applications.
- There is not much difference in hydraulic motor and hydraulic pump.
  - In hydraulic pump the rotation of the fluid is done by rotary element connected to the electric motor.
  - In hydraulic motor, the rotating element (vanes, gear, pistons) are pushed by the pressurised oil, which results into continuous rotary motion of the shaft.
- Hydraulic motors are classified on the basis of construction:
  (a) Gear motor (similar to gear pump)
  (b) Vane motor (similar to vane pump)
  (c) Swash plate piston motor (similar to swash plate pump)
  (d) Bent axis piston motor (similar to bend axis piston pump)
  (e) Radial piston motor (similar to radial piston pump)

2.9.1 Gear Motor [S-12]

- **Construction:** Similar gear pump, the motor is having two meshing gears, housing and two cover plates. The gears are mounted on two shafts and rotary motion is obtained from one shaft.
- **Working:** Pressurised oil enters from bottom port and entire chamber is pressurised. The force of oil exerts pressure on teeth of two gears which causes imbalance of forces on the two gears which in turn results in rotation of gears. The low pressure oil (used oil) goes out of upper port.

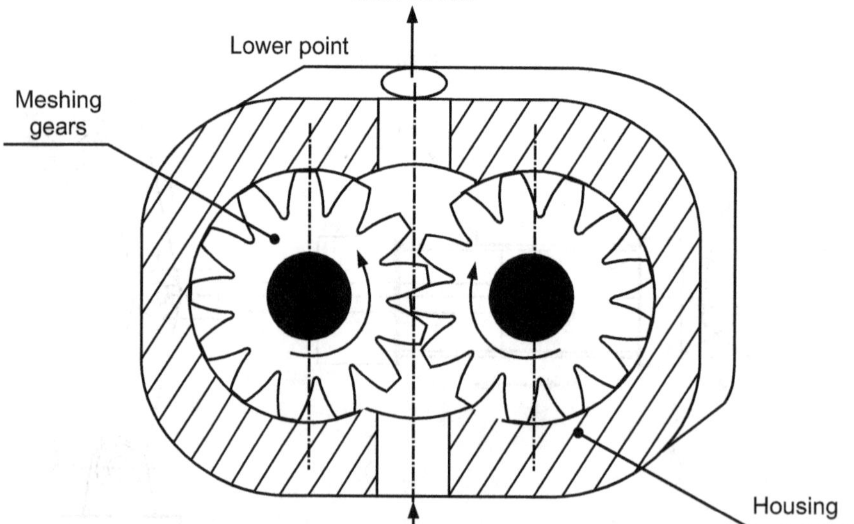

Fig. 2.37

2.9.2 Unbalance Vane Type Motor [W-12]

- **Construction:** The construction is similar to unbalanced vane type pump. The rotor is eccentric to housing. Its main parts are:
  (a) Slotted motor installed on the shaft.
  (b) Housing
  (c) Fixed cam ring to constrain outward movement of vanes
  (d) Sliding vanes

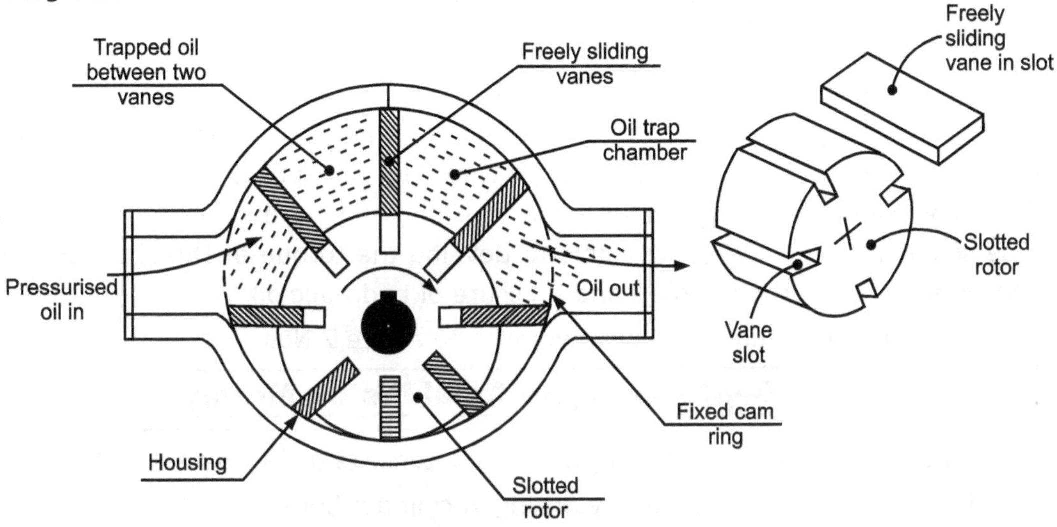

Fig. 2.38

- **Working:** Pressurised hydraulic oil enters from left port and causes the force on vanes which imbalances the rotor which then results in rotation. Due to rotation the vanes flow and through slots because of centrifugal force. In vane motor the imbalance is caused by difference in area exposed to hydraulic energy, which is due to eccentricity between rotor and housing. Imbalance results in torque on shaft.

2.9.3 Axial Piston Motors [S-10]

- Similar to axial piston pumps there are two types:
  (a) Swash Plate Axial piston Motor.
  (b) Bent Axis Piston Motor.
- The general axial piston motor is shown below. Similar to axial piston pump, the axis of all pistons are parallel to each other and are arranged on circumferencial periphery of the cylinder blocks.

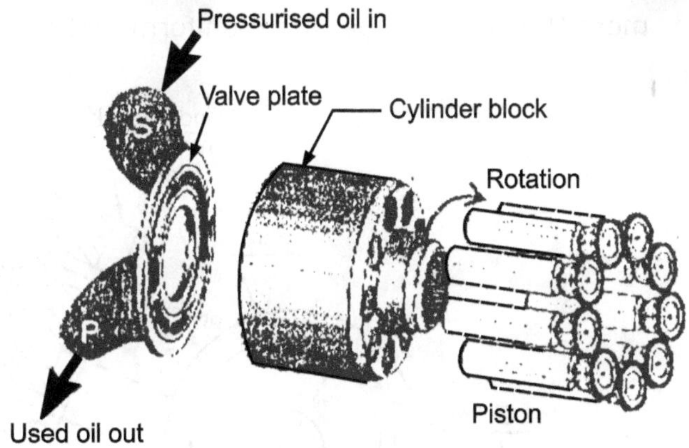

Fig. 2.39

2.9.4 Swash Plate Axial Piston Motor

- **Construction:** The motor consists of following components.
  (a) Cylinder barrel having circumferentially arranged pistons.
  (b) Pistons (minimum 3 Nos.)
  (c) Swash plate with shoe plate for piston rod connection.

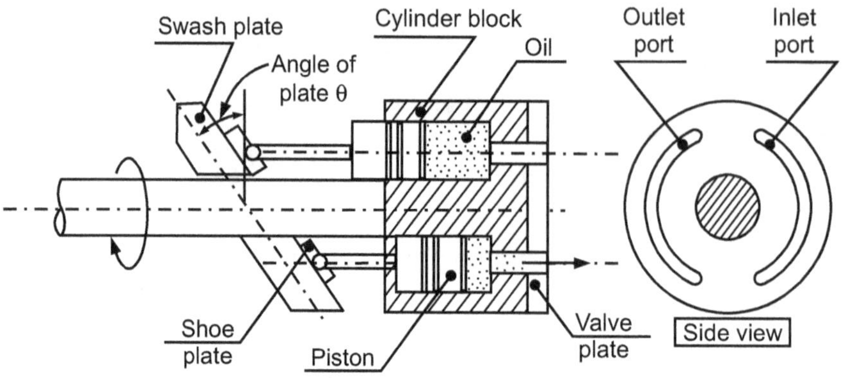

Fig. 2.40

- **Working:** The swash plate is making an angle θ with main Shaft Axis as shown. When pressurised oil comes in, it acts on piston in front of inlet port pushing the piston out of piston bores. This movement causes the piston shoe to slide across the swash plate surface. As the piston shoe slides on swash plate, it is able to develop the torque at the shaft attached to the barrel. The torque is proportional to $\angle\theta$ and pressure of hydraulic oil.

Note: For uniform motion, number of pistons must be above 7 Nos.

2.9.5 Bent Axis Type Axial Piston Motor

- **Construction:** Similar to bent axis pump, this motor consists of following components.
  (a) Cylinder block having circumferentially arranged cylinder bores.
  (b) Pistons
  (c) Driving flange similar to swash plate
  (d) Cylinder block end plate (fixed)
- **Working:** When pressurised oil comes through inlet port it acts on pistons in front of it. The oil pushes the piston out of piston bores. This movement causes the piston rod to slide on driving flange. As this piston rod slides it is able to develop torque at the shaft attached to the driving flange. The torque is proportional to $\angle\theta$ and pressure of hydraulic oil. If number of pistons are more the shaft will get more uniform motion.

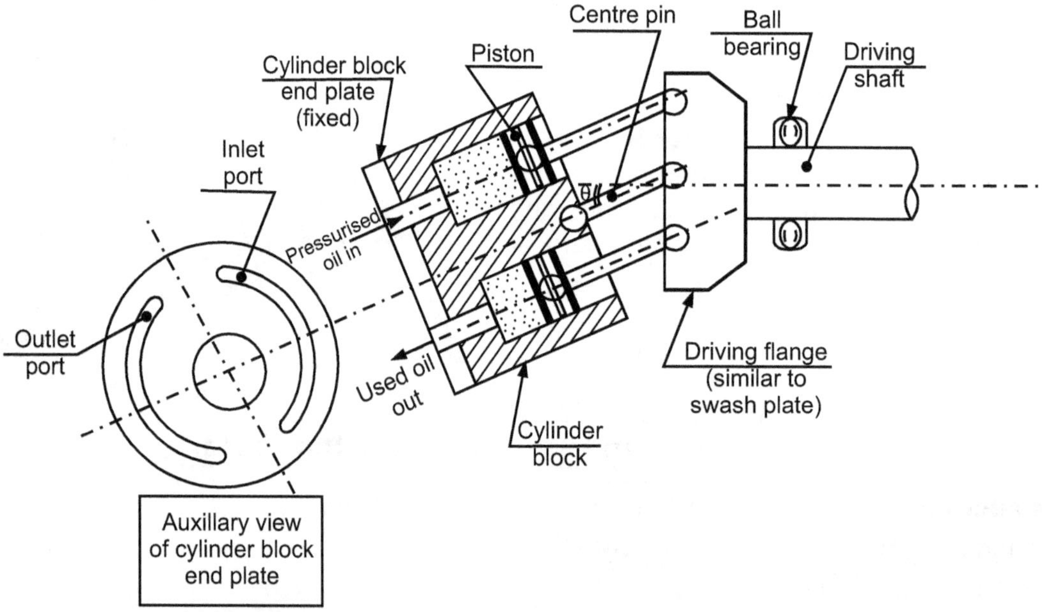

Fig. 2.41

2.9.6 Radial Piston Motor

[S-09, W-11]

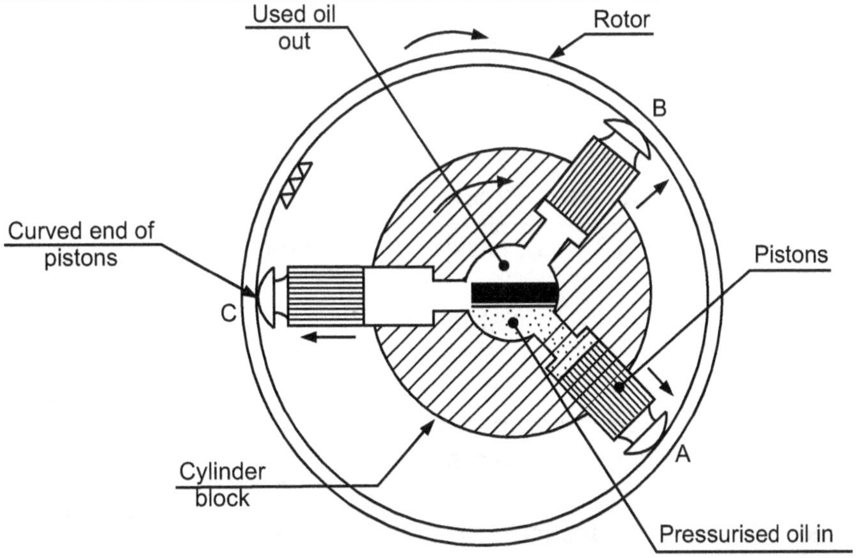

Fig. 2.42

- **Construction:** Three pistons are fitted in the cylinder block. The curved ends of pistons can rest on smooth (3 triangle finish) surface of rotor. Cylinder block and rotor are rotating members of the motor.
- **Working:** 'If fluid is introduced in the cylinder under pressure, the piston will be pushed outward'. This principle is used in this motor.
- Suppose the fluid under pressure is admitted to cylinder No. (A). The piston will move outward in its cylinder. We know that fluid i.e. hydraulic oil is incompressible; and two bodies (oil and piston) cannot occupy the same space in the cylinder. Naturally curved end of piston will slide inside the rotor with force and rotor will turn in the clockwise direction. Then cylinder B will occupy the position of A since cylinder block also starts rotating. And same cycle will start which results in rotational motion of the rotor.
- This is low speed high torque motor.

2.9.7 Symbol of Hydraulic Motor

- Symbol of hydraulic motor (irrespective of type of motor) is nearly similar to the symbol of hydraulic pump. The only difference is that the triangles shown are in opposite direction.

(a) Symbol of unidirectional hydraulic motor:

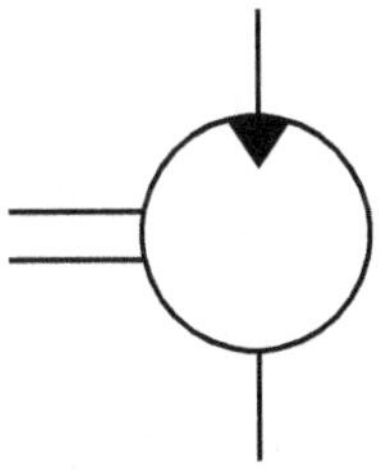

Fig. 2.43 (a)

(b) Symbol of bi-directional hydraulic motor:

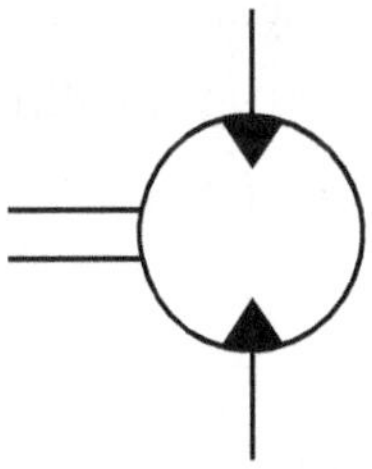

Fig. 2.43 (b)

(c) Symbol of limited rotation hydraulic motor:

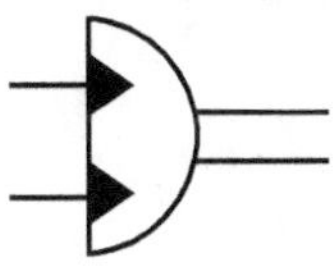

Fig. 2.43 (c)

2.10 LIMITED ROTARY ACTUATORS

- Hydraulic motors are 'full' rotary actuators. The shaft of hydraulic motor turns completely through 360°. But in maximum hydraulic circuits we need limited rotary motion (say 180°, 270° etc. but less than 360°). In such cases, we have to use limited rotary actuators, which are described below:

2.10.1 Dual Piston Type Limited Rotary Actuator OR Limited Rotary Actuator using Rack and Pinion and SA Cylinder Pair

- This actuator uses SA cylinders (2 Nos), rack and pinion mechanism and DC valve.
- **Working:** When pressurised oil is supplied through port A of cylinder 1, then its piston will move from left to right. During this movement, since rack is attached to the piston rod, it will move towards right. The pinion is meshed with rack and hence pinion will rotate to limited angle in the clockwise direction.
- When pressurised oil will be supplied through port B of cylinder 2 then piston will move from right to left and so the rack. The pinion will rotate in anticlockwise direction through limited rotation.

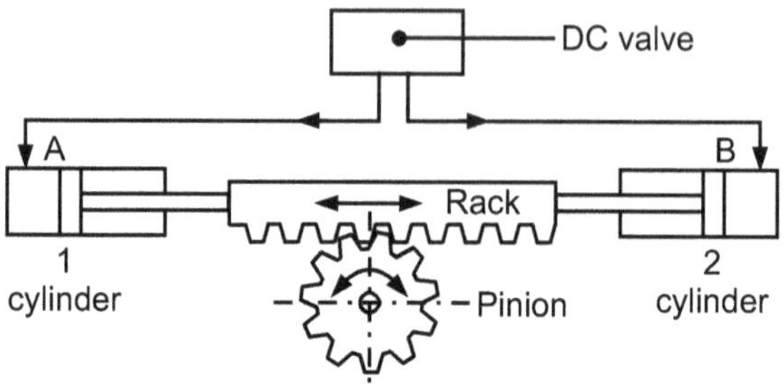

Fig. 2.44

2.11 ACTUATORS (LINEAR AND ROTARY) (PNEUMATIC CYLINDERS AND AIR MOTORS) [S-12, 15; W-13]

- Pneumatic Actuators are the devices which converts the power of compressed air into useful work. The useful work is obtained through force and motion of actuators.
- Pneumatic Actuators are broadly classified in two categories:
  (i) Air cylinders also known as pneumatic cylinders or linear actuators or reciprocating actuators.
  (ii) Air motors or pneumatic motors also known as rotary actuators.
- Now we will see all the actuators one-by-one.

(I) Air cylinders (Linear Actuators)

Air cylinders are classified on following basis:

(i) Single acting cylinder (SA cylinder).
(ii) Double acting cylinder (DA cylinder).
(iii) Diaphragm cylinder (It is SA cylinder).
(iv) Rolling diaphragm cylinder (It is SA cylinder).
(v) Turn cylinder (It is DA cylinder).
(vi) Tamdem cylinder (It is DA cylinder).

Above classification can also be plotted as under:

(a) Single Acting Cylinder
  (i) Diaphragm cylinder
  (ii) Rolling diaphragm cylinder
(b) Double Acting Cylinder
  (i) Turn cylinder
  (ii) Tandem cylinder

2.11.1 Single Acting Cylinder (SA Cylinder) [S-16]

- In this type of linear actuator, the air is admitted on one side of piston. Hence this cylinder can produce useful work in one direction only. The return movement of piston is obtained by built-in spring installed on other side of piston.

(a) Symbol of Single Acting Cylinder

- The symbol is based on construction itself. In the symbol one can easily understood that when air is admitted in cylinder then piston will move from left to right. When air stops admitting, then piston will move from right to left due to expansion of spring.

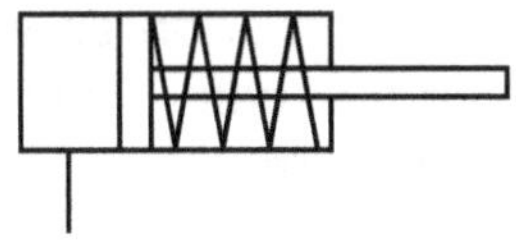

Fig. 2.45

- The actual construction of single acting cylinder is exactly similar. Fig. 2.46 shows the actual construction of SA cylinder.
- **Construction:** The single acting cylinder has following parts:
  (a) Cylinder body or cylinder barrel.
  (b) End cover with cushion packing.
  (c) Piston with piston rings and piston rod.
  (d) Return compression spring
  (e) 'O' Rings for packing.

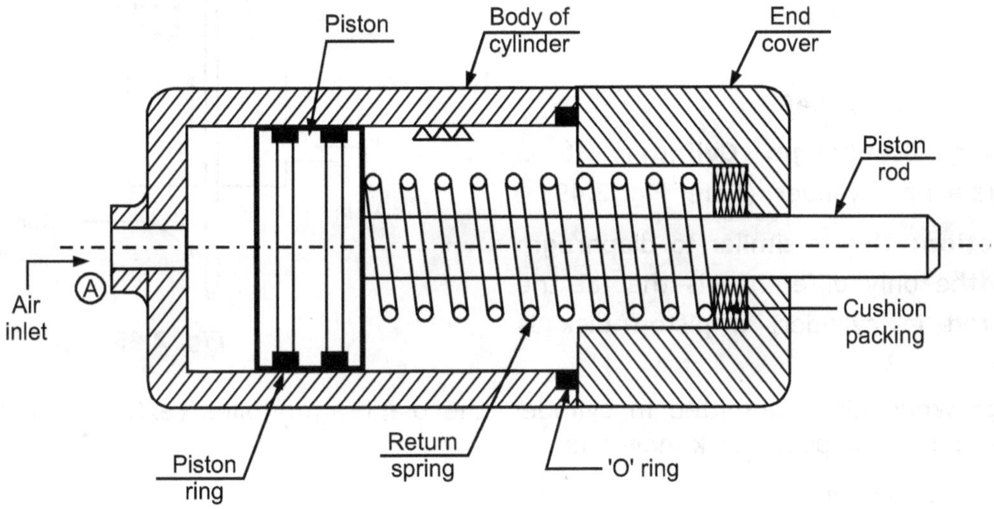

Fig. 2.46

Materials used for construction:

(a) Cylinder body or barrel: Hard drawn seamless (jointless) aluminium/brass or steel tube.
(b) End cover: Aluminium, brass or bronze or cast iron.
(c) Piston: Aluminium, bronze, cast iron.
(d) Piston rod: Ground and polished chrome plated medium carbon steel. It also is heat treated.
(e) 'O' Rings: Compounded Rubber.
(f) Spring: Spring steel.

(i) Working: Compressed air will be admitted through port 'A'. Due to pressure force piston will move towards right thereby compressing the spring. When air stops admitting, compressed spring will expand and piston will regain its original position.

(ii) Diaphragm cylinder: It is also single acting cylinder.

- **Construction:** In this actuator rubber/ plastic/(or even metal) diaphragm is used as an actuating element. This diaphragm is flexible; and is locked at the centre is locked at the centre of cylinder tube.
- **Working:** When pressurised air comes in, due to pressure force the diaphragm stretches and rod lifts-up. There is no relative motion as we have seen with piston cylinder pair. There is only stretching of diaphragm.
- **Use:** These SA cylinders are used for clamping. Also used with tools of embossing or riveting.

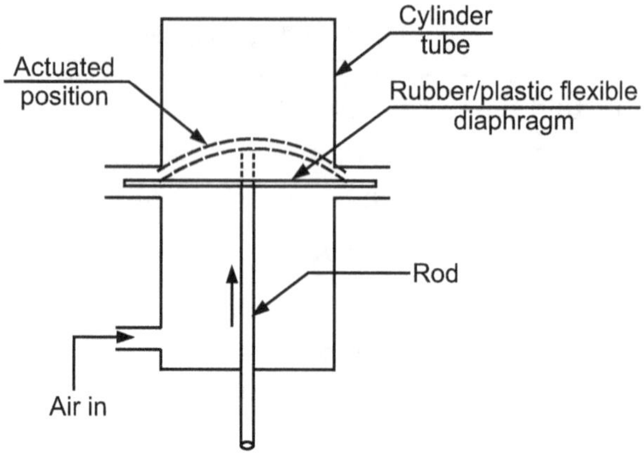

Fig. 2.47: Diaphragm Cylinder

- **Limitation:** The stroke length is very small (less than 20 mm).
- **Symbol:** Symbol is similar to construction of actuator. (Refer Fig. 2.48).

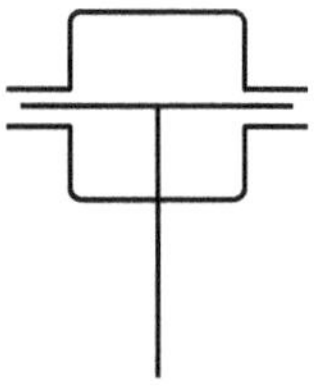

Fig. 2.48

(iii) Rolling Diaphragm Cylinder: It is single acting cylinder. (Refer Fig. 2.49).

- **Construction:** This is similar to diaphragm cylinder, the only difference is that at the end of rod in cylinder, a piston disk is attached.

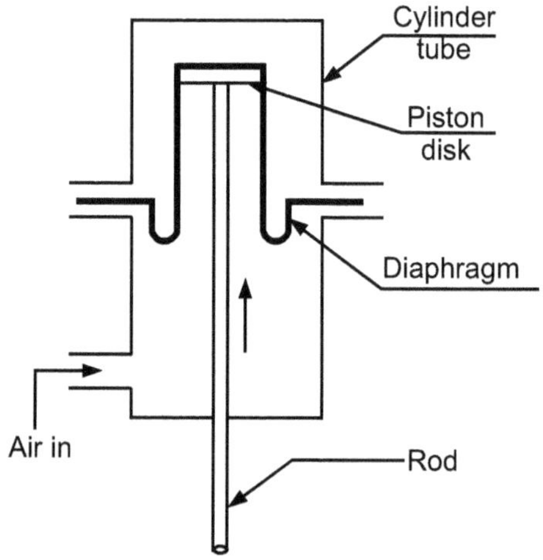

Fig. 2.49

- **Working:** When air is admitted in cylinder, the diaphragm rolles (extends) in the manner shown in figure, and piston disk moves-up.
- **Use:** Used for clamping purpose.
- **Advantage:** The higher stroke length upto 80 mm can be achieved with this SA cylinder.
- **Symbol:** Symbol is similar to its construction.

Fig. 2.50

2.11.2 Double Acting Cylinder (DA Cylinder)

[S-15]

- In this type of actuator, air is admitted on the both sides of piston. Hence this cylinder or actuator can perform useful work in both directions. There is no spring in this actuator.

Symbol of Double Acting Cylinder:

- The symbol is based on construction itself. From the symbol it is easily understood that, when air is admitted through port 'A' piston will move towards right and when air comes in through port B piston will move from right to left. (Note that do not write A and B letters in symbol).

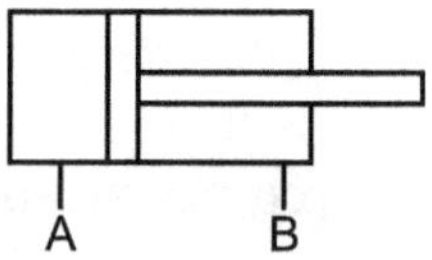

Fig. 2.51

- The actual construction of DA cylinder is as shown below:

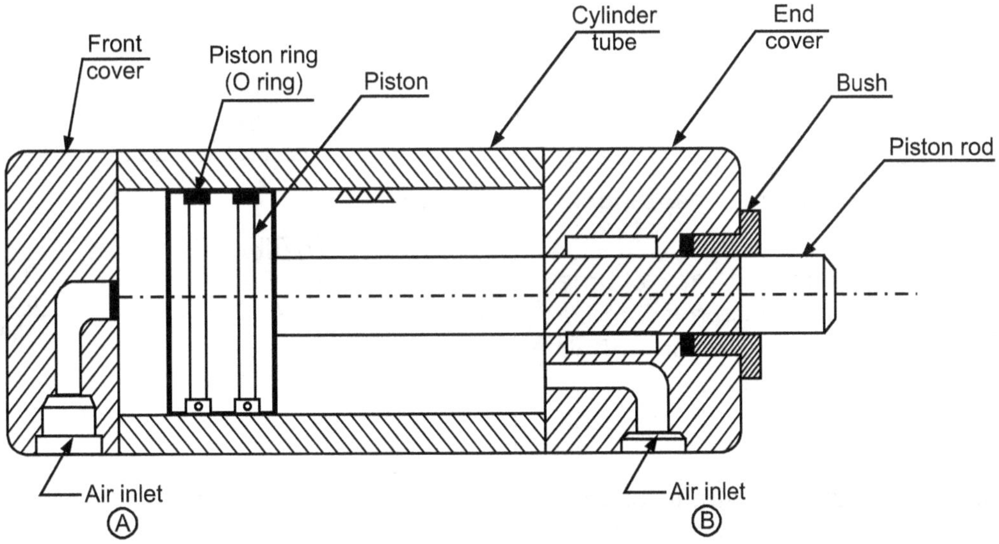

Fig. 2.52: Actual Construction of DA Cylinder

- **Construction:** Double Acting cylinder is having following parts:
  (a) Cylinder body or Barrel or Tube.
  (b) End cover with bush and built-in air-inlet port (B).
  (c) Piston with piton rings (orings) and piston Rod.
  (d) Front cover threaded to cylinder tube with in-built air inletport (A).
- **Material used for construction:** Similar to single acting cylinder.
- **Working:** Compressed air will admitted first through port (A). Due to pressure force piston will advance i.e. will move from left to right. Then airflow from port (A) will stop and flow from port (B) will start. Now piston will retract i.e. it will move from right to left. During this movement air present in the cylinder during advance stroke will move out through port (A). This movement repeats.
  (i) Turn Cylinder: It is double acting cylinder. In this type, piston rod ends is modified. The rod end is converted into rack and pinion pair through which linear motion is converted into rotary motion.
- **Working:** Air is admitted through port (A). Piston will move down. Pinion is meshed with teeth on rack. Due to downward movement of rack, pinion will move in anticlockwise direction. And we can obtain 'half turn' movement of pinion in anticlockwise direction.
- When air is admitted through port (B) piston will move-up and pinion will rotate in clockwise direction.
- We can also use worm and worm wheel pair instead of rack and pinion pair. Worm and worm wheel pair is stronger than rack and pinion.

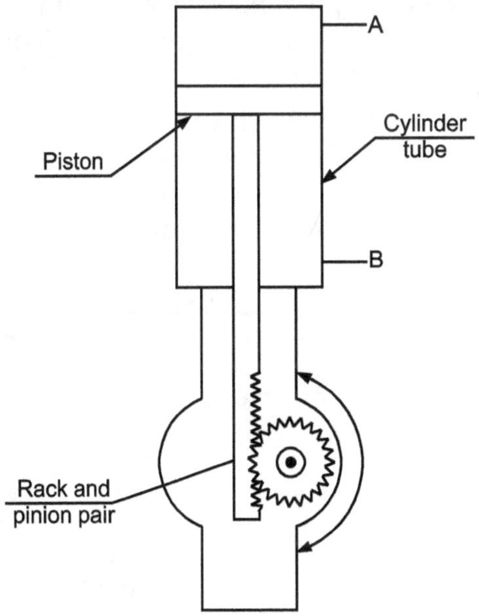

Fig. 2.53

- **Use:** This type of double acting cylinder is used in turning heavy jobs as in automatic molding line.
- **Symbol:** Same as drawn in Fig. 2.53.

(ii) Tandem Cylinder: This is special type of double acting cylinder.

- **Construction:** As shown in figure two cylinders are joined in series. There is sealed partition between two cylinders. There are two ports to each cylinder. Two pistons are fitted on one single piston rod. This arrangement is known as tandem array.
- **Working:** Air is admitted at a time through port (A) and (C). Piston will move to right with double force. Similarly, if air is admitted at a time through port (B) and (D) piston will move to left with double force.

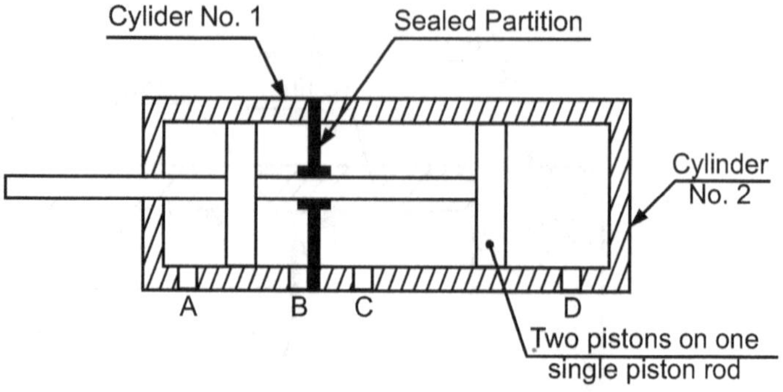

Fig. 2.54

- **Advantage:** The main advantage of this DA cylinder is that we can obtain double force with same size of cylinder.
- **Symbol:** Symbol is similar to its construction (Refer Fig. 2.55).

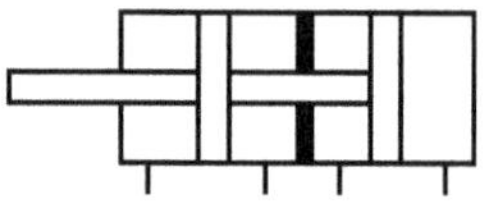

Fig. 2.55

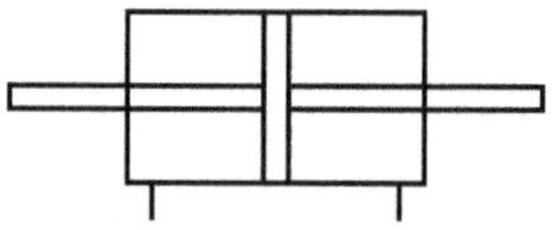

Fig. 2.56: Through Rod Cylinder

(iii) Piston Rod at Both Ends (Through Rod Cylinder): This is slightly modified double acting cylinder in which piston rod is attached on both sides of piston. The main advantage of this arrangement is that we can obtain equal force and equal speed on both sides of piston. (Refer Fig. 2.56).

- **Symbol:** As shown in above Fig. 2.56.

2.11.3 Cylinder Mountings

[W - 13]

- In every pneumatic circuit, practically we have to 'Install' or 'Mount' the air cylinder in a required fashion for smooth working of circuit and to obtain required useful motion. The cylinder mountings are listed below (Mounting is an arrangement of installing the cylinder in circuit).

1. **Foot Mounting:** This method is similar to mounting of electric motor with pump, or any other equipment like compressor etc. The foot plate is fixed to frame by nut-bolts. Mostly heavy cylinders are foot mounted for good stability. Size and number of bolts are dependent on thrust/force created in cylinder.

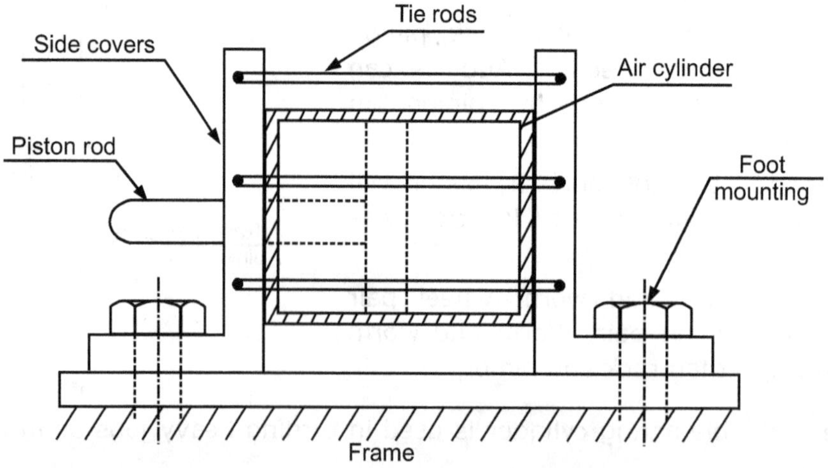

Fig. 2.57: Foot Mounting

2. **Pivot Mounting:** In many circuits, in addition to reciprocating motion of piston, the entire air cylinder body needs oscillating motion. In that case pivot mounting is used as shown.

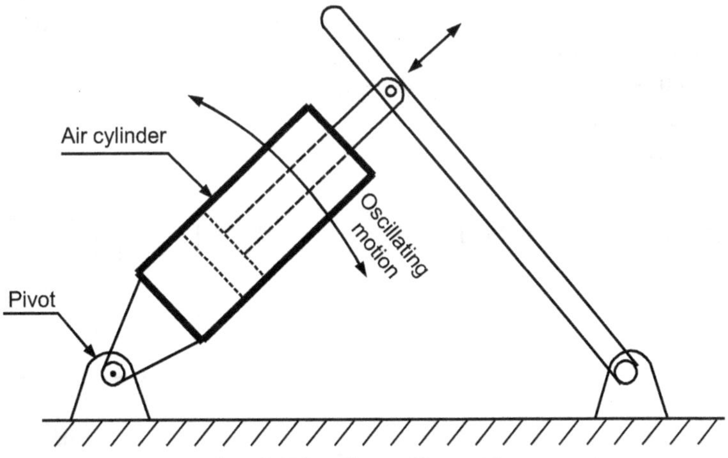

Fig. 2.58: Pivot Mounting

3. **Centre Line Mounting:** In this mounting, there is special flange bolted at the center. This special flange is then used to mount the cylinder. In this mounting, bolts are under simple shear stress or simple tensile stress.

2.11.4 Piston Force

- Consider DA cylinder having piston diameter 'D' and piston rod diameter 'd'.
  Let, P = Pressure of compressed air.

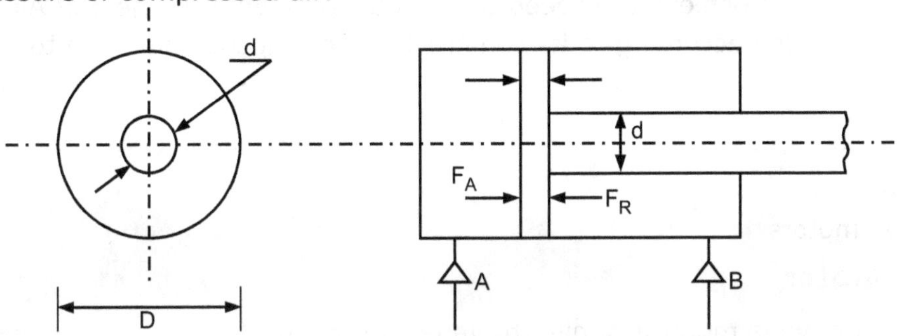

Fig. 2.59

- Note that when air is supplied through port A, piston will 'Advance' and when air is supplied through port B, piston will 'retract'.
- Now we will find out force with which piston will Advance (F_A) And force with which piston will retract (F_R).
- We know that

$$\text{Pressure} = \frac{\text{Force}}{\text{Area}}$$

$$\therefore \quad \text{Force} = \text{Pressure} \times \text{Area}$$

To find force on Piston in 'Advance Stroke'.

$$F_A = P \times \text{Area on which pressure is Applied}$$

$$\therefore \quad F_A = P \times \frac{\pi}{4} D^2$$

$$\therefore \quad \boxed{F_A = \frac{P\pi^2 D}{4}}$$

To find force on Piston in 'Retracting Stroke'

$$F_R = P \times \text{Area on which air is applied}$$

$$F_R = P \times \left(\frac{\pi}{4} D^2 - \frac{\pi}{4} d^2\right) \quad \ldots \text{ deducting cross-sectional area of piston rod}$$

$$\therefore \quad \boxed{F_R = \frac{P\pi^2}{4}(D^2 - d^2)}$$

2.11.5 Piston Speed

- In SA or DA cylinder, the piston reciprocates, that means it has linear motion (motion is straightline) and hence it has linear speed. Let 'L' is stroke length if piston is moving with linear velocity of V, then it will cover distance 'L' with in time 't'.

  Then piston speed $V = \frac{L}{t}$ mt/sec.

- The piston speed is pneumatic cylinders is dependant on:
  (a) Air pressure.
  (b) The stroke length.
  (c) Flow rate of air.

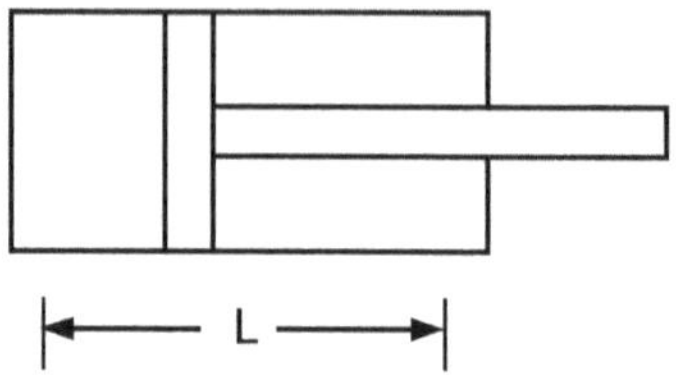

Fig. 2.60

- The average piston speed adopted in pneumatic circuits is 0.1 m/sec to 1.5 m/sec. But in very special cases and heavy applications it can go up to 8 m/sec.
- **Size of Pneumatic Cylinder:** Normal size of Pneumatic Cylinders is between 3 mm to 250 mm diameter.

2.11.6 Air Motors [W-10, 11, 16; S-11, 12]

- These are rotary actuators and are used to generate 'ROTARY MOTION' by using force of compressed air. We can achieve the speed of 10,000 r.p.m. with the help of Air Motors.
- Air motors are classified according to its construction and mechanism used to achieve rotational motion:
  (a) Vane motors
  (b) Gerotor motor
  (c) Turbine motors
  (d) Piston type motors

2.11.6.1 Vane Motor

- The construction of vane motor is similar to vane compressor.
- **Construction:** It consists of simple Vane Rotor which is having slots in which vanes (flat piece of steel) slides freely. The rotor is eccentrically located inside the stator housing.
- **Working:** When pressurised air comes in through inlet port, the pressure of air distributes equal in all directions. Since vane is sliding freely in slots of rotator, the vane comes in to way of pressurised air and air pushes the vanes so that rotor starts rotating with speed. The used low pressure air is exhausted through exhaust port. This is unidirectional motor. Since vanes are freely sliding in slots, there is possibility of leakage of air. With the help of these motors we can achieve the speeds upto 25000 r.p.m.

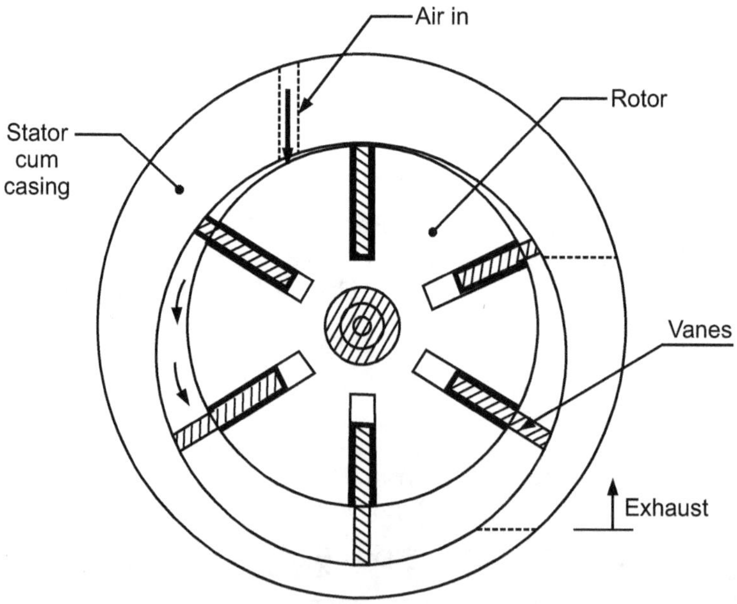

Fig. 2.61: Vane Motor

2.11.6.2 Gerotor Motor

- In this type two specially shaped gerotor blades meshing to each other are used. When pressurised air comes in, the force of air rotates one blade. Since it is meshing with other, the other blade also rotates.
- These air motor mostly used for low r.p.m. such as 200-300 rpm. Hence are not suitable for high torque applications.

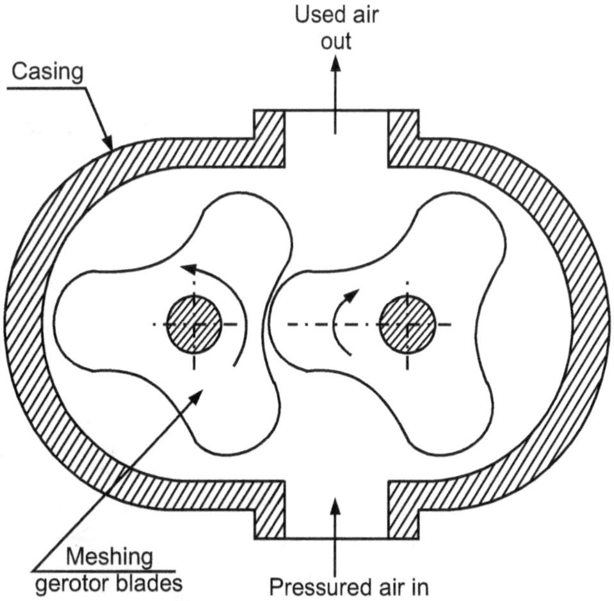

Fig. 2.62: Gerotor Motor

2.11.6.3 Turbine Motor

- In this type of air motor, light weight impeller having curved vanes is used. This pressurised air is passed through nozzle. The impact of jet will rotate the impeller. These motors are high speed low torque motors; and being simple in construction are used in many applications.

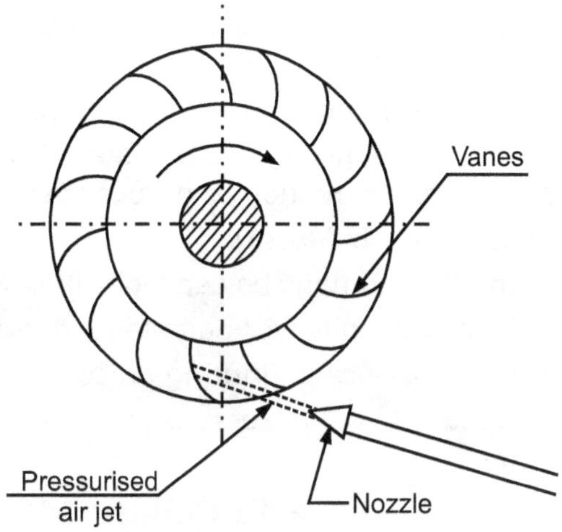

Fig. 2.63: Turbine Motor

2.11.7 Single Vane Type Limited Rotary Actuator

- This Limited Rotary Actuator consists of a stationary body, having rotor shaft at centre. This shaft hold one vane. There two inlet ports.
- **Working:** When oil comes in through port A, the force of oil pushes the vane in anticlockwise direction to obtain limited rotary motion. When oil enters through port B we can obtain limited rotary motion of shaft in the clockwise direction.

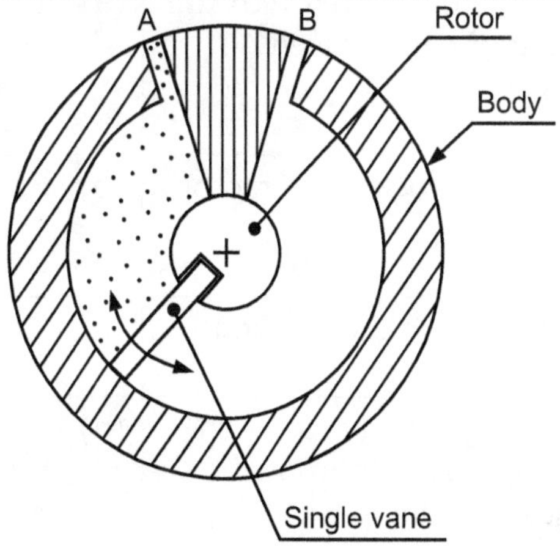

Fig. 2.64

Criterias for selection of motors in Hydraulic Circuit (or Basis of Hydraulic Motor Selection)

(a) Amount of torque needed decides the type of motor. The starting torque is higher than running torque.

(b) Speed is another important criteria for selection. When we require low speed high torque motor we have to use radial piston motor. For high speed requirement use vane motors.

(c) System pressure is another important criteria. We have to select suitable pressure to create required torque.

(d) Application of motor is also an important criteria.

(e) Power generated by motor is another criteria. For moderate power ranges Axial piston motors are preferred. Vane motors or gear motors are low power motors.

(f) Cost of motor should be moderate.

2.12 PISTON MOTOR

- There are two types of Piston motors:
  (i) Axial Piston Motors
  (ii) Radial Piston Motors
- Piston type motors may have 4, 5 or 6 cylinders. The power developed by these motors is dependent on the inlet air pressure, number of pistons, the cross-sectional area of piston, the stroke of piston and linear speed of piston.
- Axial piston motors are generally manufactured upto 3.5 HP power, while radial piston motor are upto 25 HP.
- Radial piston motors can achieve speed up to 3000 r.p.m.
- **Working:** The piston reciprocates in sequence and they actuate wobble plate. This in turn imparts a rotary motion to output shaft through gear train. (Refer 'Hydraulic motor' in topic 2).

Applications of Air Motors

(i) In all pneumatic power tools like, screw drivers, angle grinders, straight grinders.

(ii) To rotate conveyor belts in food industry.

(iii) Power device in printing press machine.

(iv) Agitators and mixers.

(v) Vibrators.

2.13 COMPARISON BETWEEN AIR MOTOR, ELECTRIC MOTOR AND HYDRAULIC MOTOR [S-11, W-11, 16]

Air Motor	Electric Motor	Hydraulic Motor
(a) Air motor rotates with the help of compressed Air.	(a) Electric motor rotates with electric current which in turn creates electro-magnetic effect.	(a) It rotates with the help of compressed fluid.
(b) This is shock proof motor.	(b) Some shock is possible during starting.	(b) The shock is less compared to other two mirrors.
(c) This motor is explosion proof.	(c) Motor can be explode or burn due to short circuit.	(c) Motor can explode due to high pressure.
(d) If continuously run then temperature of motor does not increase.	(d) If continuously run then temperature of electric motor increases.	(d) Temperature increases with respect to time.

contd. ...

(e) Speed of Air motor can be varied without using complicated mechanisms.	(e) Speed variation requires complicated circuits which increases cost of motor.	(e) Speed variation can be varied using different assembly or mechanism.
(f) The power to weight ratio of this motor is high.	(f) The power to weight ratio is low.	(f) The power to weight ratio of this motor is high.
(g) Motors are simple in design and hence not costly.	(g) Motors are complex in design and construction and hence fetches more cost.	(g) Motors are complex in design.
(h) Since the weight of motor is low, this motor can accelerate and retard quickly.	(h) Electric motor takes more time to accelerate or retard.	(h) Hydraulic motor can accelerate quickly
(i) Air motors are less efficient.	(i) Efficiency of electric motor is high.	(i) Efficiency is high if motor is leak proof.
(j) Makes more noise while working.	(j) Makes low noise.	(j) Noise is more due to rotating and reciprocating parts.
(k) High power air motors are not available.	(k) High power electric motors (500 HP/800 HP) are available.	(k) High power hydraulic motors are available.

2.14 SYMBOL OF AIR MOTOR [W-16]

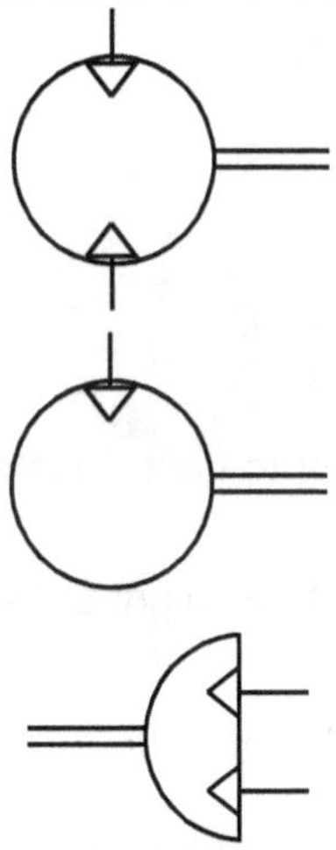

This is the symbol similar to hydraulic pump/motor. The symbol shown in 'Air motor – Bi – Directional' (can rotate clockwise or anticlockwise direction) carefully see the position of triangle. The triangle is hollow.

This is the symbol of Uni-directional air motor (It can rotate in clockwise or anticlockwise direction only).

This is the symbol of half rotation or oscillating motor.

Fig. 2.65

2.15 COMPARE LINEAR ACTUATORS AND ROTARY ACTUATORS

Linear Actuators	Rotary Actuators
(a) Linear actuators reciprocate in a cylinder.	(a) These actuators rotate about the centre.
(b) These are having linear speed measured in m/sec.	(b) These are having rotary speed measured in RPM.
(c) Used for pushing, pulling types of tasks.	(c) Used where rotary motions are required for example, straight grinders, pistol drills.
(d) Single Acting Cylinders, Double Acting Cylinders, Tandem Cylinders are its examples.	(d) Vane motors, Gear motors, Piston motors, Air motors are some of the examples.

2.16 COMPARE HYDRAULIC PUMP AND HYDRAULIC MOTOR [W - 11]

Hydraulic Pump	Hydraulic Motor
1. 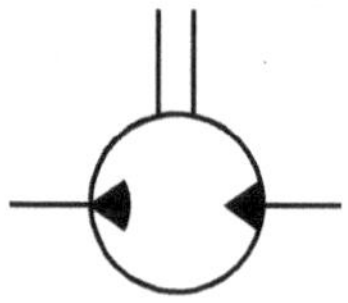 Fig. 2.66 (a)	 Fig. 2.66 (b)
2. Hydraulic pump pressurizes the oil (Medium) in hydraulic citcuit. This pressurized oil actuates the 'actuator' and mechanical advantage is obtained. The pump operates on electric motor.	2. Hydraulic motor is rotary actuator. The pressurized oil delivered by pump rotates the rotor of motor and we can obtain desired rotary motion.
3. Pumps can be classified in two broad groups. (a) Positive displacement pumps. (b) Non-positive displacement pumps i.e. Rotodynamic pumps.	3. Hydraulic motors are converting hydraulic energy in to mechanical energy (i.e. rotary motion). The hydraulic motors can be classified depending upon type of motion. (a) Unidirectional motor. (b) Bidirectional motor. (c) Limited rotation hydraulic motor.

Important Points

- Following are the types of Hydraulic Cylinders :
  - (i) Single Acting Single Piston Rod Cylinder (SA Cylinder).
  - (ii) Single Acting RAM.
  - (iii) Double Acting Single Piston Rod Cylinder (DA Cylinder).
  - (iv) Double Acting Double Piston Rod Cylinder.
  - (v) Tandem Cylinder.
  - (vi) Telescopic Cylinder.
- In DA cylinder, when oil is admitted through piston side then piston will move from left to right. This stroke is called advance stroke and piston is said to be 'Advancing'.
- In DA when oil is admitted into DA cylinder from piston rod side then piston will move from Right to Left. Then this stroke is called Retract Stroke and Piston is said to be 'Retracting'.
- In SA or DA cylinder (or any linear actuator) the piston reciprocates. That means it is having linear motion (motion in straight line) and hence it is having linear speed.
- The main advantage of this DA cylinder is that we can obtain double force with same size of cylinder.
- Word 'mounting' here means suitably installing the hydraulic cylinder in a hydraulic circuit for smooth running of circuit.
- Mounting is an important parameter both from design and maintenance point of view.
- Hydraulic motors provides rotational motion.
- Hydraulic motor converts Hydraulic energy (processed by pressurized oil) into rotary mechanical energy used in many industrial applications.
- Hydraulic motors are classified on the basis of construction :
  - (a) Gear Motor (similar to gear pump).
  - (b) Vane motor (similar to vane pump).
  - (c) Swash plate piston motor (similar to swash plate pump).
  - (d) Bent axis piston motor (similar to bend axis piston pump).
  - (e) Radial piston motor (similar to radial piston pump).

- Hydraulic motors are 'Full' rotary actuators. The shaft of Hydraulic motor turns completely through 360 degree. But in maximum hydraulic circuits we need limited rotary motion (say 180 degree, 270 degree etc. but less than 360 degree). In such cases we have to use Limited Rotary Actuators.

Practice Questions

1. Draw neat sketch of external gear pump and explain its working
2. Give classification of pump.
3. Draw figure of balanced type vane pump and label the parts.
4. What is screw pump? Explain.
5. What is swash plate? What is its use? What will happen if we change the angle of swash plate? Explain with sketch.
6. Draw characteristic curve of gear pump at constant speed.
7. Compare linear actuators and rotary actuators.
8. What is clevis mounting? Explain with neat figure
9. How hydraulic motors are classified? Enumerate their types.
10. What is limited rotary actuator? Draw labelled diagram of single vane type, limited rotary actuator.
11. Draw symbol of three types of hydraulic motor.
12. Differentiate between positive displacement pumps and roto dynamic pumps.
13. Classify positive displacement pumps and arrange them in descending order on the basis of pressure created by these pumps.
14. Sketch and explain Gerotor Pump.
15. Draw labelled diagram of any one actuators used in hydraulic system. Explain its construction and working.
16. Explain working of telescopic cylinder with neat sketch.
17. What do you mean by Non-Positive displacement pump.
18. Give the detailed classification of hydraulic actuators.
19. Explain construction and working of gear pump.
20. Enlist various factors to be considered while selecting the motor for Hydraulic circuit.
21. Draw neat figure of actual construction of single acting air cylinder and label the diagram.
22. Draw neat sketch of actual construction of double acting air cylinder and label the diagram.
23. What is rolling diaphragm cylinder? Explain
24. Compare air motor with electric motor.
25. List and explain various types of air motors.
26. Compare air motor with electric motor.
27. What is the function of actuators?
28. Differentiate between Air motor and Hydraulic motor on the basis of:
    (i) Fluid used, (ii) Material, (iii) Capacity, (iv) Application.
29. What is Clevis mounting explain with figure?
30. Draw neat sketch of Dual Piston type limited rotary actuator.
31. Explain Lobe pump with figure.

■■■

Chapter **3**

CONTROL VALVES

Weightage of Marks = 16, Teaching Hours = 12

Syllabus

3.1 Classification of Control Valve

3.2 Direction Control Valves-Check Valves 2/2, 3/2, 4/2, 4/3, 5/2, 5/3 D.C. Valves used in Hydraulics and Pneumatics

3.3 Standard Centre Positions, Methods of Actuation

3.4 Pressure Control Valves - Relief, Unloading, Sequence, Counter Balance, Pressure Reducing Valves

3.5 Flow Control Valves - Non-compensated Pressure and Temperature Compensated Valves

About This Chapter

At the end of this chapter students will be able to:

- Classify various types of valves with justification
- Sketch and explain construction of given valves
- List actuation methods of valves as per type of application
- Select appropriate control valves for given application

3.1 FLUID CONTROLLING ELEMENTS SUCH AS DIRECTION CONTROL VALVES (DC VALVES), PRESSURE REGULATING VALVES, FLOW CONTROL VALVES USED IN HYDRAULIC CIRCUITS [W-09, 16]

- Pump pressurizes the hydraulic oil. This oil flows to the hydraulic actuator. Actuator works and produces useful mechanical work. This is the general outline of the hydraulic system.
- But, one must remember that when pressurised hydraulic oil with full of hydraulic energy travels towards the actuator it must be controlled to suit the requirements of actuator (also known as consumer). So let us see 'what' to and 'how' to control this 'energetic' hydraulic oil.

  (a) Direction Control: The linear actuator creates to and fro i.e. reciprocating motion. This is depicted in following Fig. 3.1 (a).

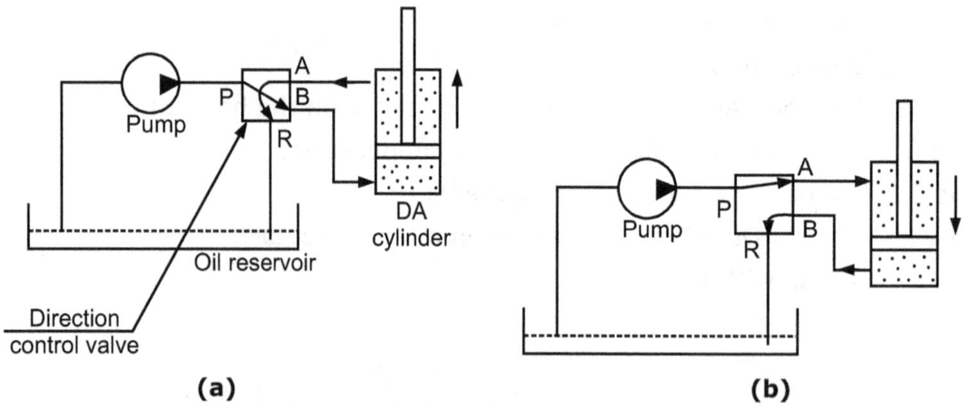

Fig. 3.1

- Hydraulic oil from pump will flow from port 'P' to port 'B'. Piston of DA cylinder will move upwards. During this stroke oil on the other side of piston (Piston rod side) will return through port 'A' via port 'R' to oil reservoir.
- In the next stroke port 'P' will be connected to port 'A' and oil will move the piston downwards. During this stroke oil present on the side of piston will return to oil reservoir through port 'B' via port 'R'.
- From above it is clear that oil must be 'directed' first through port B and then through port 'A'. At the same time port R must be connected to remaining port. All this is done by 'Direction Control Valve' or simply called as 'DC Valve'.

(b) Flow Control: Fig. 3.2 shows the double acting cylinder. When oil will enter through port A, piston will move from left to right with certain linear speed, say 6 mt/min. Suppose we have to increase this speed from 6 m/min. to 7 m/min. How we can achieve this? The answer to this question is by admitting more quantity of fluid through port A, so that piston will move faster. To admit additional hydraulic oil, we have to use valve V_1 which is known as 'Flow Control Valve'. This valve will control flow (lit./sec.) or quantity of oil.

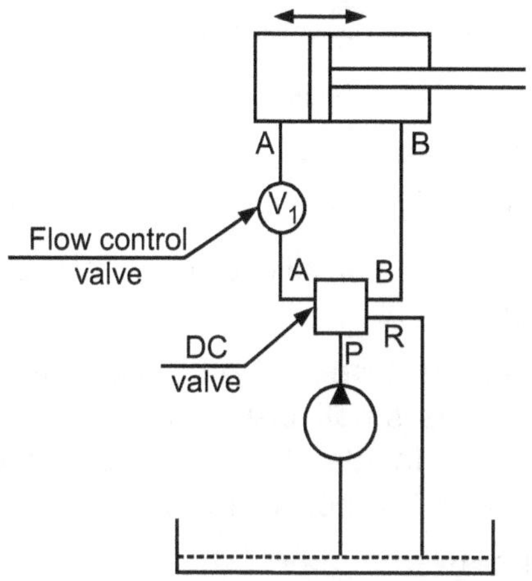

Fig. 3.2

(c) Pressure Control: The pump creates higher pressure. Suppose a piston pump in a hydraulic circuit creates 30 bar pressure. The actuator in the circuit can operate at 20 bar pressure. Then we must reduce the oil pressure from 30 bar to 20 bar for that particular actuator. The second actuator in the same circuit operates on 10 bar pressure. Then we have to reduce the pressure of oil from 30 bar to 10 bar for that second actuator. This clearly means we have to control the pressure of oil after pump to a suitable level for actuator. This is done by **pressure regulating valve.** Some pressure control valves are safety valves which do not allow the system pressure to increase beyond specific set level. These valves are called pressure relief valves.

Now we will see following valves in detail:

(i) Direction control valves.

(ii) Flow control valves.

(iii) Check valves.

(iv) Pressure control valves.

(v) Unloading valves.

(vi) Sequencing valves.

3.1.1 Direction Control Valves

[W-13]

- The name of this valve itself clarifies its function.
- This valve is like a 'Traffic Police'. It directs the flow of pressurised as well as used oil to actuator and oil reservoir respectively.

Classification of Direction Control Valves (DC Valves)

(A) Classification based on Construction:

(i) Seat or Poppet type

(ii) Spool type:
- Sliding spool
- Rotary spool

(i) Seat or Poppet Type DC Valve: In this valve a poppet or ball or similar item like plate or cone is made to seat over a specially constructed finely machined and polished seat.

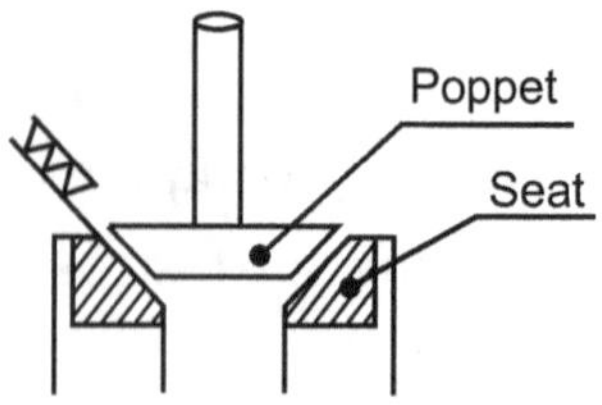

Fig. 3.3: Seat or Poppet Type DC Valve

Advantages:

(a) Suited for very high pressure applications.

(b) Very minor leakage.

Disadvantages:

(a) Not suitable for large valve sizes.

(b) Due to complicated construction, valves are costly.

(ii) Sliding Spool Type DC Valve: In this valve there is small piston inside a valve casing which slides inside the casing thereby opening or closing the ports (drilled holes) in the valve body.

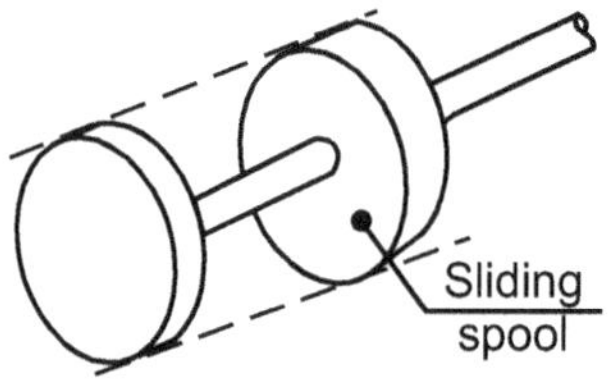

Fig. 3.4: Sliding Spool type DC Valve

Advantages:

(a) Simple in construction.

(b) Cost is low.

(c) These valves are permanently balanced.

Disadvantages:

(a) Leakage inside the valve is possible.

- Sliding spool type DC valve is most widely used valve.

(iii) Rotary Spool Type DC Valve [S-15, W-16]: Rotary spool valve consists of a rotating spool which aligns with ports in the stationary valve casing, so that fluid is directed to the required port. A/B/P/R are the ports in casing. Port 'P' is a pressure port through which pressurized oil is coming in the valve. 'R' port is the port through which used oil is returning to the oil tank. In Fig. 3.35 it is clear that:

Port P is connected to Port B (P $\rightarrow$ B) and

Port A is connected to Port R (A $\rightarrow$ R). [S-15, W-16]

Advantages:

(a) Rotary spool type DC valve is compact and simple in design.

(b) Are having low operating force.

Disadvantages:

(a) Due to rotary motion of spool, leakage inside this valve is possible.

(b) These valves are not suitable for high pressures because sometimes pressure of oil rotates the valve in unwanted directions.

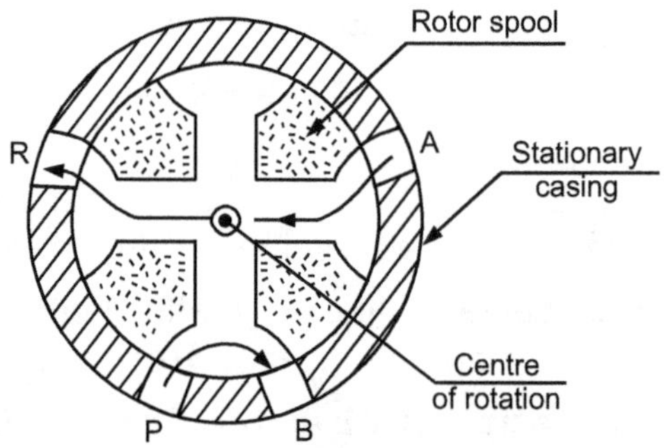

Fig. 3.5: Rotary Spool Type DC Valve

- Rotary spool valves are generally manually operated.

3.2 TYPES OF DC VALVES BASED ON NUMBER OF PORTS (Generally Spool Type) [S-09]

1. 2 × 2 DC valve
2. 3 × 2 DC valve
3. 4 × 2 DC valve
4. 4 × 3 DC valve

Now we will see these valves one-by-one in detail alongwith symbols.

3.2.1 2 × 2 DC valve (Read as Two-by-Two DC Valve) [S-09]

- The Fig. 3.6 shows 2 × 2 spool type DC valve, push button operated.

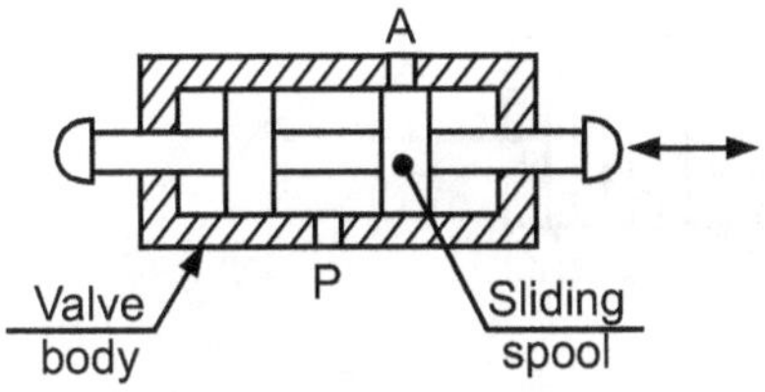

Fig. 3.6: Normal Position

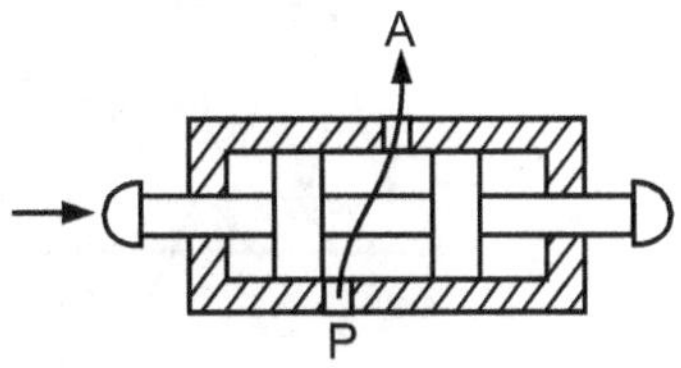

Fig. 3.7: Actuated Position

- **Construction:** It consists of sliding spool in a finely finished bore. The actuation is by push button.
- **Operation:** Through port P, pressurised oil is coming in. This oil will go out (directed) through port 'A'. In Fig. 3.7 shows port 'P' is not connected to port 'A'. Hence no flow from P to A.
- Now left push button is pressed. The spool will shift towards right and port 'P' will be connected to port 'A' and oil will start flowing from port P to port A.

Important: This valve is normally closed.

- 2 × 2 DC valve can be best compared with our domestic **water tap**. The water tap in bathroom is always closed. This is its normal position i.e. it is normally closed. When we open it, water starts flowing.
- Symbol of 2 × 2 DC valve: The symbol concept is explained below.
- The valve position is shown by squares. (Remember while drawing these squares, it should be exact square you should not draw it as rectangle).
- The direction of flow of oil is shown by arrows in these squares.

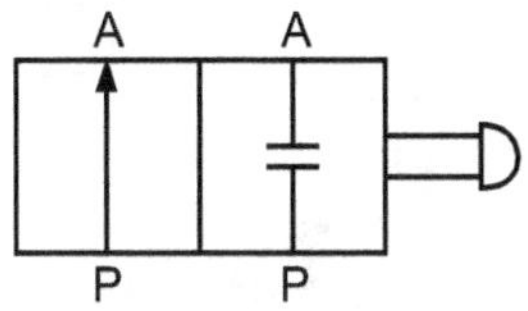

Fig. 3.8: Symbol

- Actuation arrangement i.e. push button in above case must be shown outside.
- **Remember:** At a time think of only one square i.e. one position.

3.2.2 3 × 2 DC Valve (Read as three by two DC Valve) [S-09, 12, W-09]

- Similar to 2 × 2 DC valve, in this valve also, a spool reciprocates in a finely finished bore. The spring is attached to front spool. Fig. 3.9 shows the actuated position. Push button is pressed in port P is connected to port A and no flow through port R. When spool is pressed in, spring is compressed.

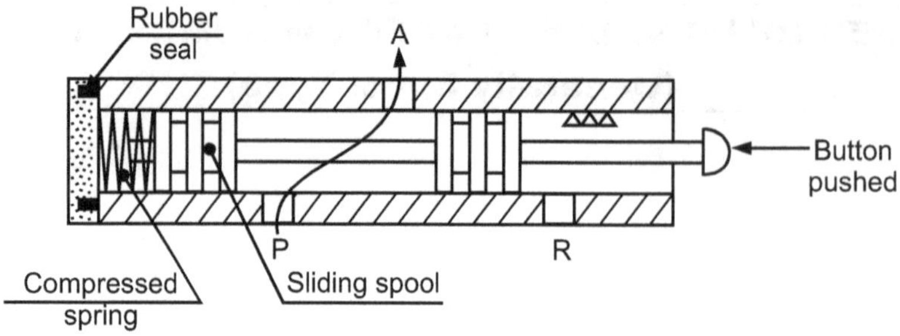

Fig. 3.9: Actuated Position

- **Now push button is released:** When push button is released, spring will expand and spool will move towards right. Now port A will be connected to port R and port P will be disconnected so that flow of used oil will start from port A to port R and then to oil reservoir.

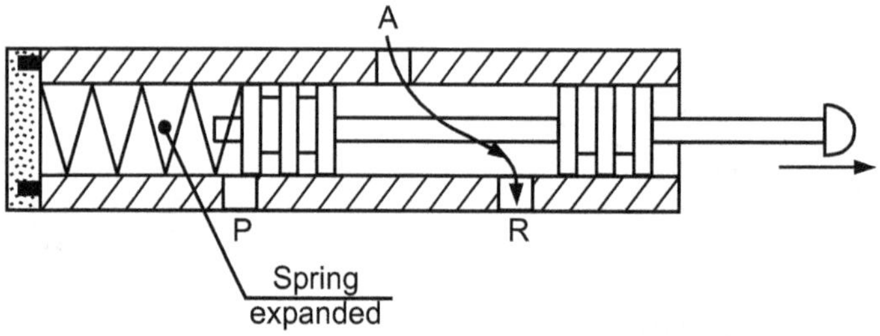

Fig. 3.10: Normal Position

- **Symbol of 3 × 2 DC valve:** 3 × 2 means 3 ports and 2 positions. Two positions mean 2 squares. Look at square (1). It shows actuated position. Port P connected to A and port R is disconnected. Similar square (2) shows normal position i.e. Port A → R, port P is disconnected.

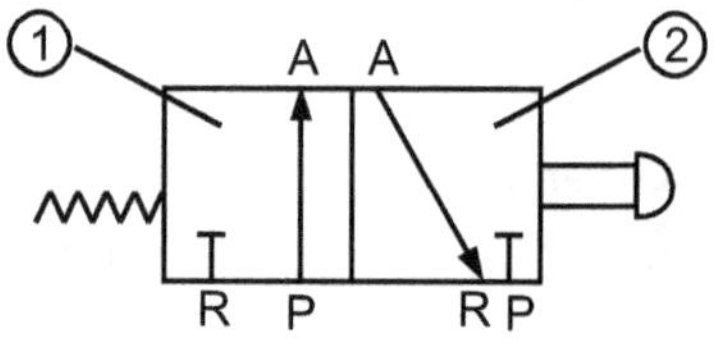

Fig. 3.11

3.2.3 4 × 2 DC Valve

(Read as four by Two DC valve can also be read as '4 way 2 position' DC valve)

[S-11, 12, 14, 16, W-13]

- 4 × 2 DC valve is shown in Fig. 3.12. It is in normal position. There are 4 ports, A and B are consumer ports (ports going to actuator). P is pressure port through which pressurized oil goes in. Port 'R' is used oil and returns to oil reservoir. The value is push button operated manually. Valve regains its normal position by spring expansion. From Fig. 3.12 it is clear that spring is fully expanded to its free length. Port P is connected to port A and port B is connected to port R. Now push button is pressed is as shown below. This is actuated position. Now sliding spool will move towards left thereby compressing the spring. Now as we can see port P is connected to port B and port A is connected to return oil port R.

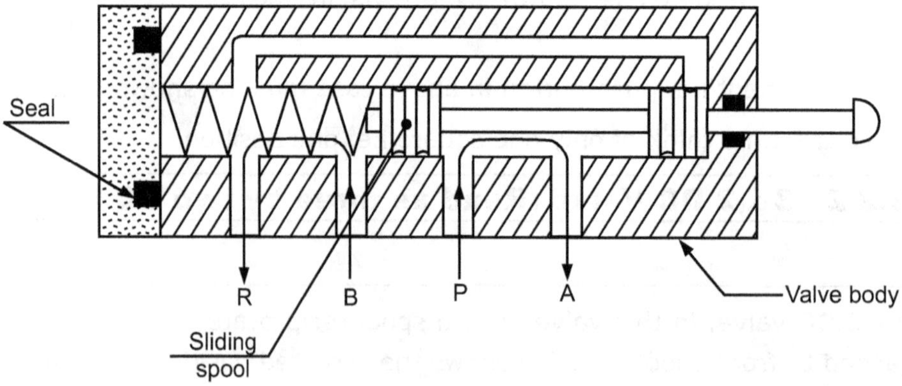

Fig. 3.12

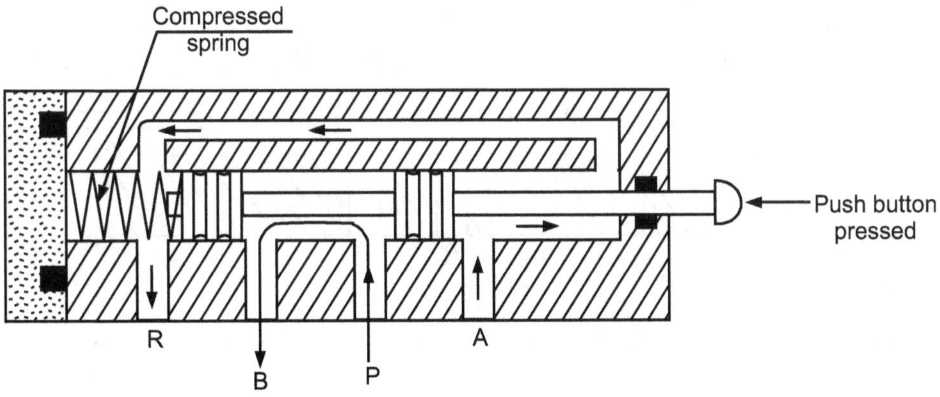

Fig. 3.13

- **Symbol:** 4 × 2 means **4 ports** and **2 positions**.
- Two positions mean two squares. Look at square (1) which shows normal position P → A / B → R.
- At square (2) we can see actuated position P → B / A → R.
- Push button and spring are shown outside.

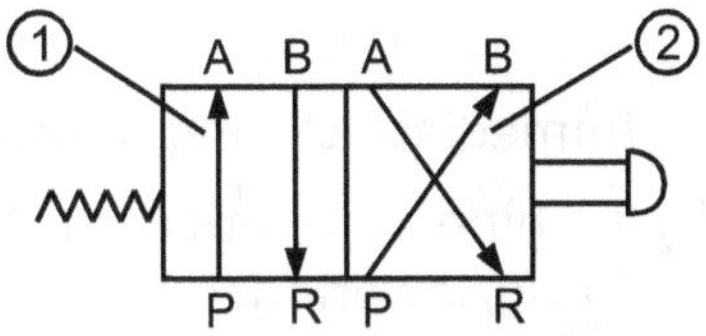

Fig. 3.14: Symbol

3.2.4 4 × 3 DC Valve [S-11, 16]

- We have seen 4 × 2 valve. This is slight modified version of 4 × 2 DC valve. Similar to 4 × 2 valve there are 4 ports (A/B/P/R). But one position is added i.e. centre position. Following two figures are showing similar positions as we have seen in 4 × 2 DC valve.

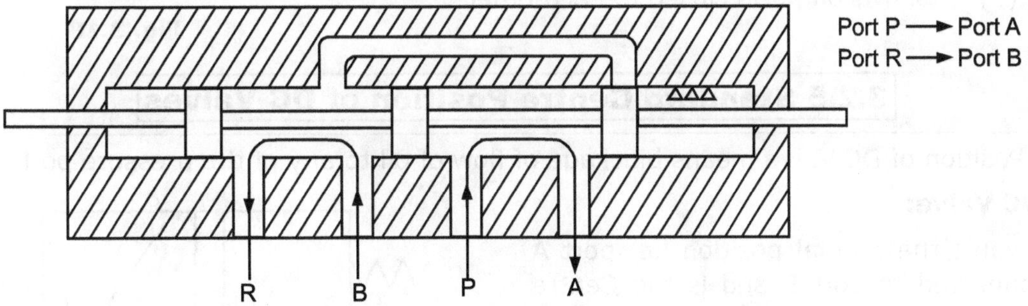

Fig. 3.15

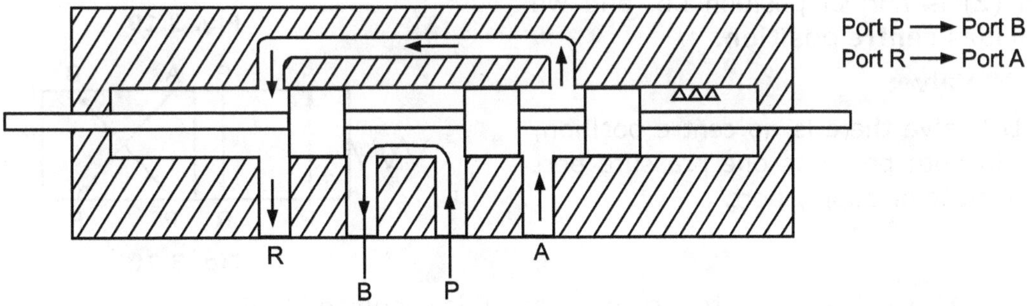

Fig. 3.16

- As we can see that, the spool of this valve is having three positions. The spool is so selected because we have to obtain third position also called as 'Closed Centre Position'. This position is shown in Fig. 3.17. We have shifted the spool in such a manner that all ports are closed to each other. No flow from port P to port A or B and no flow from port A and B to R.

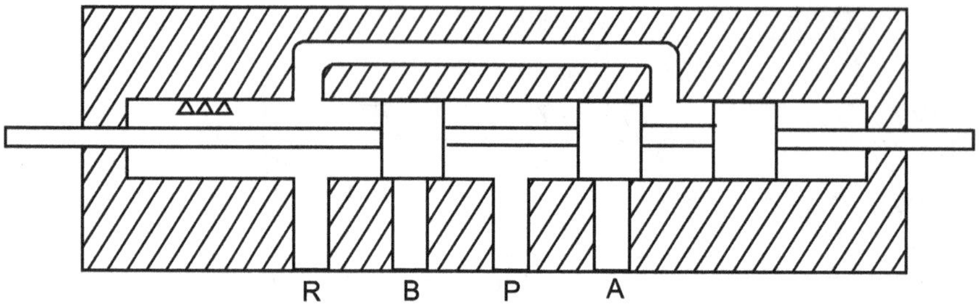

Fig. 3.17

- When DC valve attains this position, pressurised oil returns to reservoir via, pressure relief valve.

Applications of this valve:

Centre closed position DC valve is suitable for:

(a) Immediate closing of movements of actuator.

(b) 'Inching' operation in raising or lowering the heavy loads. *For example*, Dumper trolly lifting/lowering.

Symbol of 4 × 3 DC valve (Centre Closed Position):

4 × 3 means 4 ports and 3 positions. Three positions mean three squares.

Square (1): Port P → Port A

Port R → Port B

Square (2): Port P → Port B

Port R → Port A

Square (3): Shows all ports closed to each other.

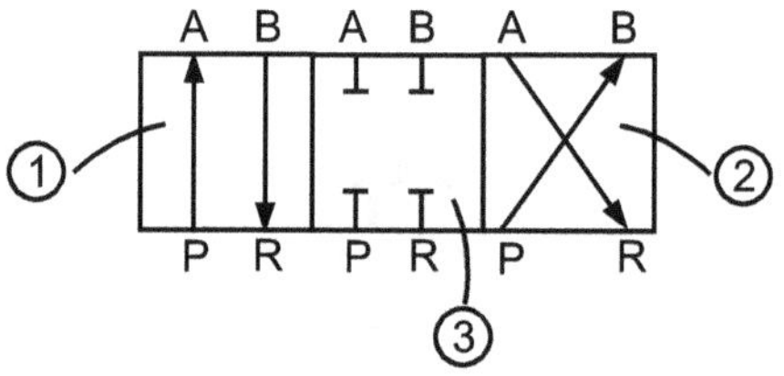

Fig. 3.18

3.2.5 Standard Centre Position of DC Valves

- Centre Position of DC Valve means blockage of flow of oil to any of the pressure port.

(I) 3 × 2 DC Valve:

- In this valve the normal position i.e. port A is disconnected to port P and is the Centre Position of 3 × 2 DC Valve.
- Position (2) is normal position i.e. and we may call it as **centre position.**

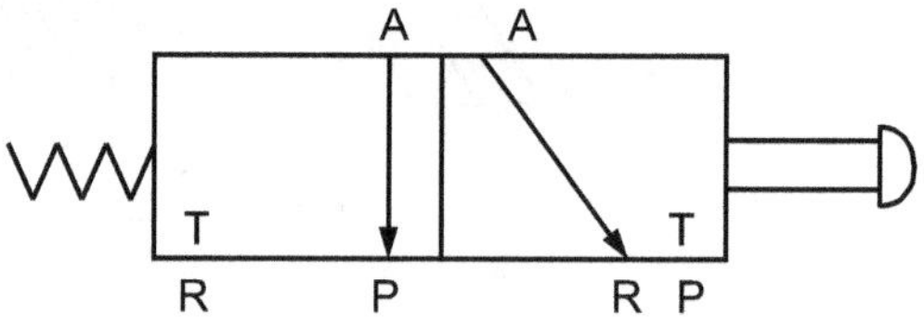

Fig. 3.19

(II) 4 × 2 DC Valve:

- In this DC valve there is no centre position, because in both positions one pressure port is connected to oil supply port.

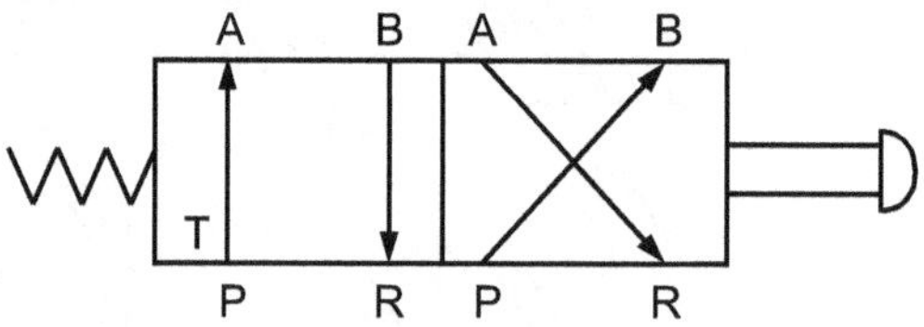

Fig. 3.20

In position (1) port P is connected to port A and B is connected to R.

In position (2) port P is connected to port P and A is connected to R.

This means in any position, ports are not blocked. So no centre position is possible in 4 × 2 DC Value.

(III) 4 × 3 DC Valve:

- In this DC Valve there are 3 position and naturally centre position decided 4 main centre positions. These centre positions are as under:

(a) Tandem Centre Position of 4 × 3 DC Valve:

Tandem Centre Position: Hydraulic DC valves with the tandem-centre position are the most commonly used 4/3 DC valve for the construction of low power hydraulic equipment with constant delivery pumps, mainly due to their energy saving feature.

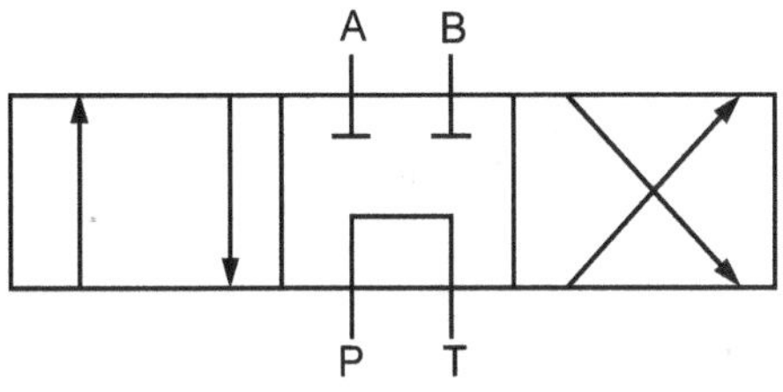

Fig. 3.21

In the tandem centre design, as shown in Fig. 3.21 above the pressure port P is connected to the tank port T and at the same time, the working ports A and B are blocked individually when the valve is in its neutral position.

Tandem centre position is 4/3 DC Valve is used in a hydraulic system for controlling the flow from the pump to the cylinder in the system. The flow can return to the system reservoir essentially at the atmospheric pressure through the internally connected ports P and T when the valve is in its neutral position. Therefore, the use of such valve in the system provides:

- The reduced build-up in the system.
- At the same time, the cylinder connected to the working ports of the valve can be stopped abruptly and hydraulically locked when the valve is brought to its centre position.
- Preventing the inadvertent of the cylinder piston.
- However, it may be noted that a cylinders tends to creep with the valve mitered due to external forces working on the cylinder and the resulting fluid leakage. Moreover, this valve is not suitable for use in a multiple actuator hydraulic system with just one pump, as the pump flow cannot be used for other parts of the circuit independently. However, the tandem centre valves can be used for the control of a series operated multiple cylinder system if the cylinders are required to operate one after another.

(b) Closed Centre Position of 4 × 3 DC Valve

Closed Centre Position: In the closed centre design of a 4/3 DC hydraulic valve, as shown in Figure above all four ports of the valve are blocked from one another in its neutral position. Consider a hydraulic system with a pump and multiple actuators.

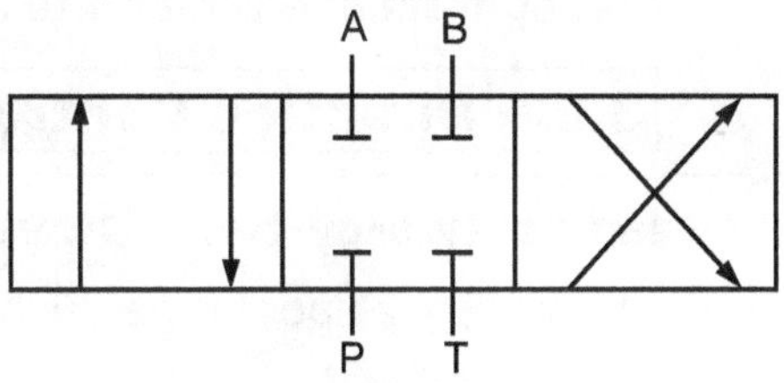

Fig. 3.22

Each of these actuators is controlled by a 4/3 DC closed centre valve. As the pressure port in each of these valves is blocked in its neutral position, the pump flow can be directed to a particular part of the circuit independently when the corresponding valve is actuated. At the same time, every actuator connected to the working, parts of the corresponding valve can be hydraulically locked when the valve is brought to its centre position.

(c) Open Center Position of 4 × 3 DC Valve

Open-center position: In the open center design of the 4/3 DC valve, as shown in figure above all four ports of the valve are connected in its neutral position.

- Consider a hydraulic system with a constant delivery pump and a cylinder controlled by a 4/3 DC open center valve.

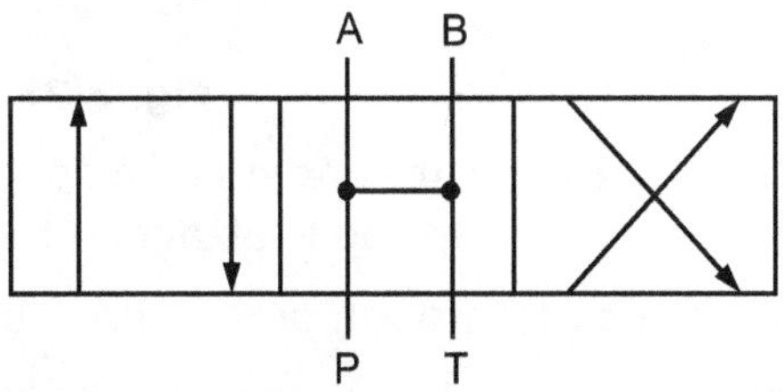

Fig. 3.23

- The flow generated by the pump can return to the system reservoir essentially at a low pressure when the valve is brought to its center position a At the same time, the cylinder can be moved freely by applying an external force.
- However, this type of valve is, usually, not recommended for the control of a cylinder because of the possibility of overshooting of its power piston while the valve is shifted to its center position.
- However this type of valves finds applications for the control of hydraulic motors.

(d) Float Centre Position:

Float Centre Position: Another commonly used configuration of the 4/3 DC valve is the float centre design as shown in Fig. 3.24.

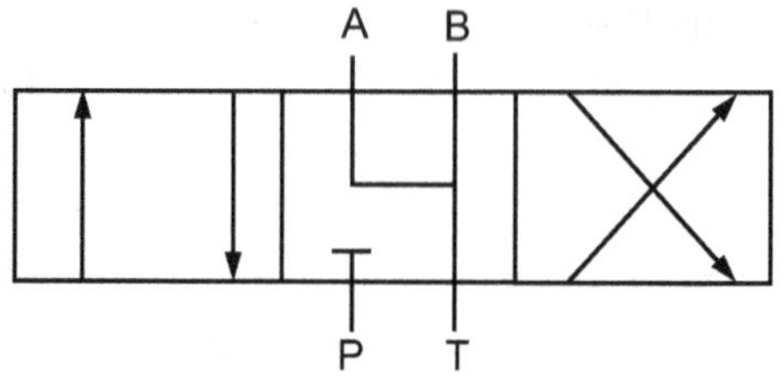

Fig. 3.24

- Consider a hydraulic system with a fixed displacement pump and an actuator controlled by a 4/3 DC float centre valve.
- The valve has both its working lines connected to the system reservoir and its pressure port blocked, in its neutral position.
- In this position, the valve allows the pressurized fluid on both sides of the system actuator to drain back to the reservoir. That is, the actuator is free to float in this position.
- Therefore, the use of this valve enables a soft shut-off while stopping the actuator.

Additional Centre Positions: Many other centre positions are also possible for the 4/3 DC hydraulic valves.

- In one type of design of the 4/3 DC valve, the pressure port P and the working port B of the valve can be blocked and the working port A can be open to the tank port T when the valve is in its centre position.
- In another design, a three position hydraulic valve can be either closed or open while crossing from one position to another.

3.2.6 Direction Control Valves (DC Valves) (Pneumatics)

1. 2 × 2 (Read as 'two-by-two') DC valve (spool type)

Fig. 3.25 shows '2 × 2 spool type push button operated DC valve'.

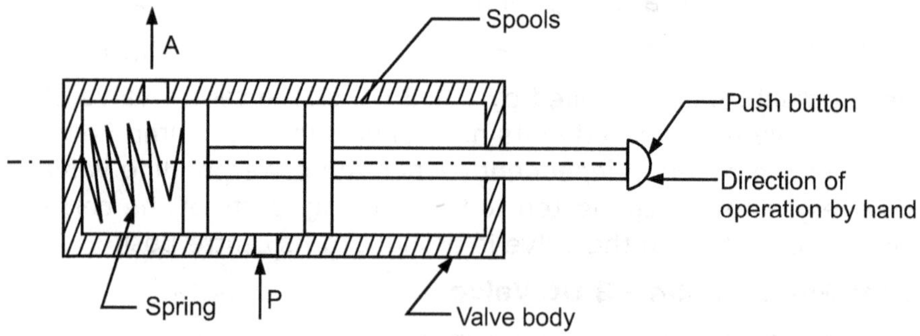

Fig. 3.25: 2 × 2 DC Valve Normal Position

- **Construction:** It consists of a spool sliding in a finely finished bore. The actuation of spool is by push button. Spring is attached to front spool.
- **Operation:** Through port P, pressurized air enters into inside bore of valve.
- When push button is pressed, spool will move forward (right to left). Spring will compress and air will move out through port A, because port P is connected to port A.

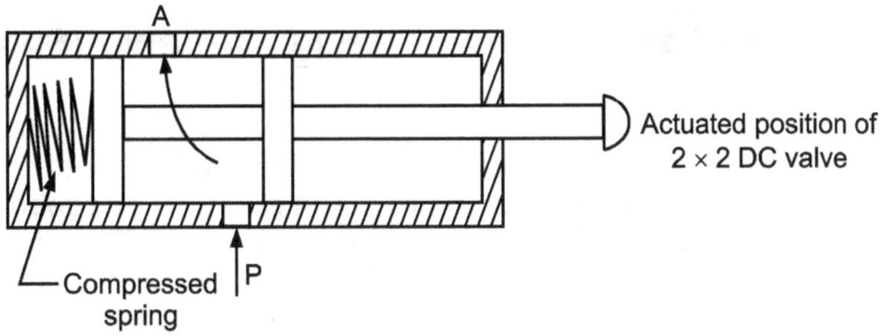

Fig. 3.26: 2 × 2 DC Valve Actuated Position

- When push button is released, spring will expand and the spool will regain its original position. Port P will be disconnected from A.
- Normal position of this valve is closed.
- 2 × 2 DC valve can be compared with ON-OFF water tap used in our house. The water tap on basin is always closed. When we open it the water starts flowing (Port P of water connects to Port A). Fig. 3.27 clarifies this concept. Remember that tap is normally closed; and it is 2 × 2 DC valve.

Fig. 3.27: Water Tap is 2 × 2 DC Valve

- **Symbol of 2 × 2 DC Valve:** The symbol generation concept can be explained as under.

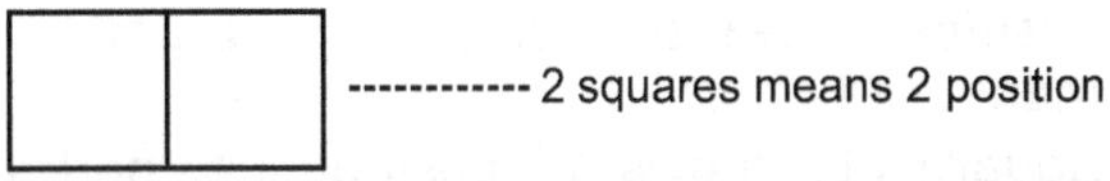

Fig. 3.28

- The valve position is shown by squares.

  In above valve, there are two positions:

  One : Air pressure port P is connected to consumer port A (On).

  Two: Air pressure port P is disconnected from consumer port A (Off).

  Hence to show 2 × 2 valve, we have to draw 2 squares (Remember: the square should be exact square having all equal sides. Student may draw squares having 10 mm (1 cm) side for suitability in drawing. Preferably use engineering plastic 'templet' to draw squares).
- The direction of flow of air is shown by 'arrows' in the squares. In square (1), Port A is connected to port P. In square (2), Port A is disconnected from port P (Refer Fig. 3.29).

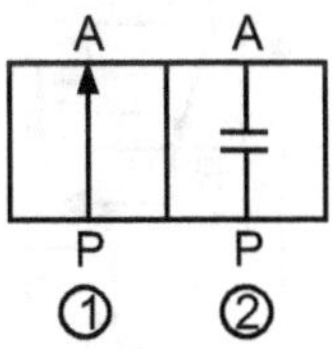

Fig. 3.29

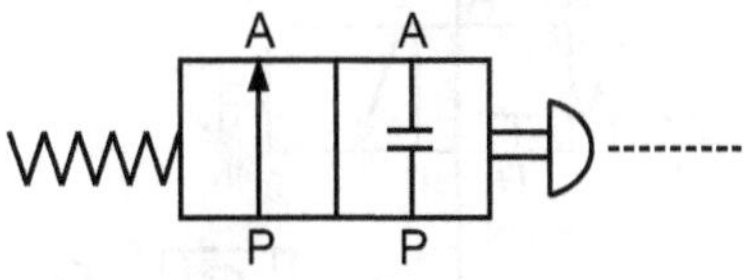

Complete symbol of push button operated 2 × 2 DC valve

Fig. 3.30

- **Remember:** At a time think of only one square and position of arrow in it.
- So now we will have to construct complete symbol of 2 × 2 DC valve. As we know from the construction, the valve is normally closed. It is operated by push button and it regains its original position by spring force. Hence we have to attach push button and spring to the two squares. So that we will get complete symbol (Refer Fig. 3.30).

3.2.7 3 × 2 (Read as 'Three-by-Two') D.C. Valve (Spool type)

[W-13, W-14]

- Similar to 2 × 2 DC valve, in this valve also, spool reciprocates in finely finished bore of valve body (machining quality is shown by two triangles). The spring is attached to front spool. In the Fig. 3.31, Port A is connected to port R. Remember that port R is exhaust port. After doing work, the air is returning from port A to port R and going into atmosphere. That means air is exhausting (Refer Fig. 3.31).

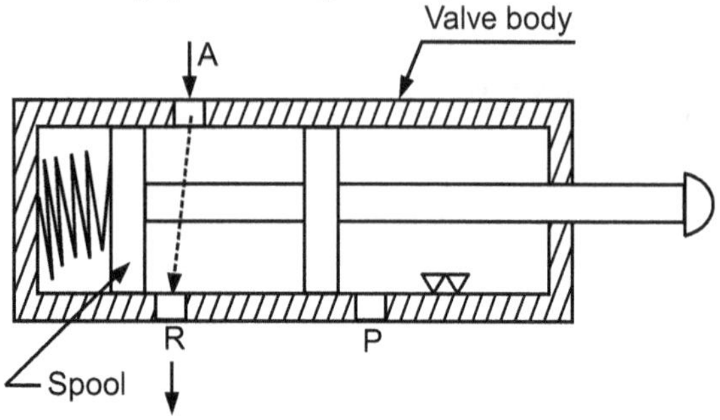

Fig. 3.31

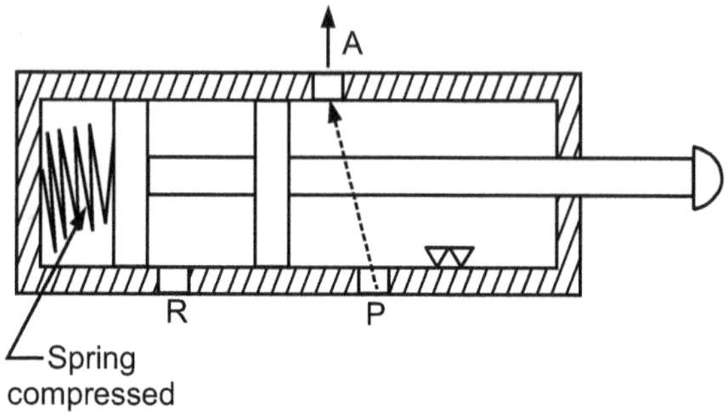

Fig. 3.32

- Now push button is operated. When push button is pressed, spool will move from right to left, thereby compressing the spring. With this movement of spool, port A will be connected to pressure port P. Compressed air will start flowing from port P to port A (Refer Fig. 3.32).
- Exhaust port R is disconnected from port A.
- Now it can be easily understood that, when push button is released, spring will expand and spool will regain its original position.

Symbol of 3 × 2 DC valve:

- Three-by-two means there will be 2 positions and 3 ports, 2 positions means 2 squares. 3 ports are shown in a specific manner.
- Square (1) shows 1st position i.e. port A is connected to port R (exhaust port) and port P is disconnected. Square (2) shows, 2nd position i.e. operated position. Push button is operated. Port P is connected to port A and port R is disconnected. In this manner symbols are formed.

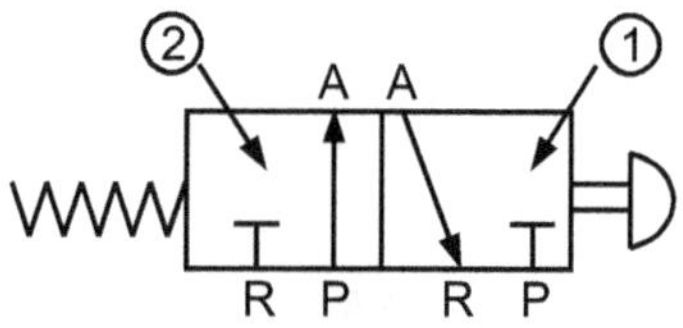

Fig. 3.33

Application of 3 × 2 DC valve:

- Let us see the applications of 3 × 2 DC valve

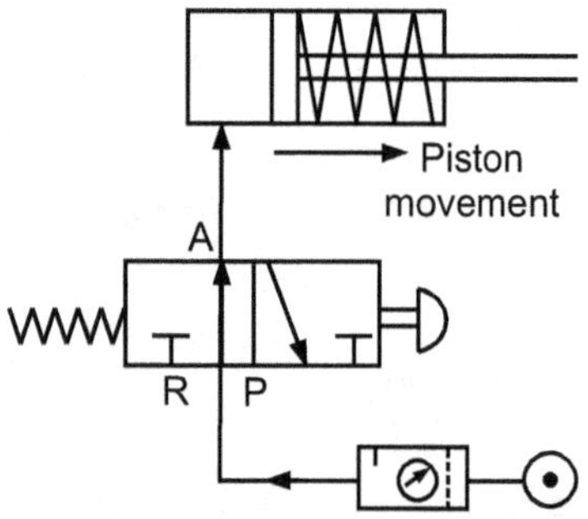

Fig. 3.34 (a)

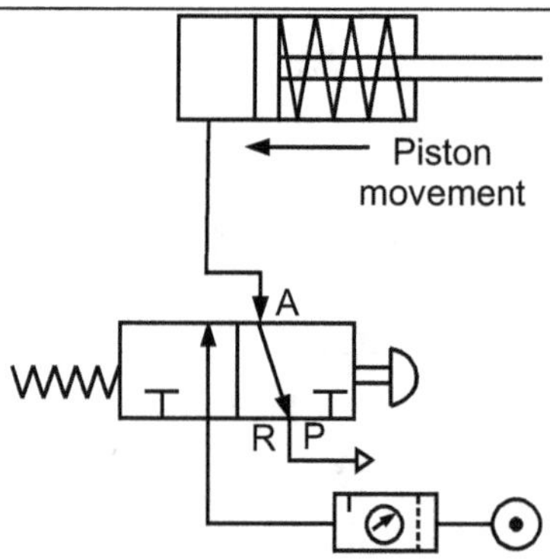

Fig. 3.34 (b)

Fig. 3.34 (a)	Fig. 3.34 (b)
3 × 2 DC valve is used here to operate single acting pneumatic cylinder (SA Cylinder) Port P is connected to port A. Compressed air will push the piston from left to right thereby compressing the spring (power stroke).	Now power stroke is complete. Push button of DC valve is released. Spring of cylinder will expand. Piston will move from right to left, there by pushing out the air out of cylinder through port A. This port is now connected to port R i.e. exhaust port. Air will return to atmosphere.

3.2.8 3 × 2 Seat Type DC Valve

- 3 × 2 DC valve can also be seat type as shown in the Fig. 3.35. There are two valve seats shown, on which cones of valve element rests. There is no spring. The valve element can move up-down. At present position shown port A is connected to exhaust port R. Pressure port P is disconnected. Now air starts entering from port P. Due to pressure force the valve element will lift-up. Due to this movement port P will be connected to Port A and Port R will be blocked by upper cone of valve element.
- The symbol of seat type valve is same as that of spool type valve.

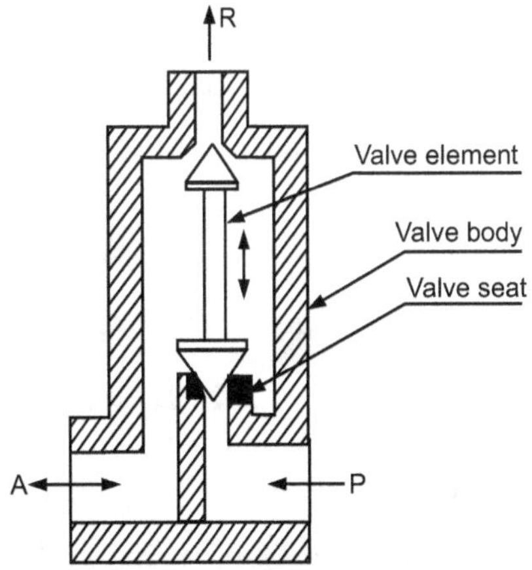

Fig. 3.35: 3 × 2 Seat Type DC Valve

3.2.9 4 × 2 DC Valve (Seat and Poppet Type)

[S-16]

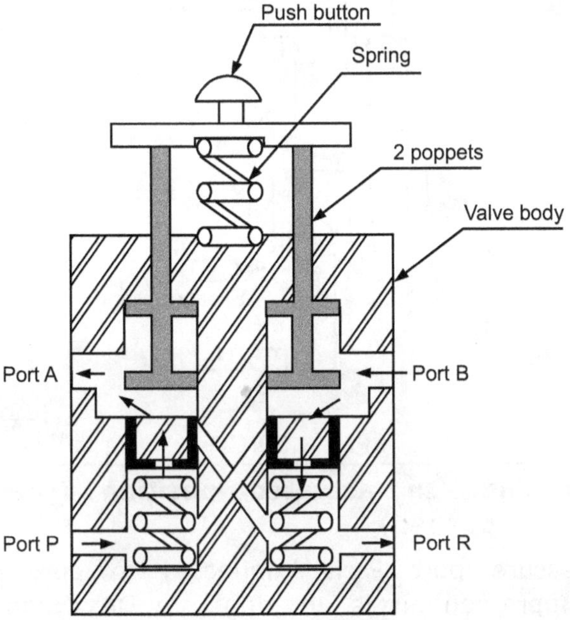

Fig. 3.36: Normal Position of Valve (Push Button NOT Pressed) of 4 × 2 Valve

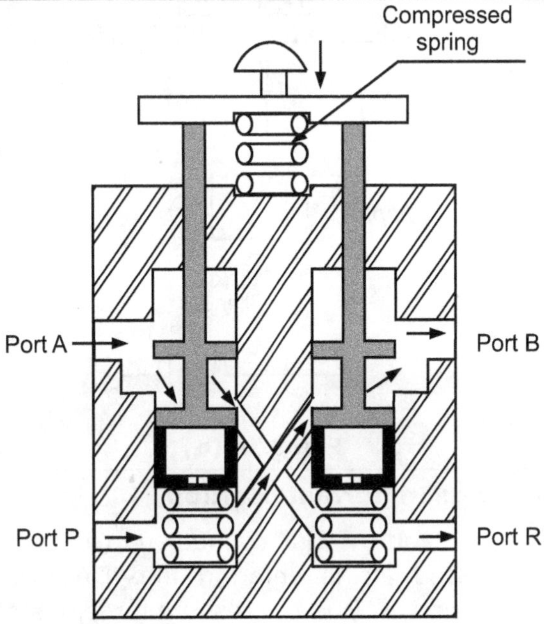

Fig. 3.37: 4 × 2 DC Valve (Actuated)

- This Fig. 3.36 shows normal position of 4 × 2 DC valve. There are two poppets. A and B are consumer ports. (Ports attached to actuator to obtain useful work).
  Port P is pressurized compressed air inlet port.
  Port R is exhaust port.
  The present position shows that:
  - Port P is connected to Port A.
  - Port B is connected to Port R.

  Now push button is pressed downwards.
  It shows actuated position of valve. See the above Fig. 3.37.
- As push button is pressed downwards, all springs will be compressed. And spool will be in such a position that:
  - Port A will be connected to port R.
  - Port P will be connected to port B.
  - When button will be released, spring below button will expand and spool will regain its original position i.e. normal position.

- **Symbol of 4 × 2 DC valve:** Four-by-two means there will be 2 positions and 4 ports. 2 positions means 2 squares. 4 ports are shown in specific manner as below (See Fig. 3.38).

  Square (1) shows – normal position of valve i.e.
  - Port P connected to Port A
  - Port B connected to Port R

  Square (2) shows – Actuated position of valve i.e.
  - Port A connected to Port R
  - Port P connected to Port B

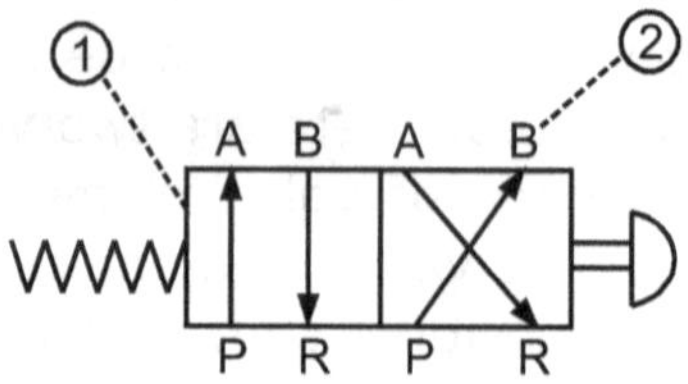

Fig. 3.38

- **Application of 4 × 2 DC valve:** This valve can be used to operate reciprocating actuator called 'Double Acting Cylinder' (DA cylinder) 4 × 2 DC valve is most widely used valve in pneumatic application/circuits.
- Observe the following Figs. 3.40 (a) and (b).

Forward stroke of DA cylinder (Normal position)	**Retract stroke of DA cylinder (Actuated position)**
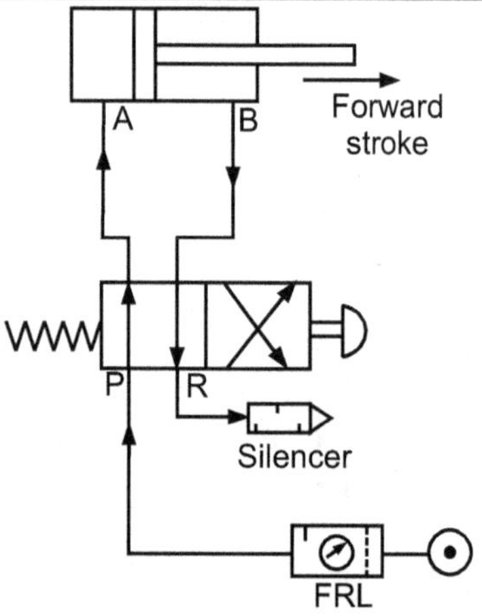 **Fig. 3.39 (a)**	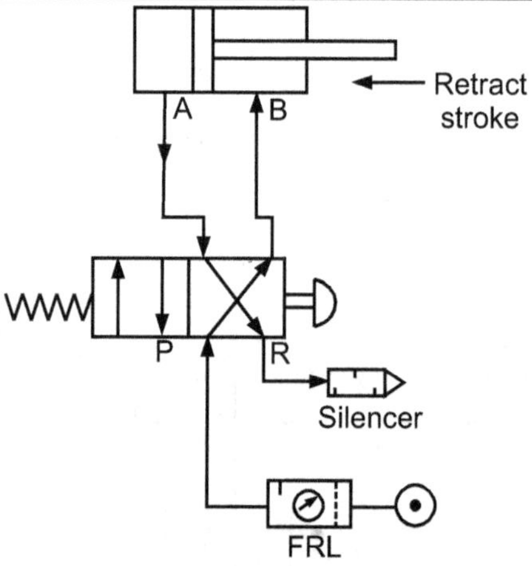 **Fig. 3.39 (b)**
Working: In normal position Pressure Port P is connected to port A. Compressed air is entering into DA cylinder through port A. Due to pressure force, piston of DA cylinder will move from **left to right**. During this movement air on other side of piston will come out through port B and will be exhausted via DC valve into atmosphere through silencer.	**Working: In actuated position (press button pushed)** Pressure port P is connected to port B. Compressed air is entering into DA cylinder through port B. Due to pressure force, piston of DA cylinder will move from **right to left**. During this movement air on other side of piston will come out through port A and will be exhausted via DC valve into atmosphere through silencer.

3.2.10 Important

What is the meaning of 'Piston of DA cylinder is forwarding'?

Ans. When compressed air is admitted in the bore of DA cylinder from piston side then, movement of piston (left to right as shown in figure) is called forward stroke, it is said that piston is forwarding. Two cases are shown in Fig. 3.40. In Fig. 3.40 (b) piston is moving from Right to left. Even then it is forwarding. Because air is admitted from piston side.

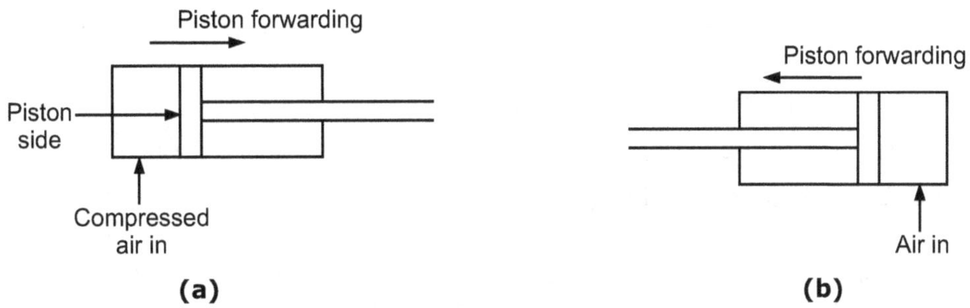

Fig. 3.40

What is the meaning of 'Piston of DA cylinder is Retracting?

Ans. When compressed air is admitted in the bore of DA cylinder from piston-rod side, then movement of piston from left to right (or right to left) is called retract movement and piston is said to be 'retracting'.

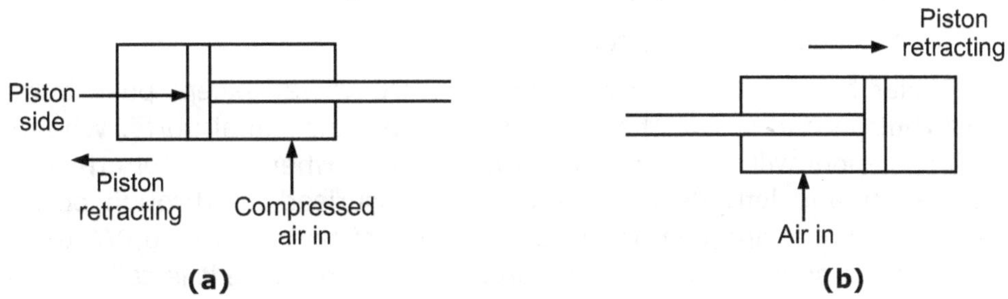

Fig. 3.41

3.2.11 5 × 2 DC Valve (Spool Type) [W-10, 13; S-15]

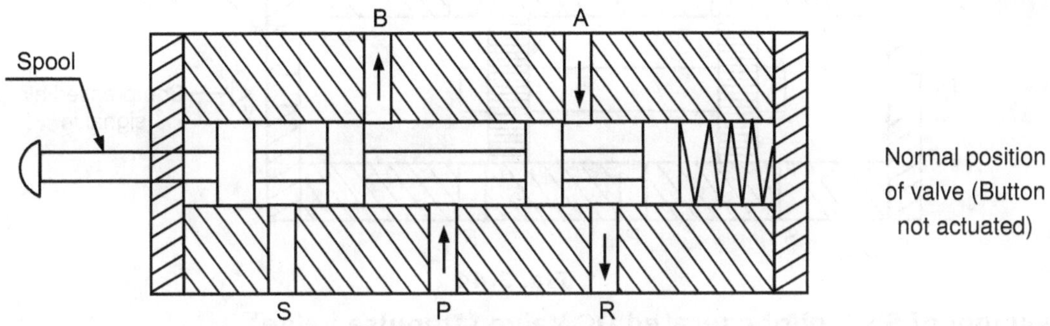

Fig. 3.42: 5 × 2 DC Valve

- Spool type 5 × 2 DC valve (push button operated) is shown in the Fig. 3.42. There is single spool. A and B are consumer ports (ports attached to actuator to obtain useful work). P is pressure port, R is normal exhaust port and S is easy exhaust port.
- The normal position shown in above figure shows that:
  - Port P is connected to port B.
  - Port A is connected to port R (Air is exhausting).
  - Port S is closed or disconnected.
- Now if push button is operated. Spring will compress due to movement of spool from left to right and then
  - Port P will get connected to port A.
  - Port B will get connected to port S.
  - Port R will close or will be disconnected.

- **Symbol of 5 × 2 DC valve:** Five-by-two means there will be 2 positions and 5 ports (A, B, P, R, S) 2 positions means 2 squares. 5 ports are shown in specific manner. (See Fig. 3.43)

  Square (1) shows – Normal position.

  Square (2) shows – Actuated position (push button is operated).

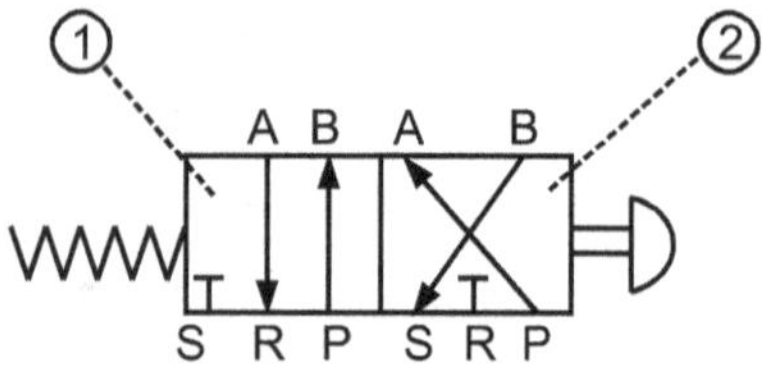

Fig. 3.43: Symbol

Note: At a time only one position (connections shown in one square) is in operation. So while studying the circuit respective one square is considered.

3.2.12 5 × 2 Pilot Operated DC Valve (Impulse Valve) [S-11]

- Pilot operated means instead of push button, the spool movement is done by compressed air. This compressed air entering into valve to operate spool of valve is called 'air signal'.
- For entry of air signal there are separate ports.
- Following Fig. 3.44 will clarify the concept.
- This valve is similar to push button operated 5 × 2 DC valve, except push button and spring. Instead of push button and spring there are x and y are air signal ports. When pressurised air enters from port x, spool will move from left to right, and when air enters through port y, spool will move from right and left. By controlling the pressurised air through port x or y we can control movement of spool and can obtain connections of ports i.e. B to P/A to R or A to P/B to S etc. The operation of spool is by impulse (force) of air and hence it is called impulse valve.

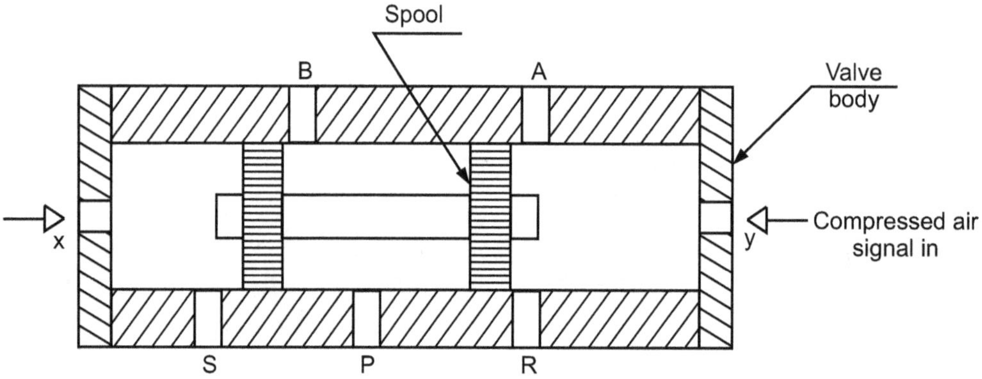

Fig. 3.44

The symbol of 5 × 2 pilot operated DC Valve (Impulse valve)

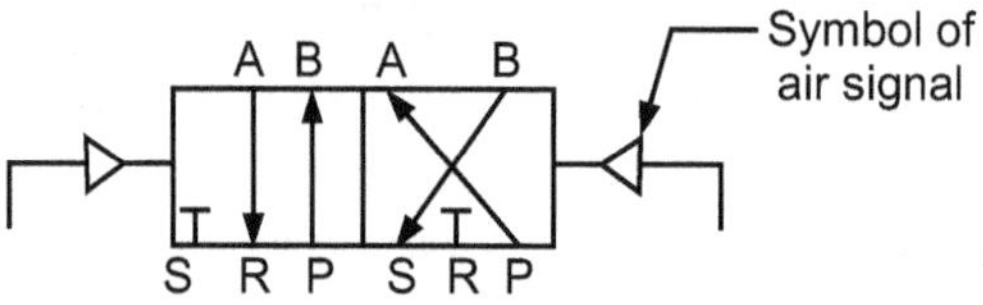

Fig. 3.45

3.3 FLOW CONTROL VALVES [W-11, S-16]

- Flow control valves (as we have seen) are used to regulate the flow of hydraulic oil in hydraulic circuit for controlling the 'SPEED' of actuators.
- For linear actuators speed is measured in m/sec.
- For rotary actuators speed is measured in r.p.m.
- **Importance of flow control :** Control of pressurised oil flow is extremely important because the rate of movement of machine element depends on rate of flow of pressurised hydraulic fluid.
- Flow control is achieved by 'throttling' or 'diverting'.

- **Throttling:** Throttling is passing the pressurised oil through small restriction like 'orifice' (hole). Fig. 3.46 (a) shows that oil flow is restricted through orifice.

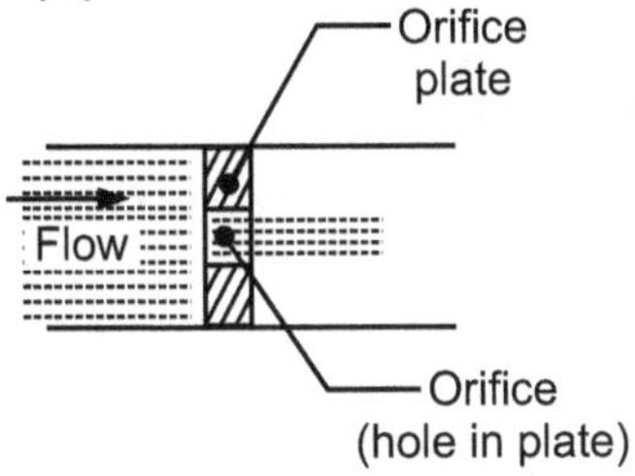

Fig. 3.46 (a)

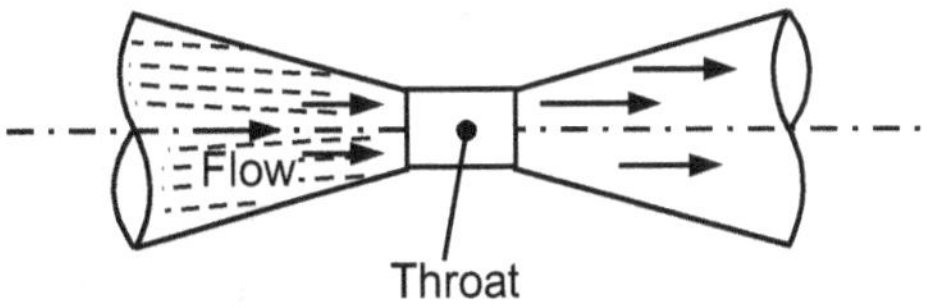

Fig. 3.46 (b)

- In venturimeter the throat of venturi does the work of throttling. (See Fig. 3.46 (b)).
- During throttling the oil looses its part pressure. But the speed of fluid increases.
- Other throttling arrangements are:

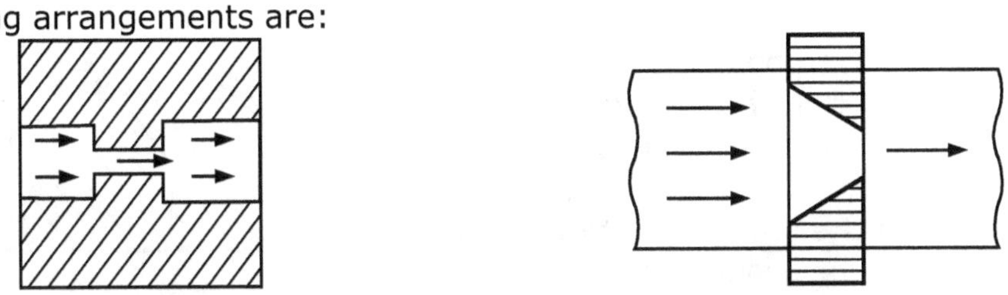

Fixed Throttle **Diaphragm Throttle**

Fig. 3.46 (c)

- **Diversion:** Diverting is by-passing the part of flow so that the actuator receives only the portion to perform its task. 'Bleed-off-Circuit' (Refer Topic 3 in this book) is the best example of diversion of flow. Fig. 3.47 shows how flow is diverted. Suppose from DC valve quantity of oil going out is say 5 lit/min. flow to 1 lit/min. is diverted. Hence actuator will receive 4 lit./min.
- **Types of Flow Control Valves:** Following are the types of flow control valves:
  (a) Needle valve (Basic two way flow control valve).
  (b) Pressure compensated flow control valve.
  (c) Non-compensated flow control valves.
  (d) Check valves.

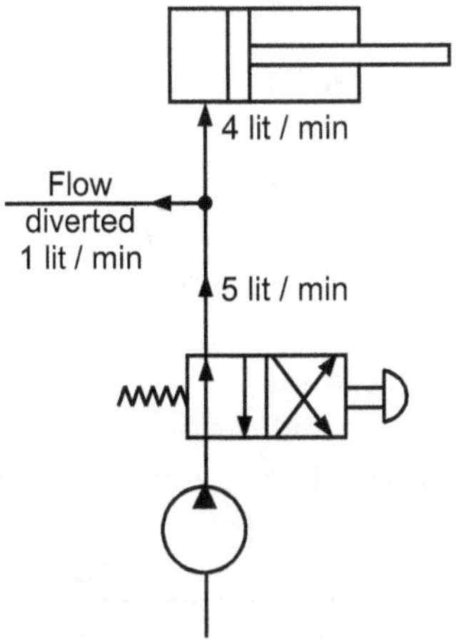

Fig. 3.47

- Now we will see one-by-one.

3.3.1 Needle Valve

[S-15]

- It has a pointed stem or needle that can be adjusted upward or downward by rotating the handwheel. Since stem is threaded we can accurately control the needle which in turn controls the outflow.
- This is the most commonly used flow control valve in many hydraulic systems.
- This valve is also used as 'stop valve' to stop or 'shut-off' the flow.

[S-15]

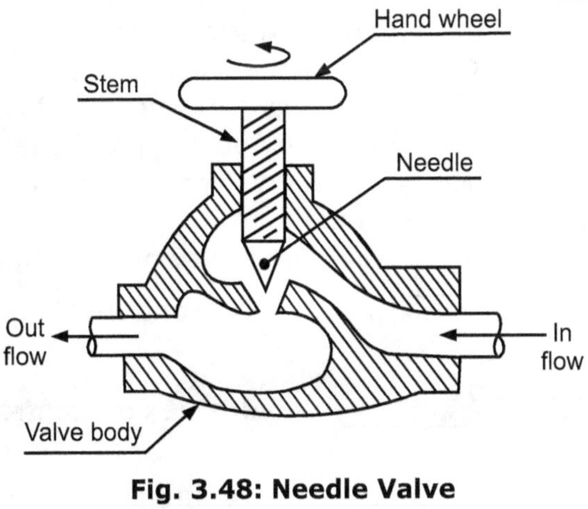

Fig. 3.48: Needle Valve

3.3.2 Pressure Compensated Flow Control Valve [W-11, W-12]

- In any hydraulic circuit there are slight variations in pressure of oil. When pressure changes, the rate of flow changes. But many circuits require constant flow regardless of input or output pressure variations in circuit. Then we must use pressure compensated flow control valve. Such a valve is depicted in below Fig. 3.49.
- **Construction:** It consists of hollow cylinder shaped poppet at the bottom of which there is fixed orifice (hole). There is spring inside the poppet as shown.
- **Working:** Pressurized oil entering through the inlet port will apply full force on the bottom of poppet and will try to compress the spring by shifting the poppet to right. The poppet will move to right and will close the outlet port. Then movement of poppet towards the right will stop.
- Now flow of oil through the orifice will start. Oil will occupy the bore of popper cylinder. This flow of oil will equalize the pressure on both ends of the poppet. The poppet will then balance.

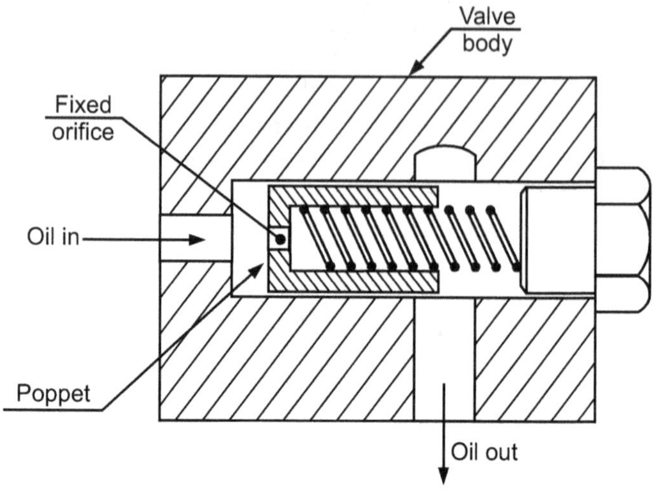

Fig. 3.49: Pressure Compensated Flow Control Valve

- During the process of poppet balancing, spring will expand and poppet will move towards left thereby uncovering the outlet port.
- A balance will automatically be established between quantity of oil through orifice and quantity of oil going out through the outlet port.
- Even if the pressure of incoming oil changes, the rebalancing will be established automatically and constant flow of oil will come out.

Applications:

- These valves are used in material handling systems. *For example*, lowering the speed of pallets where there are pressure variations.

3.3.3 Non-compensated Flow Control Valve [S-11]

- Non-compensated flow control valve that controls the flow by throttling or restricting. Needle valve we have seen is non-compensated valve.
- **Non-compensated:** It gives oil flow; which will vary if pressure increases or decreases.

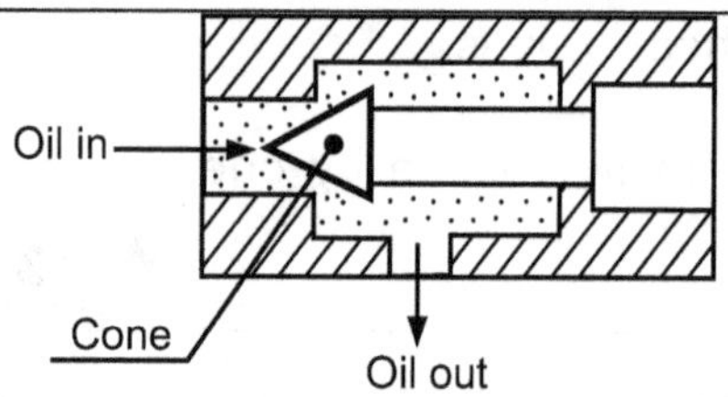

Fig. 3.50: Non-compensated Flow Control Valve

- **Non-constant flow:** A simple variable non-compensated flow control valve is shown in Fig. 3.50. The restriction is constant or variable. The flow is restricted by cone. For adjustment i.e. forward or backward movement of cone we can use screw. When pressure of incoming oil changes, the flow of outgoing oil will change. No compensation for pressure changes.
- **Applications:** These valves are used where accuracy in actuator movement and speed of actuator is not important.
- **Symbol:** Following are the symbols of flow control valves.

(a) 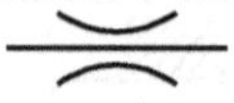 Flow control valve (Non-compensated).

(b) 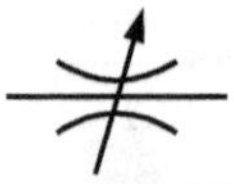 Variable flow control valve (Non-compensated).

(c) 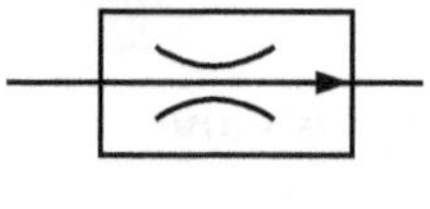 Pressure compensated flow control valve with fixed orifice.

3.4 CHECK VALVES (ALSO KNOWN AS NON-RETURN VALVES)

- **Function of check valve:** The main function of check valve is to allow the flow of fluid in one direction only. Flow of fluid in reverse direction is completely blocked or restricted. Due to this function only they are also termed as 'non-return valves'.
- In 4th semester we have studied 'foot valve' of centrifugal pump. This foot valve is non-return valve or you may call it as check valve.

3.4.1 Simple Check Valve [S-16]

- This valve consists of valve body in which valve element like cone (we can use ball or special poppet) is incorporated with specifically designed spring as shown.

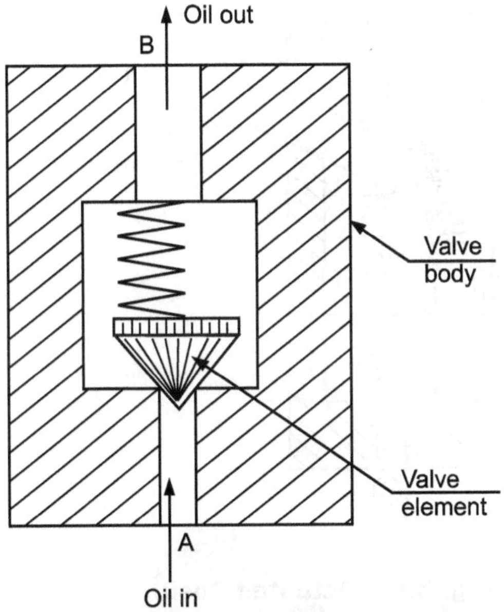

Fig. 3.51: Simple Check Valve

Fig. 3.52

- **Working:** When pressurised oil comes in through port A, it will lift up the cone by overcoming spring force and flow will start from port A to port B. When flow from A stops spring will expand and cone will block the flow. (See Fig. 3.51)
- It is easily understood that no flow is possible from port B to port A. Flow in only one direction is possible.
- **Symbol of simple check valve (Refer Fig. 3.52):** The symbol is designed based on construction itself.

3.4.2 Pilot Operated Check Valve [W-13, S-15]

- Directly operated check valve cannot be kept open because it is a safety valve. It will open when dangerous situation arises.
- If we want continuous oil flow from port (A) to (B) for some specific period then we have to modify the directly operated check valve.
- Fig. 3.53 shows the pilot operated check valve. A pilot piston is introduced below moving poppet (we can see that upper part of valve is similar to directly operated check valve). This pilot piston can move up by introducing pilot signal (it is nothing but pressurised oil) from port C.
- **Working:** As we can see in normal position shown there is no flow from (A) to (B) because the movable valve poppet has blocked the flow.
- Now pilot signal is given through port (C). This oil will push up the pilot piston upwards, thereby compressing spring (S_1). The piston rod of pilot piston will push the movable poppet in upward direction thereby compressing the spring (S_2). Now the flow from (A) to (B) will start. (Refer Fig. 3.54)

- As and when we cut-off the pilot signal the flow from (A) and (B) will continue. When pilot signal will be cut-off, springs S_1 and S_2 will expand and moving poppet will again block the flow from (A) to (B).

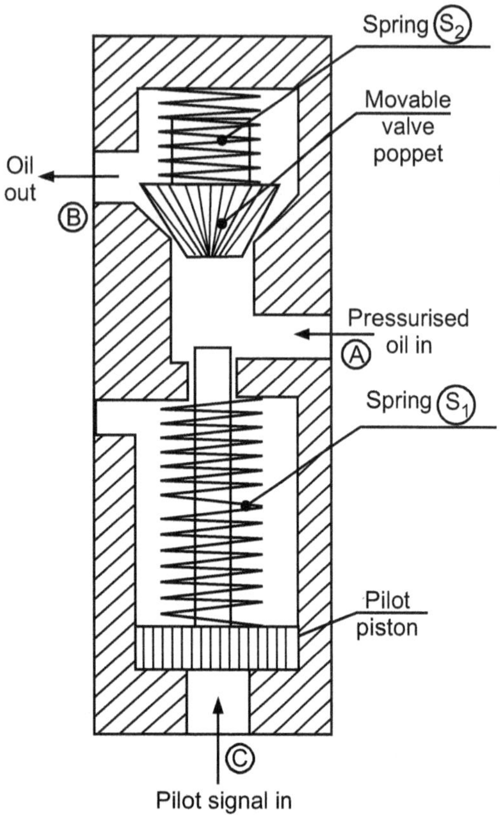

Fig. 3.53: Normal Position

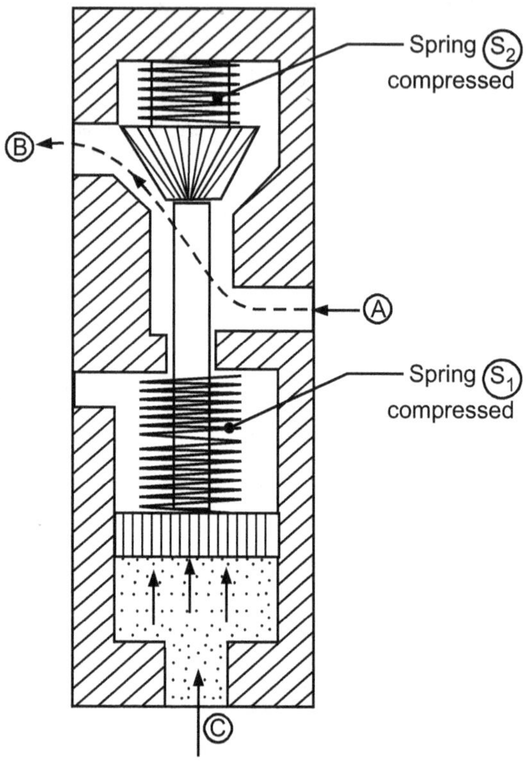

Fig. 3.54: Actuated Position

- **Symbol of Pilot Operated Check Valve**

  While drawing this symbol in hydraulic line, shows correctly pilot line. [S-15]

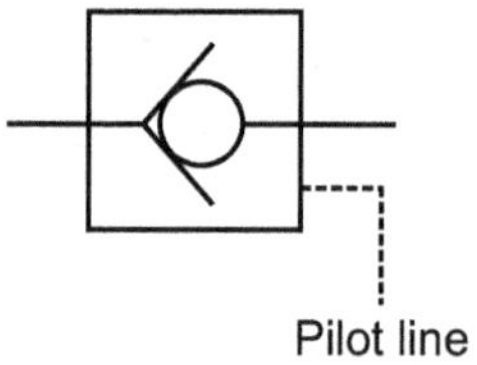

Fig. 3.55

3.5 PRESSURE CONTROL VALVE [S-11]

- Pressure control valves maintain the desired pressure level in the various parts of the hydraulic circuit.
- Functions of pressure control valves:
  (i) Limiting the maximum or set pressure limit and thus protecting the system.
  (ii) Regulating or reducing the pressure in certain portions of the circuit.
  (iii) Unloading the system pressure.
  (iv) Any other pressure related function to perform.
- We will study the following pressure control valves:
  (a) Pressure Relief Valve (Directly Operated)
  (b) Pressure Relief Valve (Pilot Operated).
  (c) Pressure Reducing Valve.

3.5.1 Pressure Relief Valve (Directly Operated) [S-11, 14, W-13, 16]

- Pressure relief valves are found in almost every hydraulic circuit. It is normally closed valve and it is connected between the pressure line (after pump) and the oil reservoir.
- **Main function of pressure relief valve:** Its main function is to limit the pressure in the system and thus to protect the individual component and hydraulic oil carrying lines from overload and danger of bursting. It is a safety valve. It takes care of safety of the hydraulic system.
- **Construction:** It consists of a valve body in which valve element (cone or ball) is fitted with spring. The compression of spring can be adjusted by pressure setting screw. There are two ports. P is pressure port connected to pump and R is oil return port connected to reservoir. (Refer Fig. 3.56)

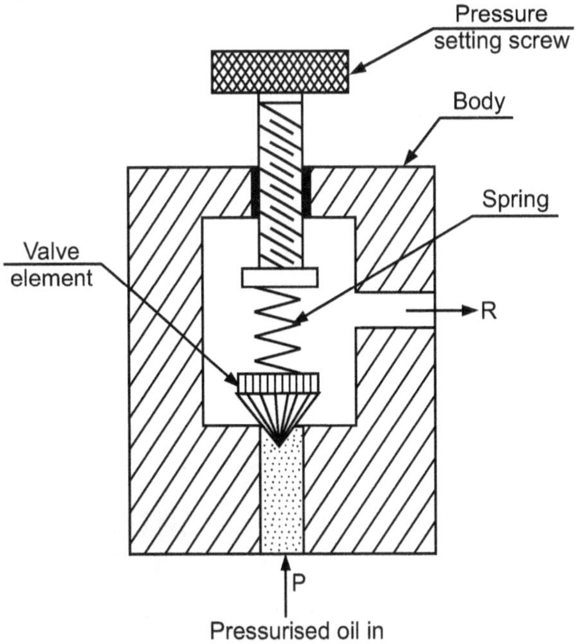

Fig. 3.56: Normal Position of Valve

- **Working:** The valve element is seated in valve body by the compressive force of spring. When inlet pressure of oil is insufficient to overcome spring force, valve will remain close. (No flow from port (P) to (R)). Hence normally this valve is closed. When pressure of input oil through port (P) increases due to some unpredicted blockages, then this increased pressure of oil will overcome the spring force and will lift the valve element in upward direction by compressing the spring. Due to lifting of valve element port P will be uncovered and flow from port (P) to port (R) will start. Since port (R) is connected to the oil reservoir, the oil will be discharged into it. This flow will continue unless and until system pressure falls below the set valve. Then again spring force will be dominating and value will close.
- The pressure level in the system is thus controlled by spring force.

Symbol of Directly operated pressure relief valve:

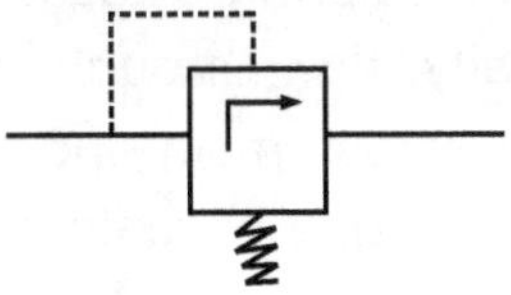

Normal Position

Fig. 3.57

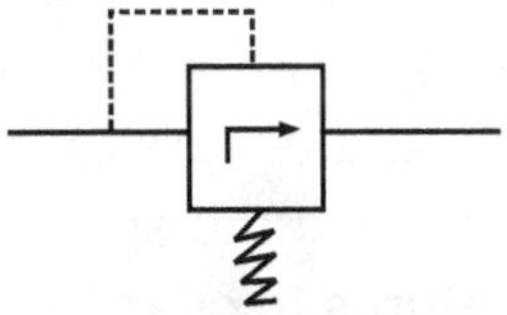

Actuated Position

Fig. 3.58

3.5.2 Pressure Relief Valve (Pilot Operated) [W-10, S-11, 16]

What is the need of Pilot Operated Pressure Relief Valve?

- When we are using high pressure oil and large flow, then we have to use heavy spring in directly operated pressure relief valve. Due to the use of heavy spring, the size of valve increases considerably. To keep the size of valve limited and to use normal size spring, the directly operated valve is modified to pilot operated valve.

Principle of Pilot Operated Pressure Relief Valve

- Pilot operated valve works on the principle of flow through orifice. It is well known that when the pressurised fluid flows through the orifice, its pressure drops. So pressure difference is set-up across an orifice. This principle is used in this valve.

- **Construction:** The upper part of valve is pilot stage and lower part is main stage. The main spool is bigger insize with bigger spring to allow more flow whereas pilot spool (cone) is smaller in size with light spring. There is orifice in main spool.

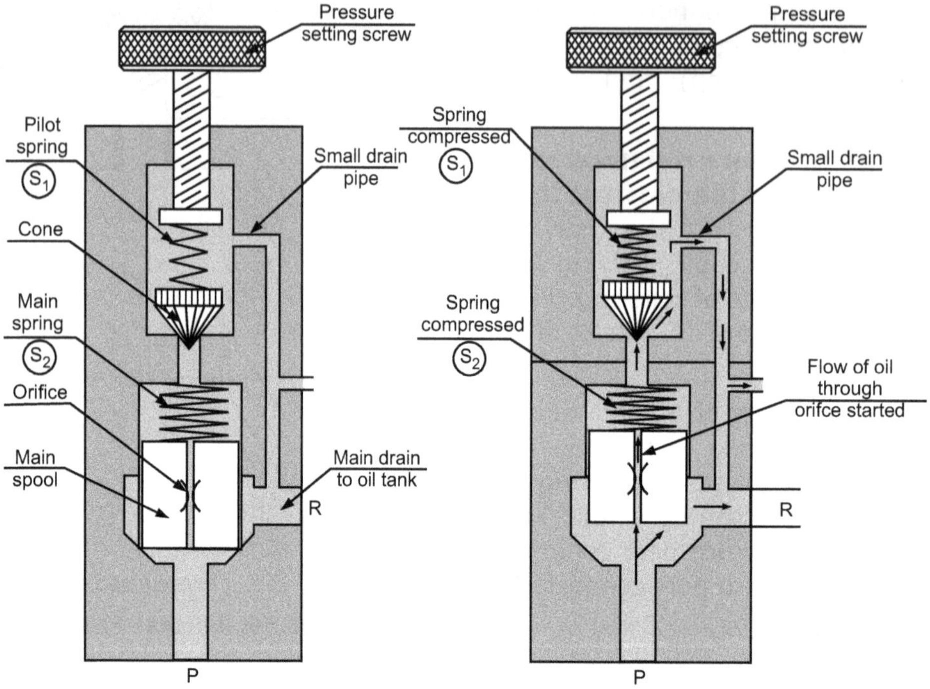

(a) Normal Position **(b) Actuated Position**

Fig. 3.59: Pilot Operating Pressure Relief Valve

Operation of Valve:

- **Normal position:** In normal position i.e. when system pressure is in constant state, the pressure on bottom of main spool and on the bottom of cone is same. There is no flow across the orifice. Hence, there is no flow from port P → R. Everything is safe.
- **Actuated Position:** Now suppose due to some problem of blockage, pressure of oil started increasing. Then due to light weight of cone and light spring attached to it, cone lifts up and small drain flow starts as shown in Fig. 3.59. Naturally, flow through the orifice is starting. This flow through the orifice creates pressure difference across main spool and automatically main spool lifts up, by compressing the spring (S_2). Due to lifting of main spool, main flow from port P to port R starts. Oil returns to oil reservoir.

Symbol of Pilot Operated Pressure Relief Valve

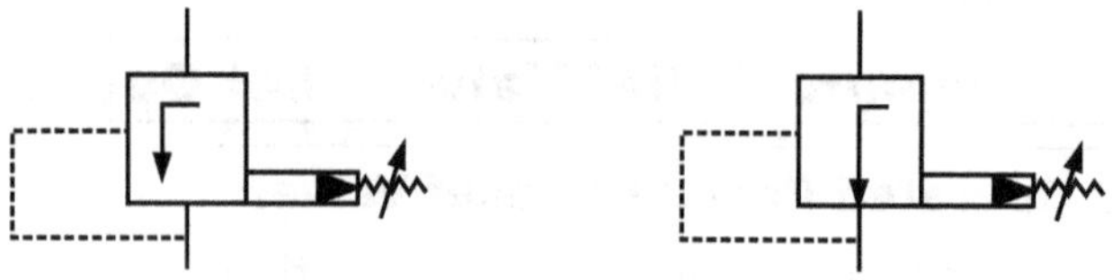

Normal Position **Actuated Position**

Fig. 3.60

3.5.3 Pressure Reducing Valve [S-11, W-11]

- The main function of pressure reducing valve is to reduce the pressure in particular branch of the circuit to different level as demanded by consumer in that branch.

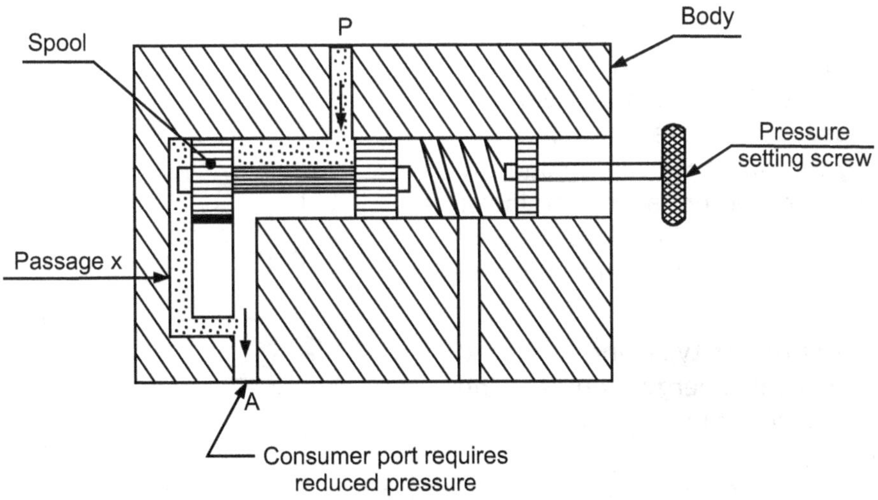

Fig. 3.61: Normal Position of Pressure Reducing Valve

- **Construction:** It consists of spool and spring enhoused in the bore of valve body. Spring compression can be adjusted by pressure setting screw. Port P is pressure port connected to pump. Port A is consumer port requiring reduced pressure.
- **Working:** As shown in normal position, port P is supplying oil to consumer port A. If the main supply pressure is below the set pressure, there will be continuous flow from P to A. Hence normally this valve is open.
- When outlet pressure rises to valve setting, then oil will flow through 'passage x' and will act on spool and spool will shift to right thereby partly closing the port A. Now only enough flow will pass through port 'A' so that consumer connected to A will receive reduced pressure.

3.5.4 Use of Pressure Reducing Valve in Circuit and Symbol [S-11]

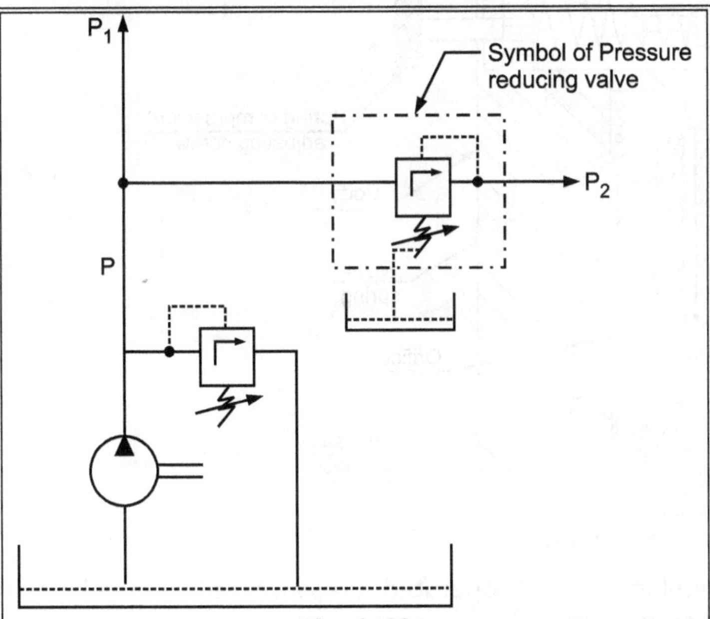

Fig. 3.62

The circuit shows that reduced pressure P_2 is available to 2nd branch of circuit. ($P_2 < P$).

3.5.5 Unloading Valve [S-16]

- **Function of unloading valve:** The power of oil in terms of pressure of flow can be given by following expression

$$\text{Power} = \text{Pressure} \times \text{Flow rate}$$

- If we allow excess flow of oil through pressure relief valve (in case of blockages) then it is wastage of 'Energy'. This can result into rise in temperature of fluid, because wastage energy generally converts into heat.

- To overcome this difficulty, additional valve called 'unloading valve' is used before pressure relief valve. This unloading valve is opened or closed by external electrical circuit. By using this valve we can unload the pressurised oil from pump at much lower pressure and there is very low wastage of energy. Naturally, there is no much heating of oil.
- Unloading valve is on-off type and operated by external electrical energy source. The hydraulic circuit using unloading valve is as under.

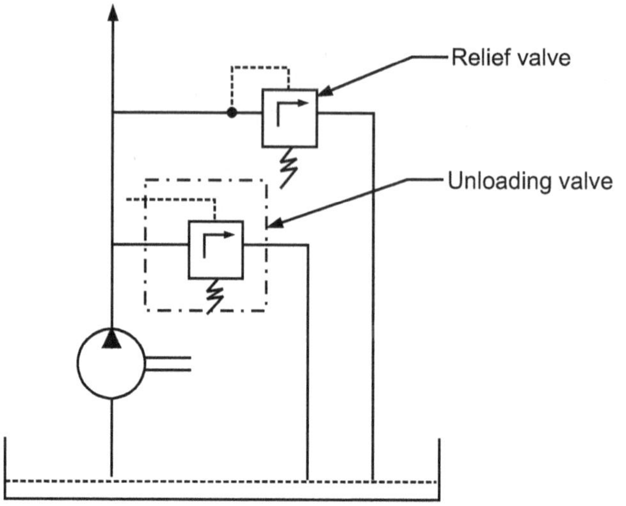

Fig. 3.63

3.6 SEQUENCE VALVE [W-09, 10, S-11, 16]

- The main function of sequence valve is to operate 2-3 operations in sequential order i.e. one after the another. Suppose we have to carry out drilling a hole in a block. Then first operation must be clamping and second operation is drilling. Third operation is declamping. These three operations must be done in a specific sequence. This task is carried out by sequence valve in the circuit. (Refer sequencing operation in Topic 5).

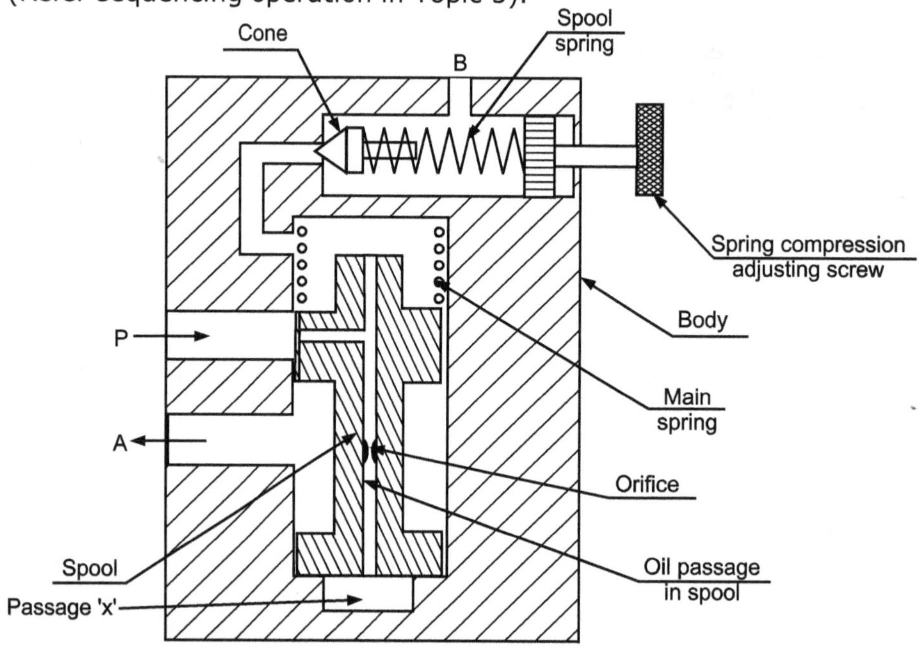

Fig. 3.64

- **Construction:** The construction is similar to 'pilot operated pressure relief valve'. It has a special spool having specifically drilled oil passages.
- **Operation:** Oil under pressure goes inside through port P. It will pass through oil passage to space 'x'. When pressure is sufficiently build-up, oil flow will start from 'x' because cone will move towards right. Due to orifice, pressure difference will set-up across the spool and spool will lift-up by compressing the main spring and flow will start from port P to port A.
- **Symbol of Sequence Valve: (Refer Fig. 3.65)** [S-15]

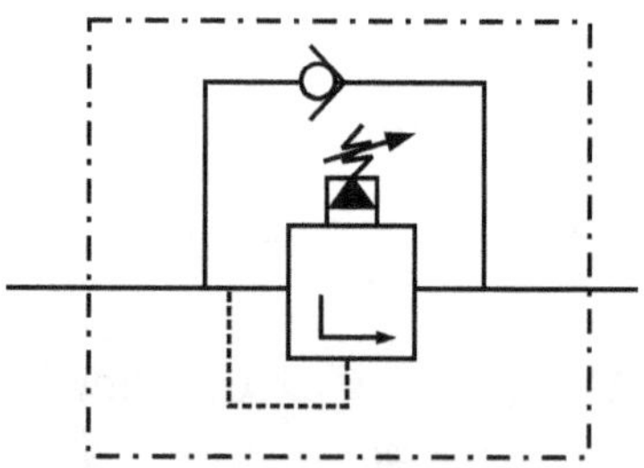

Fig. 3.65

Important Points

- Direction of flow of compressed air is controlled by direction control valve (DC valve). it acts like traffic police.
- Seat type (poppet type) and spool type (sliding spool) are types of DC valves.
- Sliding spool type DC valve is commonly used.
- 3×2 DC valve means 3 Ports and 2 position DC valve.
- DC Valve can be operated by button/by solenoid valve/plunger control/spring control/by hydraulic oil pilot signal.
- Flow control valves are used to regulate the flow of hydraulic oil in hydraulic circuit to control speed of actuator.
- Flow control is achieved either by throttling or by diverting
- Needle valve/pressure compensated flow control valve/Non-compensated flow control valve/ Check valve are the types of flow control valve.
- Check valve is also non-return valve.
- The main function of check valve is to allow the flow of fluid in one direction only.
- Check valve can also be pilot operated check valve. A pilot signal operates this valve.
- Pressure control valve maintains the desired pressure level in various parts of hydraulic circuit.
- Limiting the maximum or set pressure limit and to protect the system is main function of Pressure Control Valve.
- Pressure relief valve /pressure reducing valve are the types of pressure control valve.
- To keep the size of valve limited for higher pressures and to use normal size spring, the directly operated pressure relief valve is modified to pilot operated pressure relief valve.
- The main function of Pressure Reducing Valve is to reduce the pressure in particular branch for suitability of working of actuator in that branch.
- In case of blockages if we allow excess flow of oil through pressure relief valve then energy will be wasted. To solve this problem Unloading Valve is used. This valve is generally operated by external electrical energy.
- The main function of sequence valve is to operate 2-3 operations in sequential order i.e. one after another.

Practice Questions

1. Which are the various valves used in controlling the air in pneumatic circuit? List out the valves and explain function of each valve.
2. What is the difference between spool type and seat type DC valve?
3. Draw neat sketch of 3×2 DC valve. Explain its working.
4. How you will use 4×2 DC valve to control DA cylinder? Draw circuit and explain
5. What is impulse operation? Explain it in respect of operating 5×2 DC valve.
6. Draw neat sketch of variable flow control valve with built-in check valve and label the diagram.
7. What is the other name of check valve? Explain the function of check valve.
8. Explain working of directly operated (Poppet type) check valve with neat sketch.
9. Draw a sketch of normal and actuating position of 4/2 direction control pneumatic valve.
10. Explain with symbol and sketch pilot operated poppet valve used in pneumatic system.
11. What is the difference between spool type and seal type DC valve?
12. Explain with neat sketch the working of push button operated 5×2 DC valve used in pneumatics
13. Give classification of pneumatic control valves.
14. Explain working of directly operating check valve with neat sketch
15. State any two applications of 3×2 DC pneumatic valve with any circuit diagram.
16. Explain with neat sketch the construction of 'Non-relieving pressure regulator'.

17. Sketch and explain working of 5/2 DC pneumatic valve.
18. State the functions of pressure relief valve for pneumatic system. Explain its working with neat sketch
19. What is the use of direction control valve? Explain with sketch?
20. Compare seat type and spool type DC valves.
21. Explain with neat sketch the working of rotary spool type DC valve.
22. What is the function of flow control valve? Explain.
23. Draw neat sketch of needle valve and explain its working.
24. What is pressure compensated flow control valve? Explain with Sketch.
25. Why pressure relief valve is used in hydraulic circuit? Explain in details. Draw sketches if required.
26. What is unloading valve? Explain with circuit.
27. Explain 3×2 DC valve with neat sketch and symbol.
28. Explain rotary spool DCV with neat sketch.
29. Explain the working of sequence valve with neat sketch.
30. Draw a sketch of manual and actuating position of 4/2 direction control valve.
31. Explain with neat sketch quick exhaust valve.
32. What is pressure compensated flow control valve? How does pressure compensation takes place?
33. Draw Tandem Centre Position of 4×3 DC Valve.
34. Draw neat sketch of Float Centre Position.
35. What is the function of relieving pressure regulator? Draw neat sketch of it and explain.

■■■

Chapter **4**

COMPRESSOR, PNEUMATIC COMPONENTS AND ACCESSORIES IN FLUID SYSTEM

Weightage of Marks = 12, Teaching Hours = 08

Syllabus

4.1 Types, Construction, Working Principle of Reciprocating and Rotary Compressors

4.2 Construction, Working Principle of FRL Unit, Duel (Twin) Pressure Valve, Shuttle Valve, Quick Exhaust Valve, Time Delay Valve

4.3 Accessories: Pipes Hoses, Fittings, Oil Filters, Air Filters, Seals and Gaskets, Accumulators Heat Exchanger, Muffler

About This Chapter

At the end of this chapter students will be able to:

- Classify various types of compressors with justification
- Sketch and explain construction of given valves
- Select compressor for given application
- List various accessories required in hydraulic/pneumatics
- Selection of appropriate accessories in hydraulic and pneumatic system

4.1 INTRODUCTION

- We have seen general layout of pneumatic system in which the list of various components is made. Again we will have quick look at all components:
  (a) Air compressor.
  (b) Air receiver or air tank.
  (c) Moisture separators.
  (d) FRL unit.
  (e) DC valves and other valves.
  (f) Actuators.
  (g) Air conducting elements (accessories).
- In this chapter, we will study the various components listed above. The first one is air compressor.

4.2 AIR COMPRESSORS [W-12]

- Air compressors are the devices which gives energy to the medium of pneumatic system i.e. Air.
- The compressors can be divided in many types. But those can be broadly classified in two groups:
  (a) Positive displacement compressors.
  (b) Dynamic compressors or turbo compressors.

4.2.1 Positive Displacement Compressors [W-12]

- In this type a definite volume air is compressed. So that the volume of air reduces and naturally its pressure increases. These compressors are either reciprocating or rotary type.

- When we visit to petrol pump with our two wheeler or four wheeler we check the air in tyre of vehicle. There we can see the compressor. That compressor is reciprocating compressor. Reciprocating compressor is very similar to engine. The difference between reciprocating compressor and engine is that, engine generates rotary motion through reciprocation of piston where as in compressor we achieve reciprocation of piston by using electric motor and 4 bar mechanism (piston, cylinder, connecting rod, crank) and there by compressing the air. Output of engine is rotary motion, where as output of compressor is compressed (you may say energized) air which would be stored in air tank attached to it.
- In industrial sector, most widely used air compressor is reciprocating compressor. This compressor is also known as **'Piston Compressor'.**
- **Important:** Reciprocating air compressor is specified on the basis of working pressure of output air. This, pressure is specified in 'bar' or kg/cm^2.

  $$1 \text{ bar} \cong 1 \text{ kg/cm}^2$$

  Screw compressor: This is rotary type positive displacement compressor. This compressor is also most widely used compressor in industrial sector similar to reciprocating compressors.
- 'Piston compressors' are used when high pressure (above 10 bar) and relatively low volumes (below 10,000 m^3/hr) are required. But when medium pressures (below 10 bar) and relatively high volumes (around 10,000 m^3/hr) are required then rotary i.e. screw compressor is adopted.
- Other positive displacement compressors are:
  1. Diaphragm compressors.
  2. Lobe compressors (Roots Blower).
  3. Sliding vane compressor.

4.2.2 Dynamic Compressors or Turbo Compressors [W-12]

- In this type of compressors, rotating impellers impart velocity and pressure to the flow of air being handled.
- If large volume of air with low pressure is required then dynamic compressors are adopted.

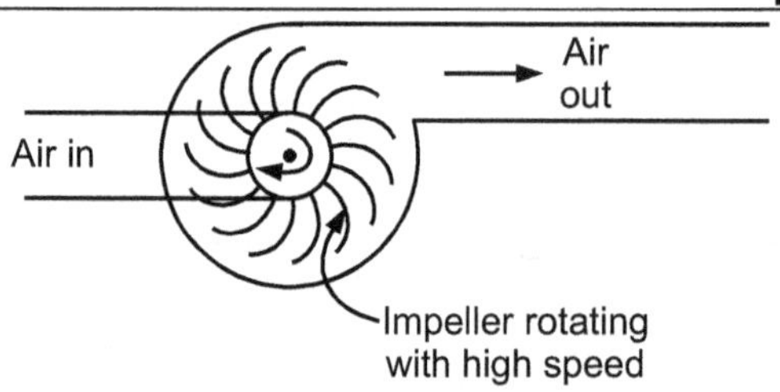

Fig. 4.1

- Fig. 4.1 shows centrifugal blower which draws air (similar to suction of water in centrifugal pump) and forces it out due to centrifugal force created by rotation of impeller. ('Hair Dryer' used by barber in hair cutting saloon/beauty parlor is a dynamic compressor).

4.2.3 Classification of Compressors [S-10, 12, W-10, W-12, W-14]

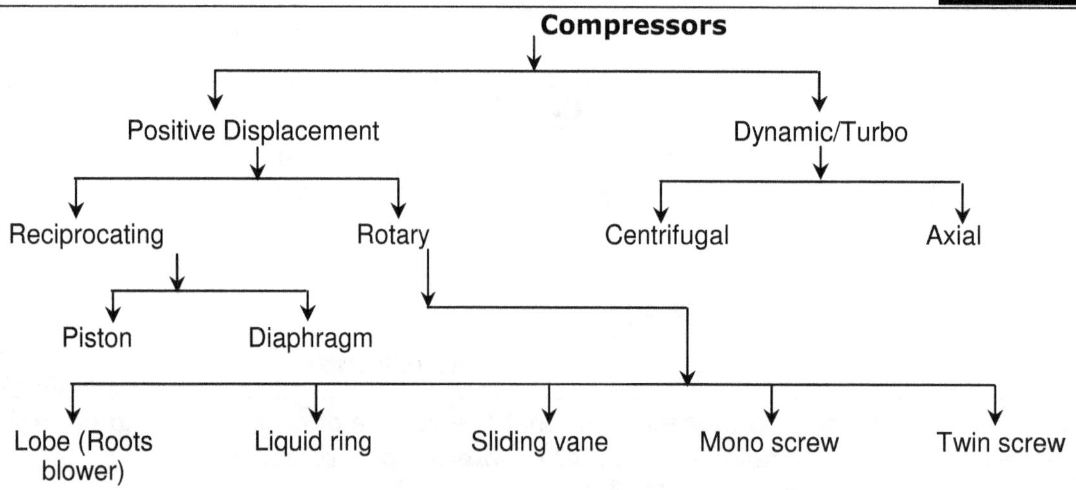

4.2.4 Classification based on Geometry and Operational Characteristics or Features [W-12]

Geometry or Operational Characteristics	Classification
Action	(a) Single acting reciprocating air compressor (b) Double acting reciprocating air compressor
Stages	(a) Single stage (b) Double stage (c) Multi stage
Position of cylinder	(a) Verticle (b) Horizontal (c) V type
Drive of compressor	(a) Motor driven (b) Engine driven (c) Gas/Steam turbine drive
Cooling medium	(a) Air cooled (b) Oil cooled (c) Water cooled
Portability condition	(a) Portable compressor (b) Stationary compressor

- The compressors you observe on petrol pump (puncture repair shops), cycle shops are 'single stage or two stage, portable reciprocating air compressor'.
- 'ELGI' is famous brand in reciprocating compressors in the market.
- 'Atlas-Copco' is another gaint in screw compressors manufacturing.
- 'Ingersol-Rand' is another famous brand in reciprocating as well as screw compressors.
- The compressor of jet engine used on modern air crafts is gas turbine driven.

4.2.5 Positive Displacement Compressors [W-12]

- As stated earlier, we will study following compressors:

  (a) Piston compressors: (i) Single acting.
  (ii) Double acting.
  (iii) 2 stage single acting.

  (b) Screw compressors: (i) Mono screw.
  (ii) Twin screw.

  (c) Rotary Vane compressor (Sliding Vane).

  (d) Twin Lobe Compressor (Roots Blower).

4.2.6 Single Acting Reciprocating Air Compressors [W-11]

- This is piston compressor in which there are following parts:

  (a) Cylinder.

  (b) Piston and piston rod, connectivity rod.

  (c) Crank and crank case.

  (d) Two valves (one suction valve and one delivery valve).

  (e) Inlet and outlet ports.

- **Construction:** In its simplest form single acting reciprocating air compressors uses 4 bar mechanism i.e. cylinder – piston – piston rod-crank. There are 2 valves which are spring operated. The cylinder is having fins from outside to increase the surface area so that it will dissipate the heat and will be cooled earlier. The crank is driven by electric motor/ engine/turbine.
- **Working:** Piston reciprocates in cylinder. Fig. 4.2 shows suction stroke. In this stroke the piston is at bottom dead centre position. Due to downward movement of piston vacuum is created in cylinder and air sucks in by operating suction valve. The cylinder is now full of atmospheric pressure air.
- Now piston starts moving-up. During this movement suction valve closes. Air gets compressed and delivery valve opens. The compressed air moves out of delivery port and goes to air receiver. In this compressor compression of air takes place on one side of piston only.

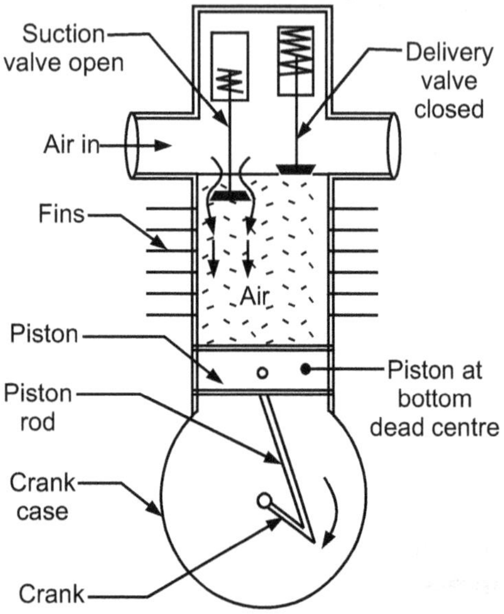

Fig. 4.2: Suction Stroke

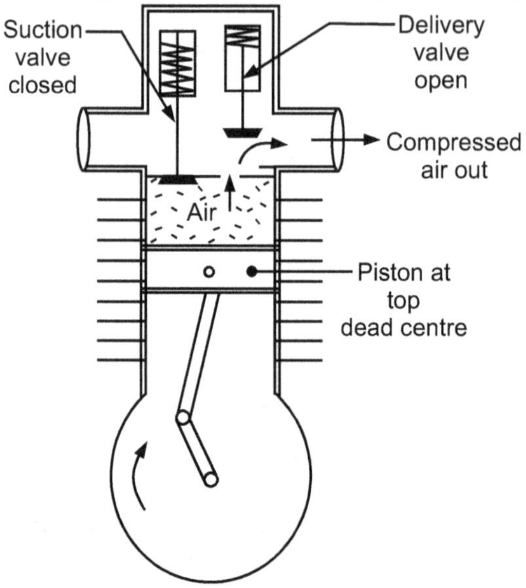

Fig. 4.3: Delivery Stroke

Important:

- In this compressor, air supply to receiver is pulsating because compressed air delivers only in delivery stroke. Hence this type of compressor is having limited applications.
- But it is widely used in 2 stage or 3 stage or multistage reciprocating air compressors.

4.2.7 Double Acting Reciprocating Air Compressor [W-11, 16, S-15]

- Double Acting Reciprocating Air Compressor, is similar to 'double acting reciprocating pump'. It is comprised of following parts:
  (a) Cylinder.
  (b) Piston and piston rod and connecting rod.
  (c) Crank and crank case.
  (d) Two suction valves and two delivery valves.
  (e) One inlet port and one outlet port.
- **Construction:** In its simplest form, double acting air compressor uses four bar mechanism. There are 4 valves (2 suction valves and 2 delivery valves) shown at A, B, C, D in Fig. 4.4. There are cooling fins similar to single acting compressors. The crank rotates on electric motor/engine/turbine.
- **Working:** In this compressor, compression of air takes place on both side of piston. When crank rotates, the piston starts reciprocating.
- When piston comes down and attains, 'Bottom Dead Centre Position', the air comes in through port A due to vacuum created due to downward movement.
- When piston start moving upward, the air starts compressing. When piston attains, 'Top Dead Centre Position', the stroke is complete and air is fully compressed which goes out through delivery valve B to air receiver.

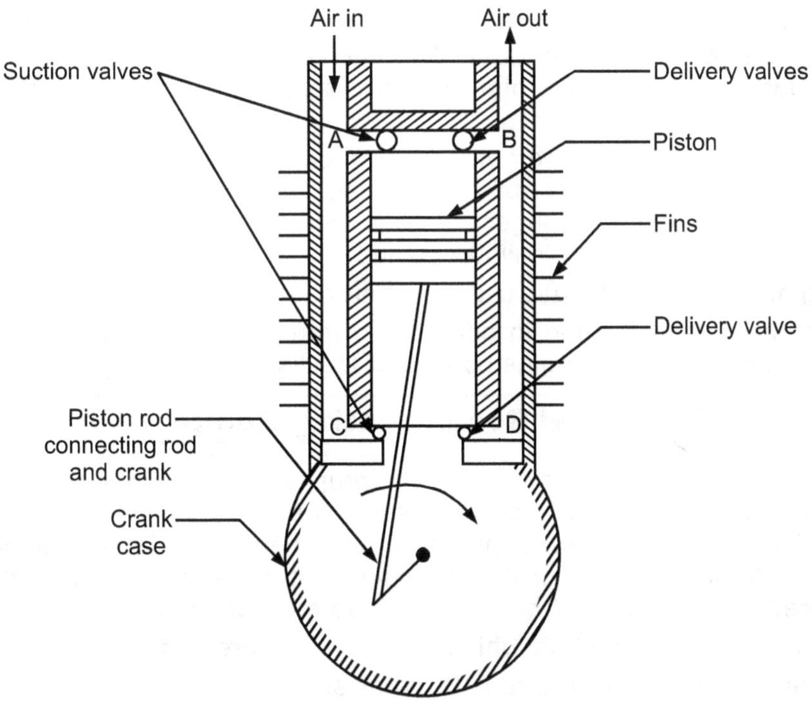

Fig. 4.4: Double Acting Reciprocating Air Compressor

- During this upward movement, the vacuum is created on other side (piston rod side) of piston. Suction valve 'C' open and air comes in.
- When piston starting comes down, this air which came through valve 'C', gets compressed and compressed air goes out through delivery valve 'D' to air receiver. In this downward movement air comes in through valve 'A' and entire cycle repeats.
- **Important:** In this compressor, the air supply to air receiver is more smooth than single acting compressor, because in each half stroke compressed air is delivery to air receiver.

4.2.8 Two Stage Reciprocating Air Compressor with Intercooler [S-09]

- Two stage Reciprocating Air Compressor using intercooler is shown in Fig. 4.5. This type of compressor is used extensively in all sectors of engineering.
  - Atmospheric pressure air comes in to 1st cylinder (1st stage). This is single acting reciprocating compressor. The pressure of air is increased by compressing this air. This is low pressure stage (For example, pressure increases from atmospheric to say 3 bar).

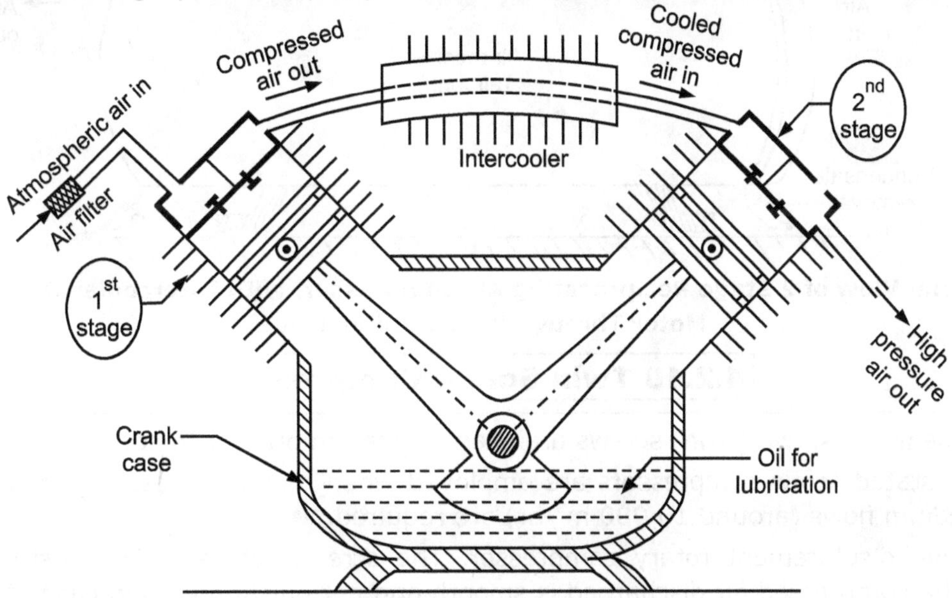

Fig. 4.5: Two Stage Reciprocating Air Compressor with Intercooler

- The compressed air then enters to suction side of 2nd cylinder (2nd stage). This is also a single acting reciprocating compressor. The pressure of air is further increased in this stage (say from 3 bar to 7 bar). This is high pressure stage. The high pressure air then goes to air receiver.

- **Intercooler:** In first stage, the air temperature increases due to reciprocation. Intercooler cools the air. It also reduces the volume of compressed air and there is energy saving in second stage. Intercooler is a heat exchanger. It is a cylinder having cooling fins on the outside surface. Sometimes the intercooler is water cooled.
- 2 staging/3 staging/or multistaging is used of pneumatic pressure upto 15 bar.
- When staging is more than 3, then in addition to intercooler, the cylinders are also cooled by water cooling system (similar to engine cooling of trucks).

4.2.9 Actual Reciprocating Compressor [W-11]

- The compressor shown in Fig. 4.6 is a common compressor used in industry. The two stages of compressors are mounted on air receiver itself. The drive is given to compressor through electric motor. The compressor is reciprocating and hence similar to reciprocating pump, its speed is low. The standard speed of AC induction motor, as you know is either 960 r.p.m. or 1440 r.p.m. To reduce the speed, V-belt pulley drive mechanism is used.
- A fan is attached to compressor shaft which cools the compressor.
- For protection purpose protective guard is fitted as shown.
- **Important:** When pressure in air receiver falls below set pressure i.e. required pressure, then compressor automatically starts. It starts supplying compressed air to receiver. When sufficient i.e. set pressure is attained the compressor motor stops. The arrangement is done to save power.

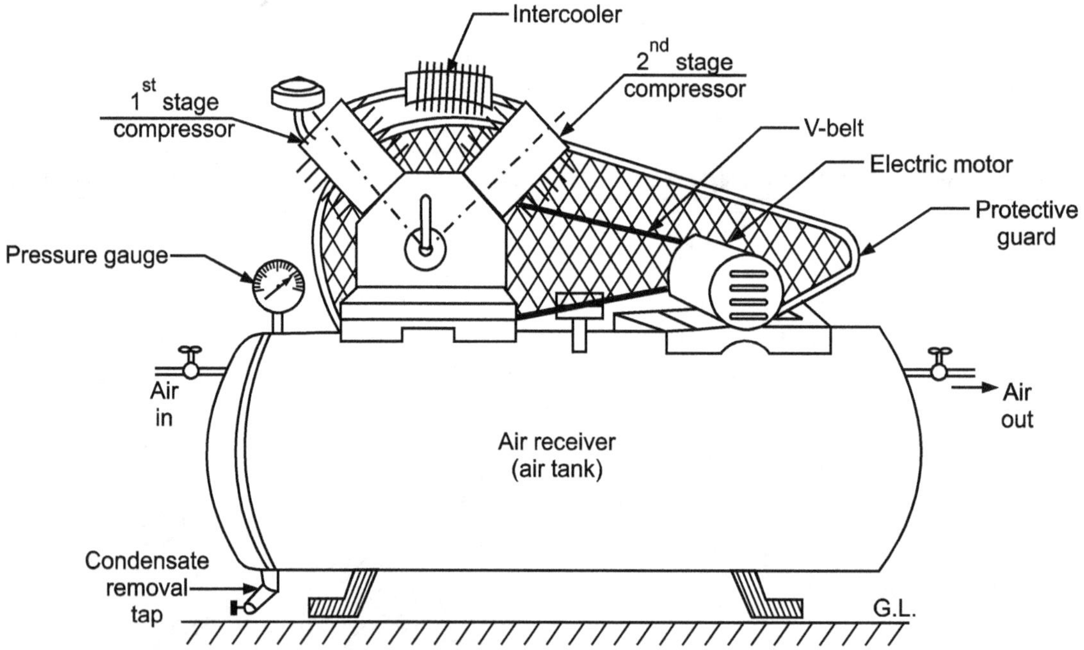

Fig. 4.6: Pictorial View of 2 Stage Reciprocating Air Compressor, with Intercooler-Driven by Electric Motor Through V-Belt Pulley Drive

4.2.10 Twin Screw Compressor [S-10]

- As the name implies, two (twin) screws are used in this compressor to compress the air.
- As already stated screw compressors are employed when medium pressure air (below 10 bar) having medium flows (around 10,000 m^3/hr) are required.
- It is positive displacement rotary compressor. The screws rotates with constant speed and naturally the compressed air discharged is smooth and not pulsating (as in case of reciprocating compressors).

- On the basis of rotation arrangement twin screw compressors can be divided in two categories:
  1. **Dry screw rotary compressor:** In this type both screws are rotated by external timing gears fitted on shaft of each screw and meshing with each other. There is no lubrication required between two screws. Hence it is called dry screw compressor.
  2. **Wet screw rotary compressor:** In this type one screw is driving screw. i.e. motor is coupled to this gear. Other screw is meshing to it and hence it is driven gear, which rotates because of contact with driving gear.

     This system require oil for lubrication of screw meshing surfaces. The lubricating oil is mixed with compressed air. This oil needs to be removed by special oil separation units before air reaches the pneumatic circuit or consumer.
- **Important:** The construction of both twin screw compressors is same.
- The screw compressors are also having air receivers to store compressed air.

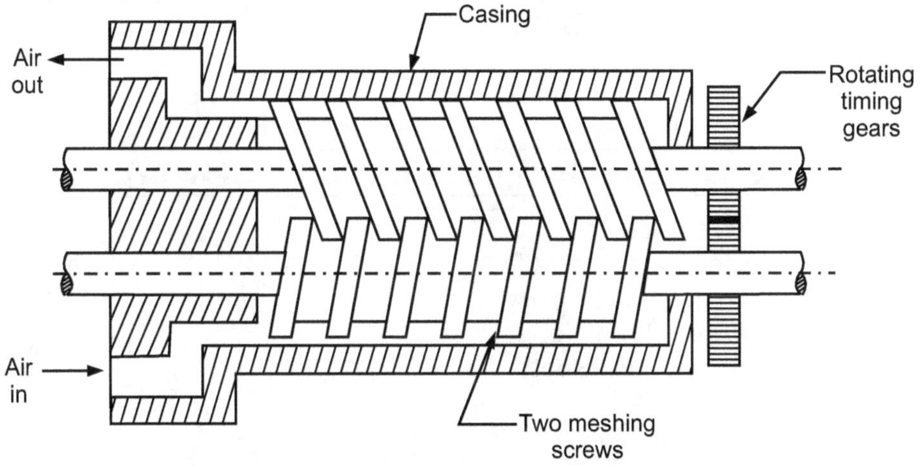

Fig. 4.7: Schematic Diagram of Dry Screw Rotary Compressor

- **Construction:** It consists of two intermeshing screws with 0.05 mm clearance. The timing gears are fitted on shafts of each screw. There are two ports. One for entry of atmospheric pressure air and other for exit of compressed air.
- **Working:** Drive is taken by two gears, which rotates both screw at a time. As screws rotates, air is drawn into the housing, trapped between the screws and carried along the discharge port, which then goes to air receiver.
- The Figure of Wet Screw Rotary Compressor is almost same as above. The only difference is that there are no timing gears. One shaft is coupled to electric motor or shaft of engine or turbine. Rest is same.

4.2.11 Rotary Vane or Sliding Vane Compressor [W-13]

- A single stage rotary vane compressor can deliver pressure upto 3 bar (much lower than reciprocating or screw compressor).
- It is +ve displacement compressor.
- Its output is smooth (not pulsating) because it is rotary compressor similar to screw compressor.
- The construction is similar to that of hydraulic vane pump.
- **Construction:** The compressor consists of simple vane rotor which is having slots in which vane (flat piece of steel) slides freely. The rotor is eccentrically located inside stator housing. The rotor is coupled to motor.
- **Working:** When rotor starts rotating, all vanes flows out from centre due to centrifugal force and touches the inside surface of casing. Due to rotating, partial vacuum is created in compartment 'A'. So air rushes in this compartment. The trapped air is carried towards delivery port. Due to reduction in area air gets compressed and goes out.

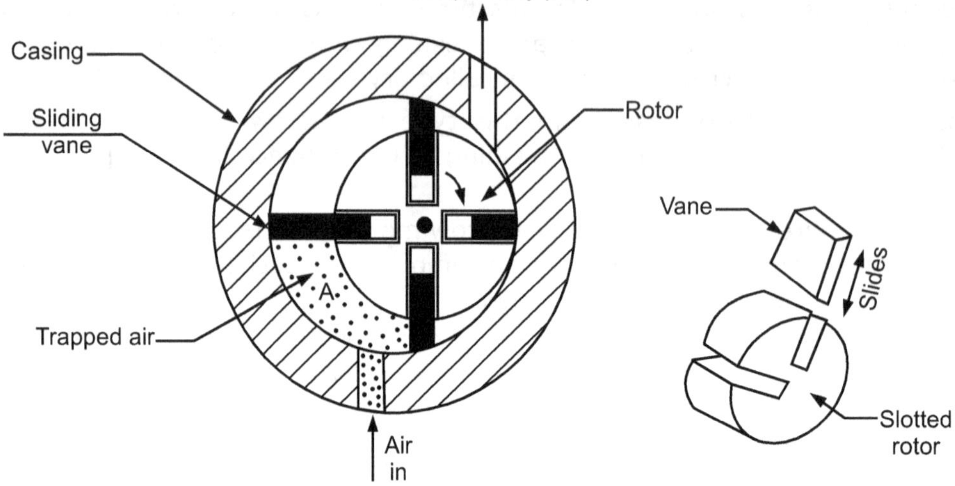

Fig. 4.8

4.2.12 Lobe Compressor (Roots Blower)

[W-13]

- This is also a rotary positive displacement compressor. In this compressor instead of impeller two 'lobes' are used. Hence it is called lobe compressor or 'Twin lobe compressor'. It is also known as 'Roots Blower'.
- This compressor can generate very low pressure air i.e. upto 1-2 bar. But the delivery volume of air is very high (more than 12000 m^3/hr.).
- **Construction:** In its simplest form it consists of typical shaped casing and two meshing lobes. Similar to dry screw compressor, both lobes are driven by timing gears hence there is no question of lubrication.

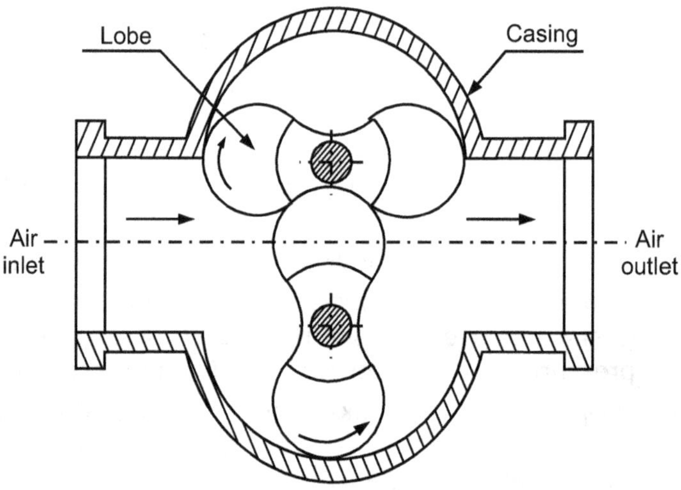

Fig. 4.9

- **Working:** When lobes start rotating with high speed, the air starts transferring from suction side to the delivery side. The air is trapped between casing and outside surface of lobes. Both are machined with close tolerances. Hence, the compression of air is very minor. The compression ratio of this compressor is just 1.7 (not even 2)

$$\left[\text{Compression ratio} = \frac{\text{Volume of Inlet air}}{\text{Volume of Outlet air}}\right].$$

4.2.13 Dynamic or Turbo Compressors

[W-13]

- We will study following compressors under this head:
  (a) Centrifugal type compressor.
  (b) Axial type compressor.
- **Centrifugal type compressor:** This compressor is like centrifugal pump, but only it handles air instead of water. Centrifugal compressor is dynamic compressor.
- In case of burners used in oil fired furnaces, only large quantity of air is required. There is no requirement of pressurized air.
- In case of air ventilation of air curtains used in cold storages, large quantity of air is required. In such cases dynamic compressors are employed.
- Centrifugal compressor is also known as blower. Its impeller rotates at very high speed (50,000 r.p.m.).

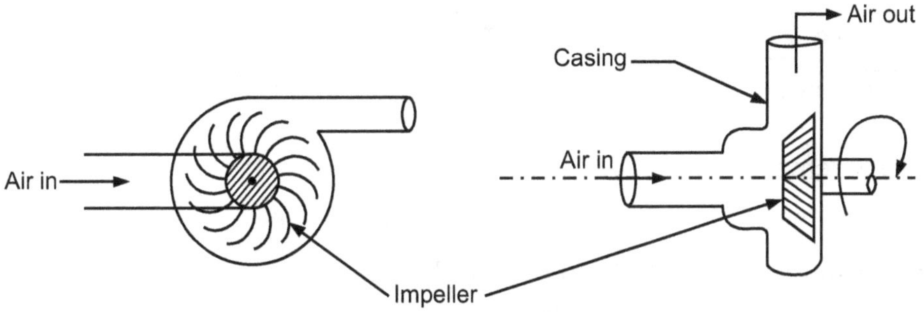

Fig. 4.10: Centrifugal Compressor

- **Construction:** The casing of this compressor is similar to casing of centrifugal pump (small blower casing is made-up of plastic while, big blowers casings are made-up of MS Sheets). The impeller is enhoused in casing and is coupled to motor through speed increase mechanism (such as belt pulley).
- **Working:** As impeller rotates with high speed, it imparts kinetic energy to air. This high velocity air passes through special shaped casing (similar to scroll casing of centrifugal pump), which increases its pressure. The pressurized and high velocity air comes out of discharge port.
- **Important:** This air is not stored in air tank. It is directly used.

4.2.14 Axial Type Compressor [W-13]

- This is also dynamic compressor used for high discharge of air upto 50000 m^3/min. Pressure of air is very minor.

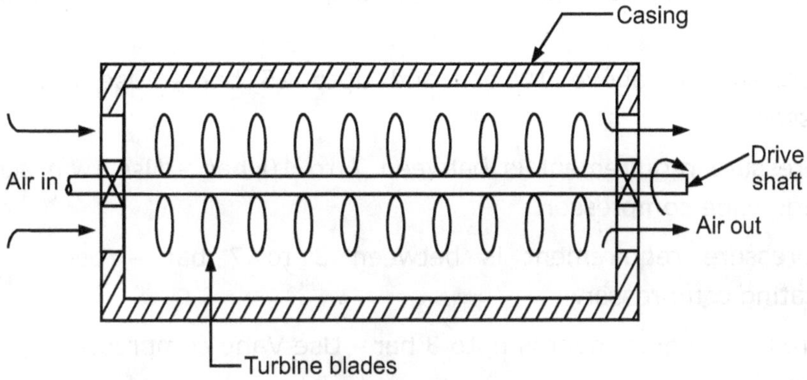

Fig. 4.11

- **Construction:** A series of blades are fixed on shaft and the assembly is enhoused in casing. Shaft is coupled to motor through belt pulley.
- **Working:** As shaft rotates with high speed, large volume of air is transferred from input side to output side. The pressure achieved at discharge point is just 1 bar.

4.2.15 Symbol of Compressor

- Similar to electric circuits, there are symbols for all elements used in pneumatic circuits.
- Let us see the symbol of compressor.
- The symbol drawn is for compressor, irrespective of type of compressor. The compressor is motor driven because (M) symbol shown is for motor. If letter 'E' is written then it is engine driven. If letter 'T' is written then it is turbine driven.
- The triangle shown is empty because it indicates the handling of air.

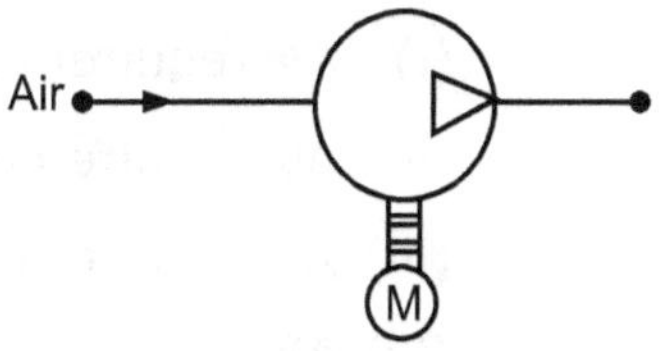

Fig. 4.12

4.2.16 Selection of Criteria for Compressor

[S-10

- Following are the basic criteria (factors) which needs to be considered while selecting type of compressor:

  (a) Pressure requirement of pneumatic circuit.

  (b) Capacity or volume of air required by circuit or consumer.

  (c) Compressor configuration (space availability).

- Now we will discuss the factors in details:

  (a) Pressure Requirement: We know that the pneumatic systems are used when pressure requirement is below 10 bar (10 kg/cm^2). Following broad criteria is used for selection of compressor based on pressure ratio requirement $\left(\text{Presure ratio} = \dfrac{\text{Output pressure}}{\text{Input pressure}}\right)$.

  (i) For pressure Ratio 10 ⇒ Use ⇒ Reciprocating compressor.

  (ii) For pressure Ratio 8 ⇒ Use ⇒ Vane compressor

  (iii) For pressure Ratio 3 ⇒ Use ⇒ Screw compressor

  (iv) For pressure Ratio 2 ⇒ Use ⇒ Lobe compressor

  (v) For pressure Ratio 1 to 2 ⇒ Use ⇒ Lobe compressor

  We can have criteria of selection based on output pressure (Pressure of Compressed Air) also:

  (i) When pressure is above 10 bar and upto 15 bar – Use multistage reciprocating compressors.

  (ii) When pressure requirement is between 7 to 10 bar – Use twin screw compressor or multistage vane compressor.

  (iii) When pressure requirement is between 3 to 7 bar – Use single or two stage reciprocating compressor.

  (iv) When pressure requirement is upto 3 bar – Use Vane compressor.

  (v) When pressure requirement is between 1 to 2 bar – Use lobe compressor.

  (b) Volume of Air: Compressor compresses the air and this compressed air is stored in air receiver. The air under pressure is supplied to pneumatic circuit through air receiver. In case of lobe or axial or centrifugal compressors no air tank is used. The air is directly supplied through compressor.

  Air requirement is in thousand cubic meters per hour or even per min. Following are some broad guidelines for selection of compressor on the basis of air requirement:

  (i) Air requirement below 5000 cu. mt./hr – Reciprocating compressors.

  (ii) Air requirement 10,000 cu. mt./hr – Screw compressors.

  (iii) Air requirement above 10,0000 cu. mt./hr – Vane compressors/lobe compressor.

  (iv) Air requirement about 5000 cu. mt./hr – Centrifugal or Axial compressor.

  (c) Compressor Configuration: Sometimes there is limited space for compressor installation and we need high pressures. Then we have to use V-type cylinder configuration of reciprocating compressor. This is a geometrical aspect of selection. In any factory, compressor is installed in a separate room because it makes noise.

4.3 COMPARE SINGLE ACTING AND DOUBLE ACTING COMPRESSORS

Single Acting Compressors	Double Acting Compressors
(a) Used for small capacities only.	(a) Used for large capacities.
(b) Compressed air delivery is in one stroke only.	(b) For same speed and cylinder volume air delivery is double.
(c) More compact and less costly.	(c) Cost is more.
(d) They are normally air cooled.	(d) For big size they are water cooled.
(e) Use for light duty work.	(e) Used for heavy duty work.
(f) Limited applications.	(f) Most widely used compressor.

4.4 COMPARE RECIPROCATING COMPRESSOR AND ROTARY COMPRESSOR

Reciprocating Compressor	Rotary Compressor
1. These are positive displacement compressors.	1. These are also positive displacement compressor. But centrifugal and Axial compressors are dynamic compressor.
2. Piston compressors and Diaphragm compressors are Reciprocating compressors.	2. Screw compressors, Vane compressors, Lobe (Roots blower), Liquid ring are the types of rotary compressors.
3. Reciprocating compressors especially piston compressor (using 4 bar mechanism) are used when very high pressure air is required (above 10 bar)	3. Screw compressor is used when air pressure requirement is below 10 bar but delivery air volume requirement is high.
4. Most popular compressor in Industrial sector is reciprocating compressor.	4. Screw compressor is also popular in industry.
5. It is less costly than screw compressor.	5. It is costly than reciprocating compressor.

4.5 COMPARE AIR RECEIVER AND OIL RECEIVER

Air Receiver	Oil Receiver
(a) Used to store compressed air.	(a) Used to store oil which is at atmospheric pressure.
(b) Its pressure vessel closed from both ends.	(b) It is MS fabricated box. It is not pressure vessel.
(c) Air compressor is fitted above the air receiver.	(c) Hydraulic power pack comprising of shut-off valve, pump, pressure relief valve is fitted above oil tank.
(d) It is verticle or horizontal vessel.	(d) No question of verticle or horizontal orientation.

4.6 AIR RECEIVER OR AIR TANK [S-12, 15]

- The next component of any pneumatic circuit after compressor is Air Receiver. Generally air compressor and air tank is an essential pair in any pneumatic circuit. The main functions of air receiver are:

(a) It provides reserve storage of compressed air.

(b) It smoothens out the pulsation in flow from reciprocating machine.

(c) It helps to cool the warm compressed air.

(d) It helps to condense the moisture in air.

Schematic Diagram, Construction Details of Air Receiver:

- Air receivers are either horizontal or verticle.
- Big capacity receivers are generally verticle. Due to high storage capacity, they can accommodate compressed air from 2-3 compressors. This means that there is common air storage tank between 2-3 compressors.
- Air receiver is a pressure vessel and is made-up of minimum 5 mm thick MS plate. The thickness of plate increase with increase in storage capacity. The cylinder is closed by 'dish ends' from both sides.

[S-10

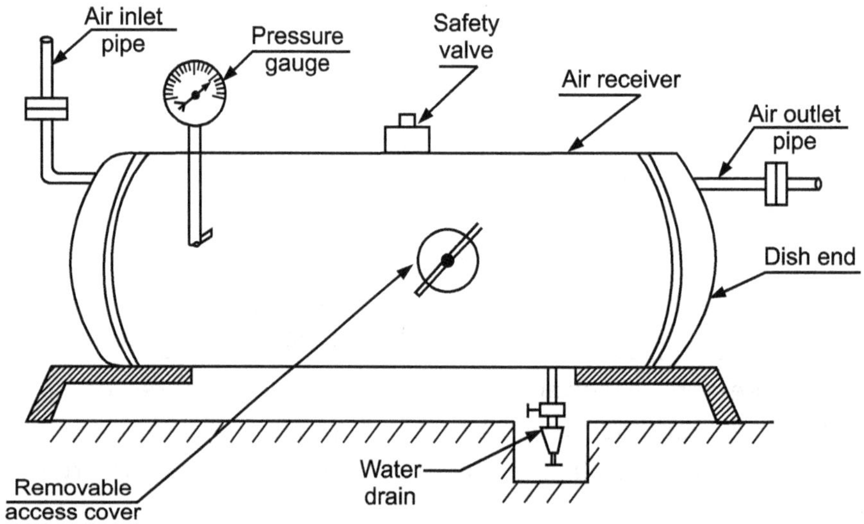

Fig. 4.13: Air Receiver

- Fig. 4.13 figure shows essential features of air receiver. Pressure gauge (Bourdon's Pressure Gauge) shows the pressure of air inside the tank. Safety valve is operative when pressure in tank rises the 'danger level' due to some fault. Air coming from compressors is generally warm (not hot) and the large surface area of air receiver dissipates the heat and air gets cooled. Moisture present in air also condenses. This condensed moisture in the form of water can be removed from water drain tap.
- There are two pipe connection of air tank. One is inlet pipe and other is ofcourse outlet pipe.
- The compressed air is stored in receiver. When tank is 'full', the air pressure is shown on pressure gauge. This is required pressure. Then compressor must stop. After use of air stored in tank, the pressure starts dropping. When it drops to certain value, then compressor must start automatically. This control of compressor can be done by using 'Pressure Switch'.

4.6.1 Symbol of Air Receiver or Air Tank

[S-12]

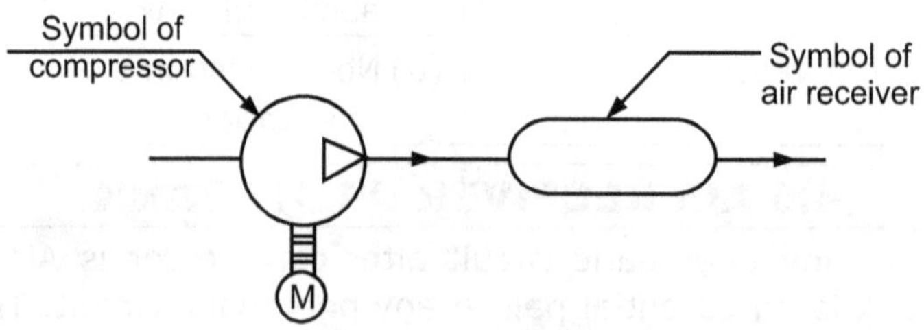

Fig. 4.14

4.7 MOISTURE SEPARATORS AND AIR DRYERS [W-13, W-14]

1. Air that is sucked by compressor is atmospheric air and we all know that it contains water particles called 'moisture'. Amount of moisture in a given volume of air is termed as 'Humidity'. It is expressed in percentage. Humidity of regions near sea shore i.e. Mumbai, Goa, Kokan is 60 to 70%, where as humidity of regions away from sea shore i.e. Sangli, Osmanabad, Nagpur is ranging between 15 to 25%. Hence, we call the air of Sangli/Nagpur as 'dry air' and that of Goa as 'Sticky air'. So it is natural that if we reduce the temperature of air, the moisture in air condenses (water) and this water is removed by taps.

2. Compressed air may be passed through chemical dryers. These are 'hygroscopic chemicals' (hygroscopic chemicals are having a special property of absorbing moisture from air. Table salt, is hygroscopic chemical. In rainy season, the humidity increases, and if salt is kept open to atmosphere, it absorbs moisture and becomes wet).

   The compressed air is passed over these dryers. They absorb moisture present in air and air dries.

   Note:

   o Some hygroscopic chemical dryers (desiccants) are solid but when they absorbs moisture, they becomes liquid. These types of dryers are known as 'Deliquescent Dryers'. These are salt based and urea based driers.

   o But some hygroscopic chemical dryers are solid but even after absorbing moisture they remain solid and in same shape. An interesting point to note that, these driers are reusable. The absorbed water can be removed and can be used again and again. These driers are known as 'Adsorbant Driers'. Activated Alumina and Silica gel are Absorbent Driers.

3. Another but very rarely used way is to heat the compressed air. Due to heating of compressed air the moisture evapourates. When we are operating a pneumatic circuit in 'Dry Regions' then we will have to take less efforts to remove the moisture from air, but if the region is near sea shore, then we will to have take extra efforts to remove the moisture because of the water constraints are more in that air.

   o What will happen if we do not remove the moisture from air ?

   o As we know the amount of moisture present in air depends on the:

     (a) Nearness of region to sea shore.

     (b) It is also related to compression ratio.

     (c) Moisture containts increases at lower pressures and higher temperature.

   If we do not remove the moisture from air, then water particles will carry along with compressed air. The air enters into pneumatic equipments like, air motors, air cylinders. There, these particles will create corrosion, which will hamper the smooth running of actuators/equipments and will also reduce the life of actuators/equipments.

   Hence we must remove the moisture and make the air 'dry'.

4.7.1 Ways of Reducing Moisture

- There are following ways of reducing moisture i.e. Drying of Air:

  o Compressed air is passed over chilled pipes. The chilling effect is achieved by using refrigeration circuits. When temperature of compressed air is reduced to its dew point temperature the moisture will condence. Another way is to heat the air. But hot compressed air will heat the pneumatic equipments and circuits and will reduce the life.

  o The moisture can also be reduced by installing 'water traps' at the end of each pipeline or where the branching of pipe is taking place. Let us see what is 'Water Trap'.

4.7.2 Water Trap [W-10]

- **Construction:** It consists of water trap drum to which inlet and outlet air pipes are attached. There is water tap and on-off valve to remove the collected water in drum.
- **Working:** The compressed air containing moisture enters into drum. To go out of drum, the air has to change its direction in 90°. During this abrupt change in direction the heavy moisture particles drops at bottom due to inertia and the water is collected at bottom. This water needs to be removed after regular intervals.

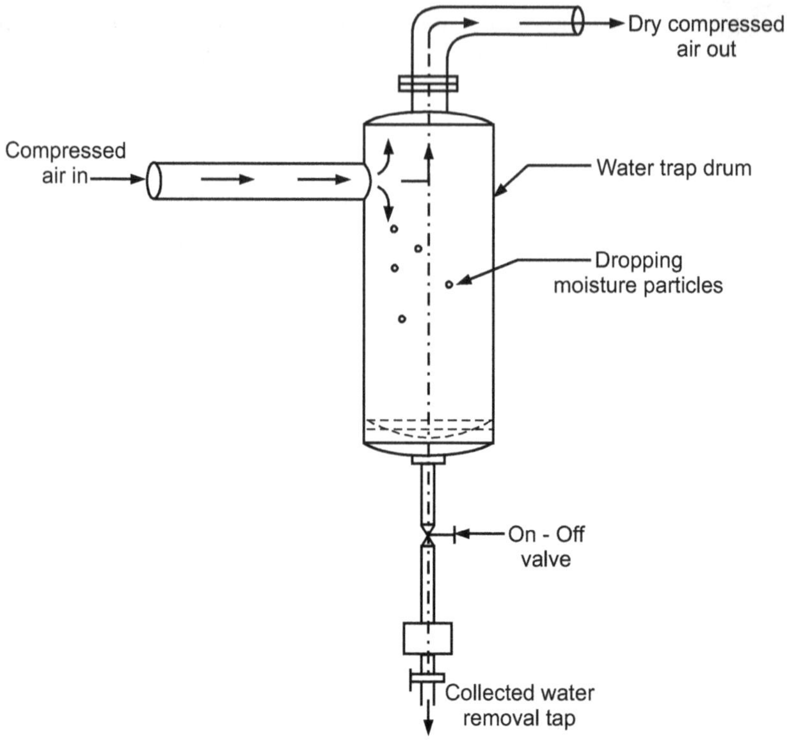

Fig. 4.15: Water Trap

4.8 FRL UNIT [W-10, 12, 13; S-15, 16]

F – Filter

R – Regulator

L – Lubricator

- These three units are called service unit or FRL unit. These three units are in bunch and are installed on pneumatic pipe lines. Compressed air passes through these three units one-by-one.
- First air enters into 'F' i.e. Filter. In this unit air gets filtered. The micron and submicron particles (very very small like dust particles) are arrested in filter and air gets cleaned.
- Filtered compressed air then enters in to 'Regulator'. This unit regulates the pressure of air required by pneumatic mechanism or circuit. Suppose the pressure of compressed air is say 8 bar. The circuit demands 3 bar pressure. Then regulator is used to reduce the pressure from 8 bar to 3 bar. Regulator is a pressure reducing valve.
- The regulated pressure air then enters into 'Lubricator'. In lubricator a thin film of oil or very fine oil droplets are mixed with air. The oil particles are then carried with air to pneumatic circuits and lubricates the components of circuits (actuators, valves etc.) for smooth running and to increase the life of components.

4.8.1 Symbols of Filter/Regulator/Lubricator and their Composite Symbol [W-13, W-14]

(a) Symbol of Filter:

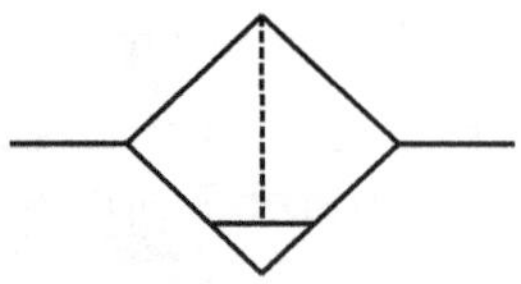

(b) Symbol of Regulator:

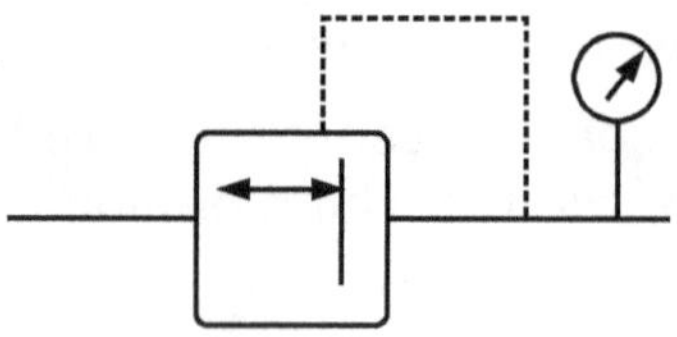

(c) Symbol of Lubricator:

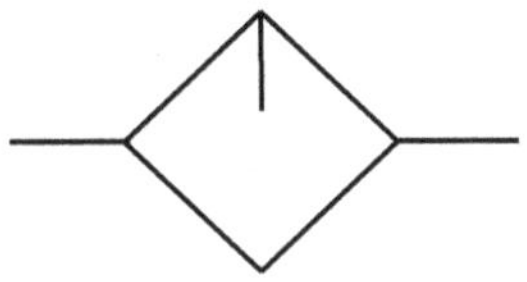

(d) Combination of FRL

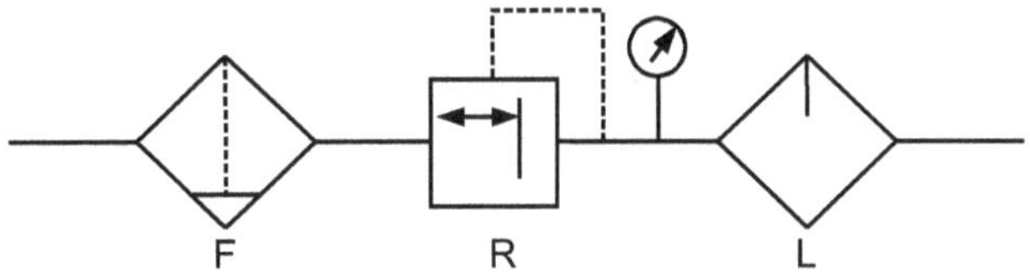

(e) Composite Symbol Used in Pneumatic Circuit

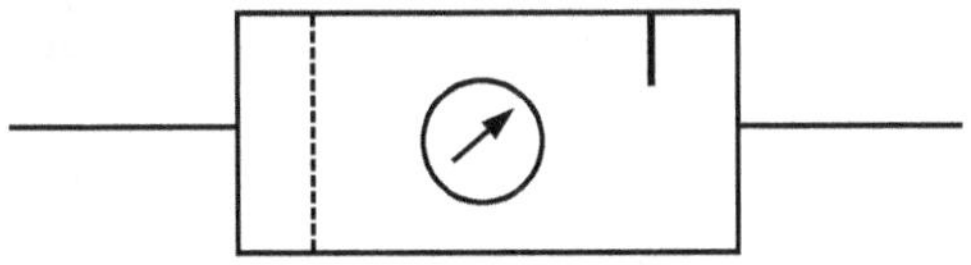

- From above symbol you can easily make out that one line or figure from each unit is taken to form composite symbol. Dotted line on left is taken from symbol of filter. Pressure gauge symbol at center is taken from Regulator and small verticle line on right is taken from symbol of lubricator. Three boxes are combined in Rectangle.
- Now let us see, the construction and working of filter/regulator/lubricator.

4.8.2 Air Filter

[S-11, 15; W-12]

- The construction of simple air filter is depicted in following schematic diagram.
- **Construction:** Air filter consists of plastic bowl, attached to body of filter. The filter element called 'cartridge' is also attached to body at the centre. Both cartridge and bowl can be removed as and when required and can be reattached. There is water tap to remove collected condensate (water) at the bottom of bowl. The air inlet and outlet are created in body of filter.
- **Working:** The compressed air which is un-filtered enters through port (A) and comes into bowl. It has no alternative than to enter into filtering element. The special zig-zag passages created in filtering element, arrests, the micron and sub-micron particles and clean air goes out through port (B).

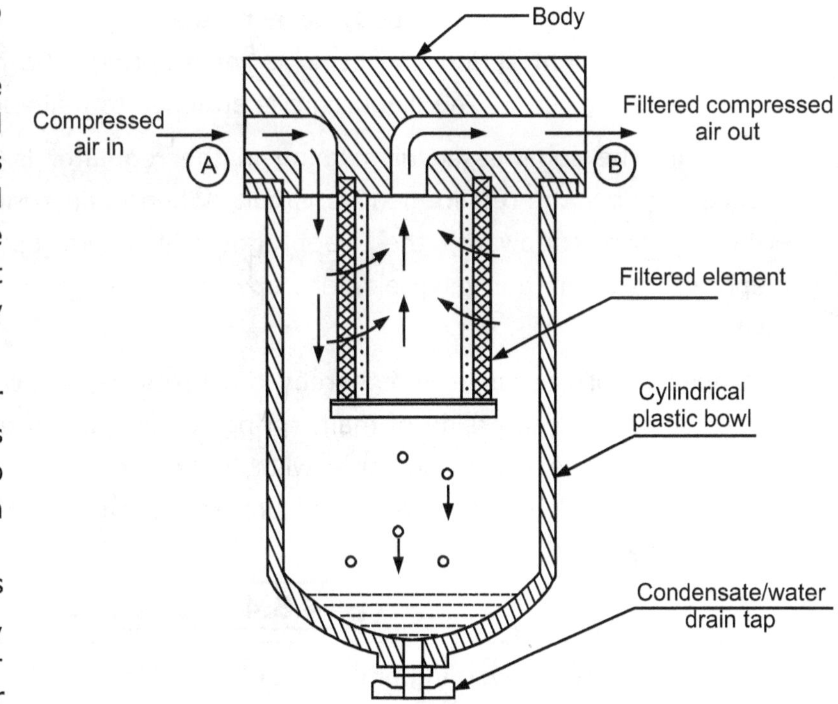

Fig. 4.16: Air Filter

- **Filter element:** It is sintered (powder metallurgy process) cylinder and uses metal-wool as a filtering material.
- The filter also acts as water trap. When air enters into filter bowl through port (A). It suddenly changes its direction. During this movement the moisture particles present in air drops down due to inertia (moisture particles are heavy and they cannot change the direction with air). The moisture is collected at bottom of bowl.

4.8.3 Pressure Regulator [W-12]

- The construction of simple regulator is depicted in Fig. 4.17.

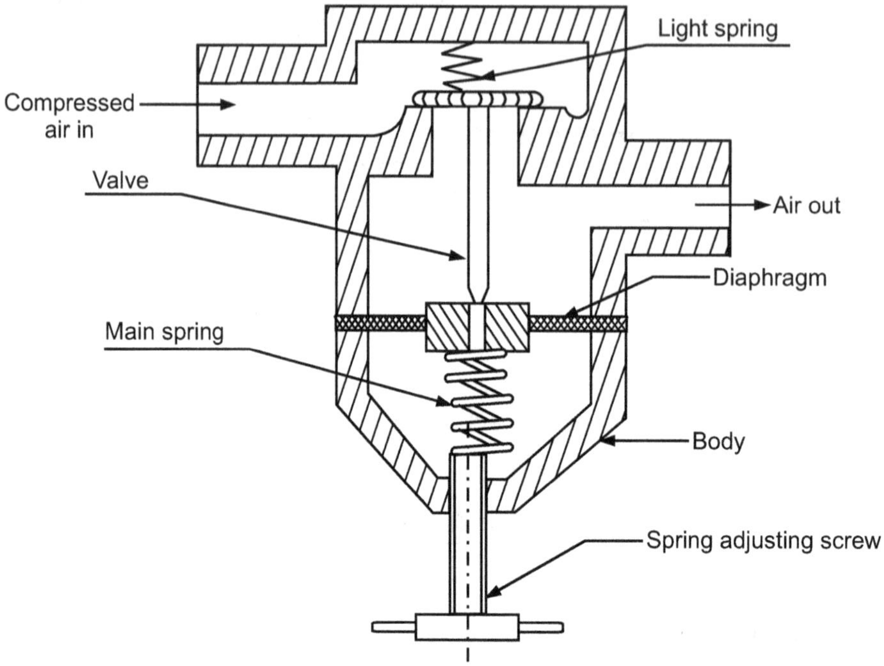

Fig. 4.17: Regulator

- **Construction:** The metallic body of a regulator has 'inlet' and 'outlet' port. There are two springs one is light spring and other is main spring. The main spring can be compressed by spring adjusting screw. The diaphragm is attached to main spring.
- **Working:** The compressed air coming in the regulator is having high pressure. We have to reduce it as per the requirement of circuit. When main spring is compressed by moving spring adjusting screw upward, the diaphragm will move upward. This upward movement of diaphragm will lift the valve element upward, creating an opening to allow the air flow from inlet to outlet.
- The opening of the valve and thereby the pressure of air flowing through it, will be directly proportional to compression of main spring. When the spring compression is less then amount of opening will be small and there will be slight pressure reduction. If spring compression is more then there will be more pressure reduction. Hence regulator is actually a pressure reducing the valve.

4.8.4 Lubricator [W-11, W-12]

- The construction and working principle of lubricator is depicted in following diagram.

Working of Lubricator:

- First we will see the working principle of lubricator:
  - As we know, the lubricator is a device which mixes mist or fine spray of lubricating oil into compressed air. This air with oil particles, enters into pneumatic equipments and oil particles lubricate the sliding/rotating parts/pairs of equipment. This arrangement enhances the life of equipment. Hence lubricator is 'must'.

How to obtain fine spray or automized spray of lubricant oil ?

- This spray is obtained by 'Venturi Effect'. The details of this effect is given below:

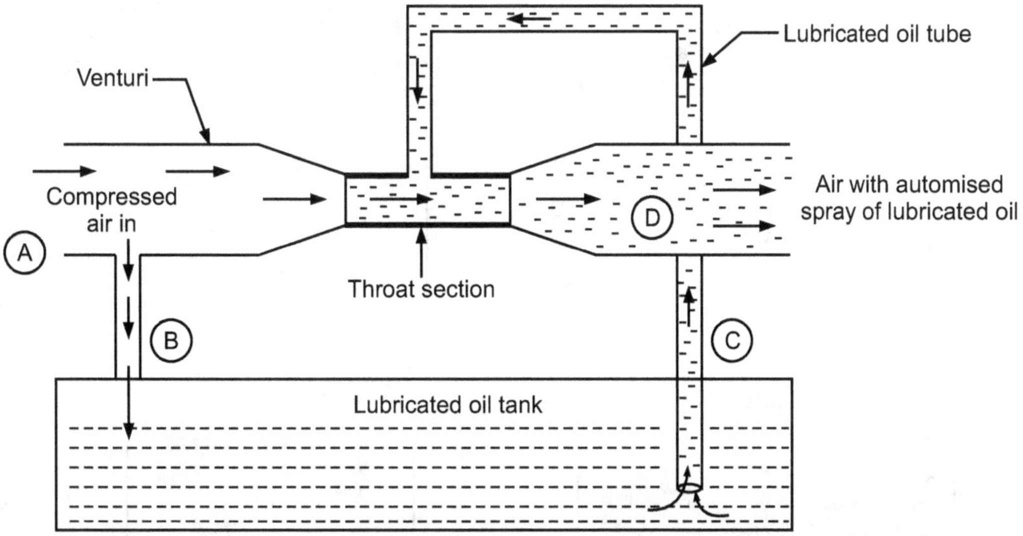

Fig. 4.18

- The arrangement shows only principle of working of lubricator (The actual lubricator is having different shape). When compressed air enters through port (A), part of it goes through tube (B) and exerts pressure on surface of oil. When compressed air passes through throat section, the air pressure drops and lubricant oil from pipe (C) rushes to throat section in the form of spray. Air goes further, crosses throat section and enters into venture portion (D). The pressure again builts-up due to diverging portion and air + oil droplets goes out.

4.8.5 Actual Construction of Lubricator

- Fig. 4.19 shows actual construction of lubricator based on the principle of venturi effect discussed earlier.
- **Construction:** It is having transparent plastic bowl, through which we can see the level of lubricating oil. Venturi is located in the body of lubricator. There are two ports in body, one for dry 'air in' and other for lubricated 'air out'. Oil pick-up tube is immersed in lubricant oil.
- **Working:** Dry air comes in with pressure. The vacuum is created at the throat section. The lubricating oil rushes to throat section through oil pick-up tube and fine droplets of lubricant oil mixed with dry air. Air plus oil goes out to circuit.
- **Actual FRL unit:** FRL unit is like three 'close friends'. They stand 'hand in hand' in every pneumatic circuits. Fig. 4.20 shows FRL units actually used in circuits.

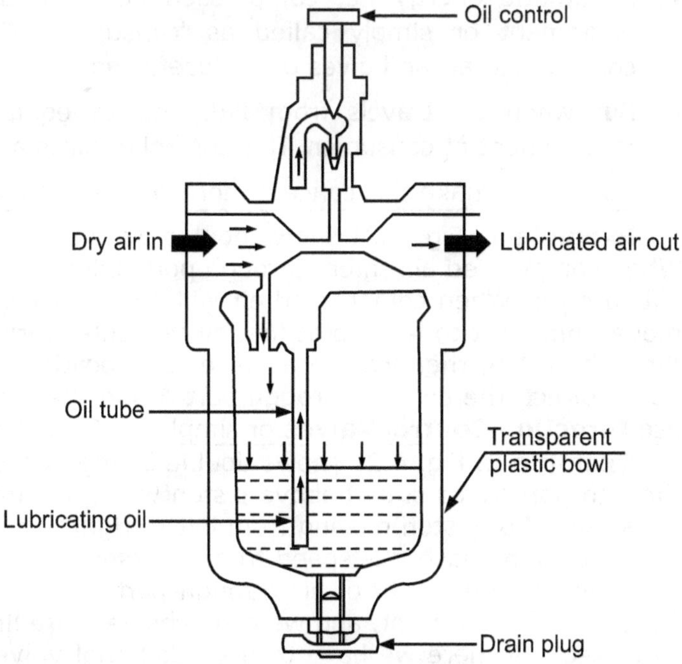

Fig. 4.19

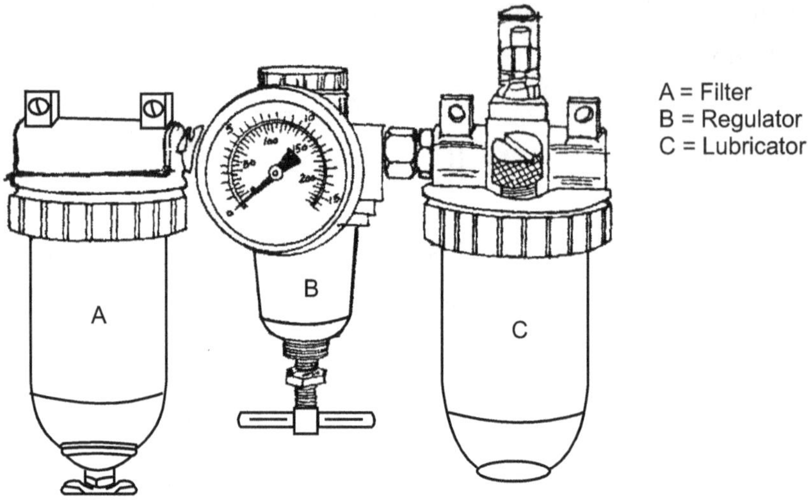

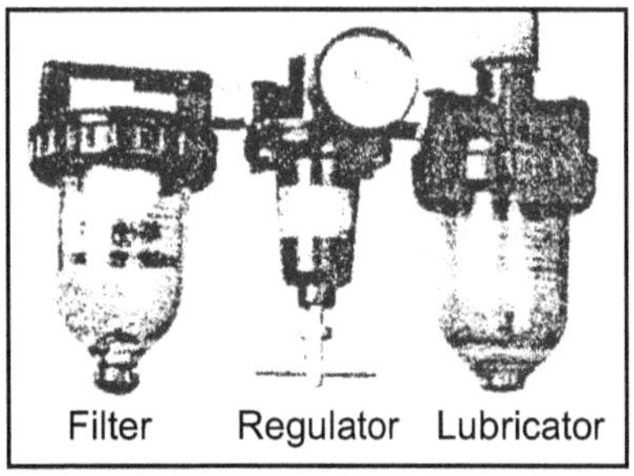

Actual Photograph of FRL Unit (Traditional Design)

Actual Photograph of Modern (Sophisticated Design) FRL Unit

Fig. 4.20

4.9 FLOW CONTROL VALVES/DIRECTION CONTROL VALVES/PRESSURE CONTROL VALVES AND OTHER VALVES

- Pneumatic energy i.e. compressed air from air receiver and FRL unit goes to pneumatic equipment or simply called as 'consumer'. Consumer (also known as an actuator) uses compressed air and gives us an 'useful work'.
- But when air travels from FRL unit to equipment, it needs to be controlled to suit the requirement of consumer. This control of air is a function of various pneumatic valves.
- Now let us make this concept more clear by studying following examples:

Example 1: Fig. 4.21 shows double acting cylinder. (It is actuator). When compressed air enters through port (inlet) A piston will move from left to right. When compressed air will enter through port B, piston will move from right to left. So at first air will enter through port A and then through port B, then we can achieve reciprocating movement of piston. So to 'direct' the air first through port A and then to port B, we have to use **Direction Control Valves** or simply called as 'DC valves'.

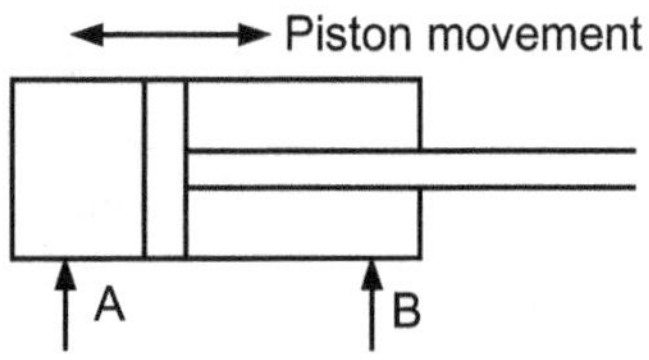

Fig. 4.21

Example 2: Fig. 4.22 shows double acting cylinder. When air enters through port 'A' as stated above piston will move from left to right. The linear speed of piston moving from left to right in say 3 m/sec. Suppose we have to increase this speed from 3 m/sec to 5 m/sec, then we have to admit more quantity of air through port 'A' so that piston will move faster from left to right, and we can achieve more linear speed of piston (5 m/sec). So here we have to use additional valves which will control the **quantity of air** entering into cylinder. This valve is known as **Flow Control Valve**.

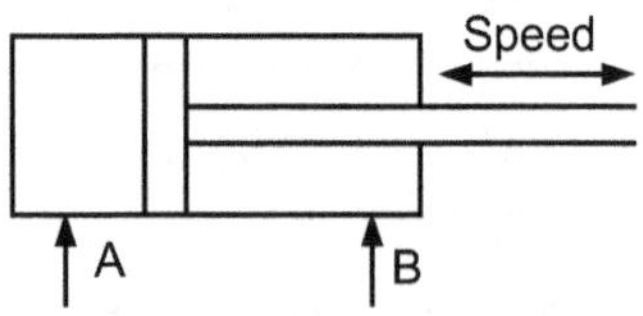

Fig. 4.22

Example 3: Consider two actuators (air cylinder or air motor) No. (1) and (2) shown below. The air pressure developed by air compressor is say 8 bar. Actuator (1) operates on 4 bar pressure and Actuators (2) operates on say 5 bar pressure. So for Actuator (1) we have to reduce the pressure of air from 8 bar to 4 bar and for Actuator (2) we have to reduce the pressure of air from 8 bar to 5 bar. So we have to control pressure of air. We must introduce another additional valve to make changes in pressure. This valve is known as **Pressure Control Valve.**

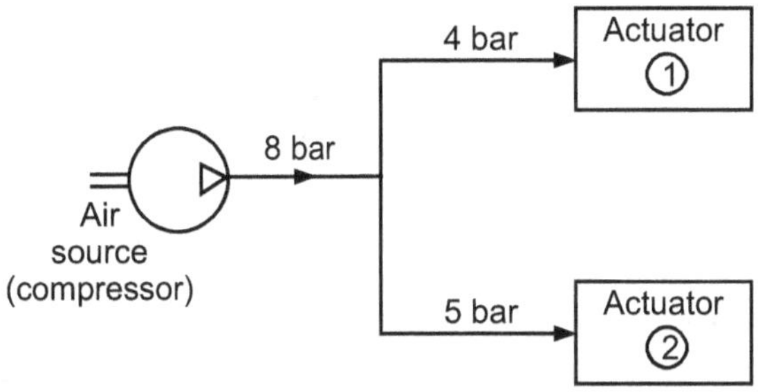

Fig. 4.23

4.9.1 Classification Pattern of Valves used in Pneumatic Circuits

[W-10, W-14]

- Now we will see the classification pattern of valves.
- As stated above valves in pneumatic circuit are divided into three main categories. Each valve will be either spool type or seat type.

  (a) Direction Control Valves (Spool/Seat).

  (b) Flow Control Valves (Spool/Seat).

  (c) Pressure Control Valves (Spool/Seat).
- Note that spool or seat type is concerned with the construction of valve. Irrespective of its construction, each valve will perform the function for which it is made.
- **Spool Type Valve:** Spool is an element similar to piston and piston rod; which slides in a closely finished bore of cylinder. The wear of these elements is uniform and are most widely used.
- **Seat Type Valve:** You must have seen the values of IC engines. It is exactly a similar valve. The valve element (either poppet/cone or ball), sits on specially machined and finished seat. The wear of these valves is not uniform, hence used on very selective basis.

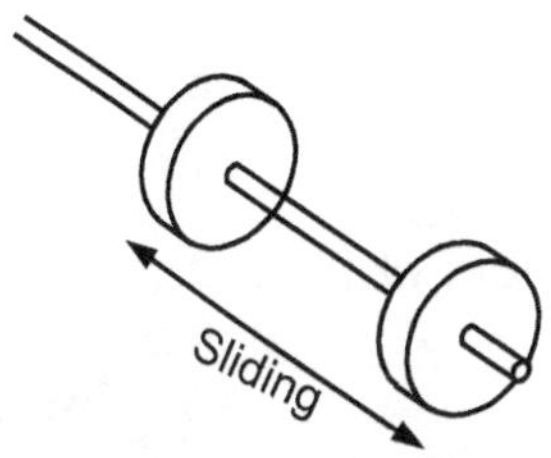

Fig. 4.24: Spool

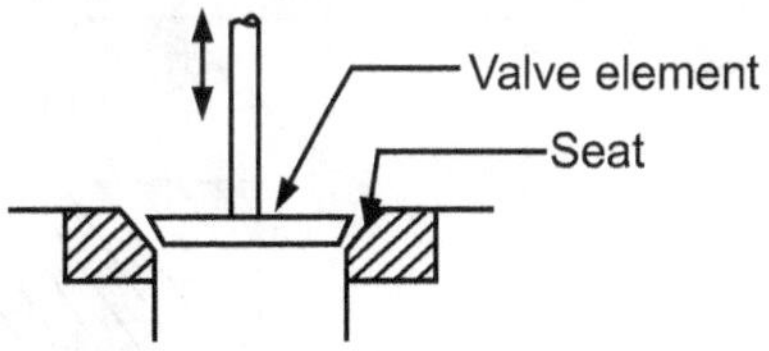

Fig. 4.25: Seat Type

Classification of Flow Control Valves:

1. Fixed type flow control valve.
2. Variable flow control valve with built-in check valve.
3. Check valve (Directly operated).
4. Check valve (Pilot operated).
5. Double check valve (shuttle valve).

Classification of Pressure Control Valves:

1. Relief valve.
2. Pressure regulator (Pressure reducing valve).
3. Non-relieving pressure regulators.
4. Relieving pressure regulators.

4.9.2 Flow Control Valves

Fixed Type Flow Control Valve

- As we have seen that by controlling the flow of air admitting in actuator we can control the speed of actuator (linear speed in case of linear actuator or Rotary speed in case of Air motors).

Fixed type flow control valve is shown as below:

- In this valve a special restricted passage is created. This passage is called throttle or flow restrictor passage (It is similar to throat of venturimeter). Suppose if 3 lit./min. compressed air comes into the valve, due to restricted passage entire 3 liters will not go out. Quantity less than 3 lit./min. will go out, and hence flow will be controlled.
- This type of flow control valve is very nearly used; or used on very selective basis because, we cannot vary the outgoing flow (outflow is fixed). Refer following Fig. 4.26.

Fixed type flow control valve:

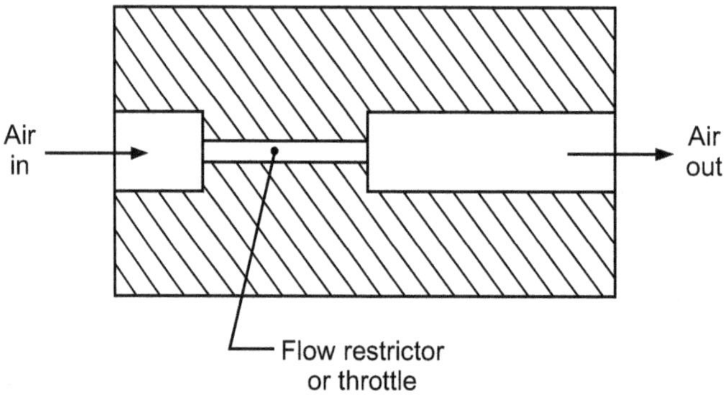

Fig. 4.26: Fixed Type Flow Control Valve

- The symbol of fixed type flow control valve is as shown in Fig. 4.27.

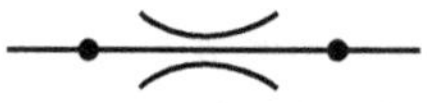

Fig. 4.27

4.9.3 Variable Flow Control Valve with Built-in Check Valve

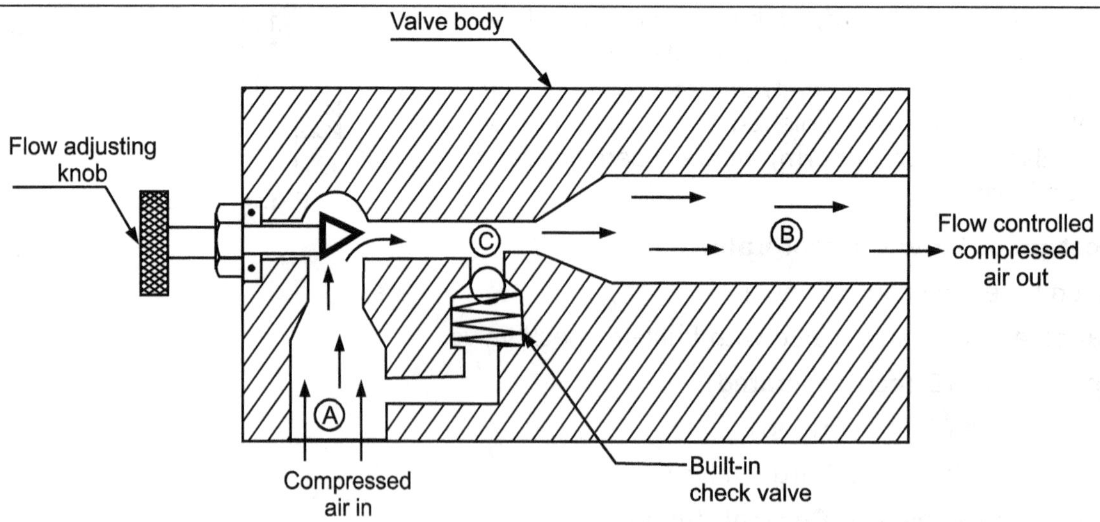

Fig. 4.28

- The internal construction of valve is shown in the figure. A built-in check valve (Also called non-return valve or unidirectional valve) is incorporated in this valve.
- **Working:** It is clear from the Fig. 4.28 that, air comes in through port (A) and goes out through port (B). The flow can be reduced or increased by rotating 'Flow Adjusting Knob', in clockwise or anticlockwise direction.

(a) Function of Built-in check valve:

- Air going out from port (B) which is connected to consumer. If due to some chocking/jamming, the consumer has stopped working; then air will not flow further and back pressure starts building. Now air starts accumulating and flow will try to reverse i.e. from port (B) to (A). But there is continuous supply of air from (A) to (B). That means, air flowing in reverse direction i.e. from (B) to (A) needs to be vented out. Look at port (C). Reverse Air will act on ball and will push down the ball by compressing the spring. Due to this, the reverse flow air will flow from (C) to (A) and back pressure will be subsized.

(b) Why check valve is called unidirectional valve ?

- When air flows from (A) to (B), some air will try to flow towards port (C). But this air will push the ball in upward direction, thereby locking the port (C). In addition to this the spring force will also keep the port (C) locked. Hence there cannot be flow from (A) to (C). The flow is possible only from (C) to (A). Hence this valve is actually to safety valve because it vents out back pressure (If this back pressure increases beyond certain limits there is possibility of damaging the consumer/pipelines/valves). i.e. flow from (C) to (A). And because flow cannot be possible from (A) to (C) it is called unidirectional valve.

(c) Symbol of variable flow control valve with built-in check valve:

- The symbol is slightly modified by addition of an arrow to the symbol of fixed type flow control valve. Arrow represents its variable flow. The symbol of built-in check valve is similar to its construction.
- Now let us study, the application of 'Variable flow control valve with built-in check valve'.

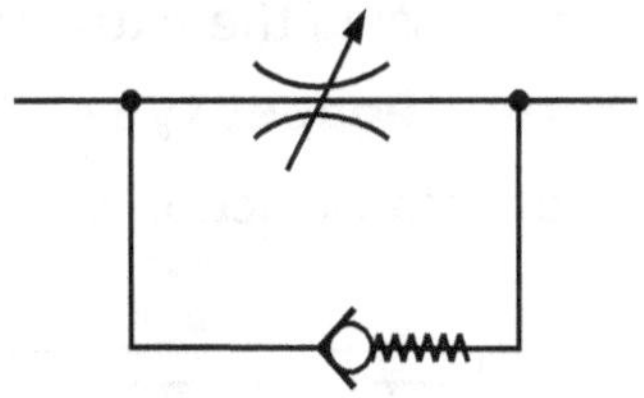

Fig. 4.29

- Let us use this valve for controlling the forwarding speed of piston of Single Acting Cylinder (SA Cylinder). (Refer Fig. 4.30)
- SA cylinder is shown in the figure. The piston moves from left to right because air is being admitted through port A (This is forward stroke). The linear speed of piston can be controlled by flow control valve. When push button of DC valve is operated, Port A will be connected to port R. Port P will be disconnected and piston will move from right to left due to expansion of spring. If suppose spring does not expand due to failure of spring, then air will be accumulated in cylinder and piston will remain on right side. Here, the check valve will operate automatically and excess air in cylinder will be by-passed through check valve to avoid damage of SA cylinder.

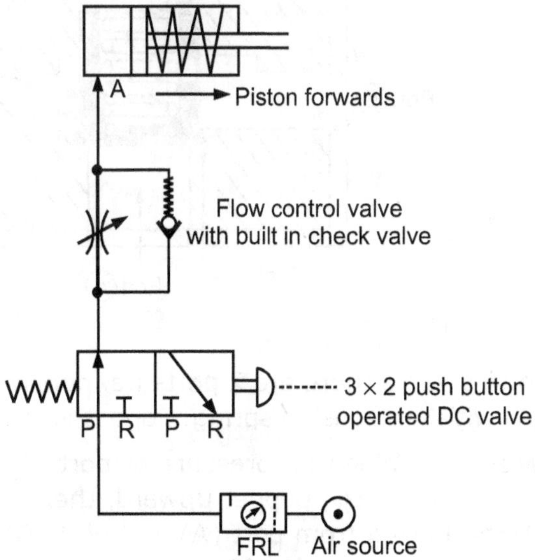

Fig. 4.30

4.9.4 Directly Operated Check Valve

[S-12, W-16]

- As we have seen that, check valve is a 'safety valve' because it allows the flow in one direction. This valve can be best compared with safety valve of domestic pressure cooker.

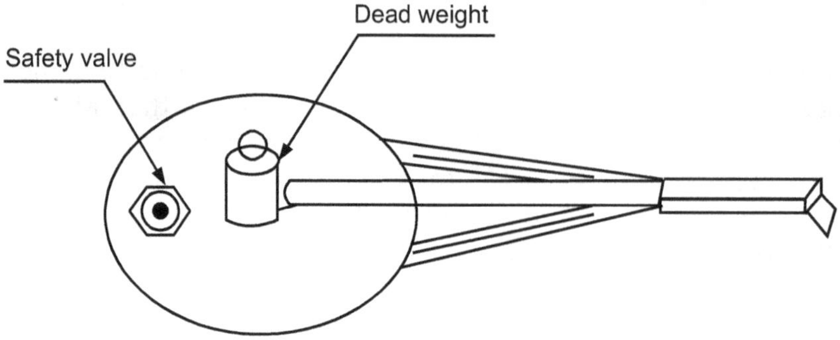

Fig. 4.31

- We know that, steam formed inside cooker pot, comes out form the opening below dead weight. If this opening is chocked, then steam starts accumulating in pot. After sometime steam superheats. Temperature increase tremendously. Then the special material below safety valve gets melted and hole is created at the centre of safety valve. Through this hole steam vents out. This safety valve is also called as fusible plug. Because it fuses (melts) at particular temperature.
- Except melting, the work of check valve is to protect the pneumatic systems and circuits by blowing or by passing the extra air.

Directly operated (Poppet type) Check Valve: [S-12]

- Fig. 4.32 shows construction of Directly Operated (Poppet type) check valve:

[S-12]

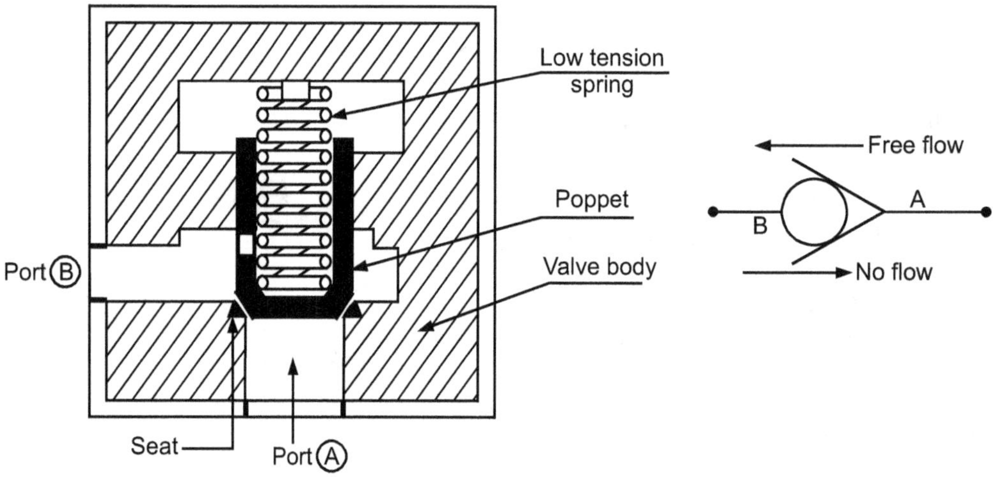

Fig. 4.32

- In this valve there are 2 ports i.e. port (A) and Port (B). A moving poppet is installed above port (A) with low tension spring. Poppet is rested on seat.
- **Working:** When air pressure at port (A) starts increasing beyond rated limit then, the force of air will move the poppet upward, there by compressing the spring. When poppet moves-up, air starts flowing from port (A) and port (B). Flow from port (B) at (A) is not at all possible. When air pressure at port (A) normalizes, the spring will expand and poppet will regain its original position.

4.9.5 Pilot Operated (Poppet type) Check Valve [W-09, S-16]

- Pilot operated means, as we have seen, the air pressure (called air signal) is utilized in this valve for special operation.
- Directly operated check valve, cannot be continuously kept open i.e. it acts as a safety valve. But if we want continues supply of air from port (A) to (B) for some particular period of time, then we have to modify this directly operated check valve in following manner.

- In the modified version of directly operated check valve, a small piston is introduced below moving poppet. (As we can see that, upper part of the valve shown in above figure is same as that of directly operated check valve). This piston can move-up by pressurised air signal.
- **Working:** When air signal is operated, pressurised air will be admitted and small piston will move upward by compressing spring S_2. Piston rod of piston, will in turn push the poppet upward by compressing spring S_1. And air flow form (A) to (B) will start. Unless and until we supply air signal to small piston, there will be air flow from (A) to (B). When air signal is cut-off, spring S_1 and S_2 will expand and poppet and small piston will regain its original position.

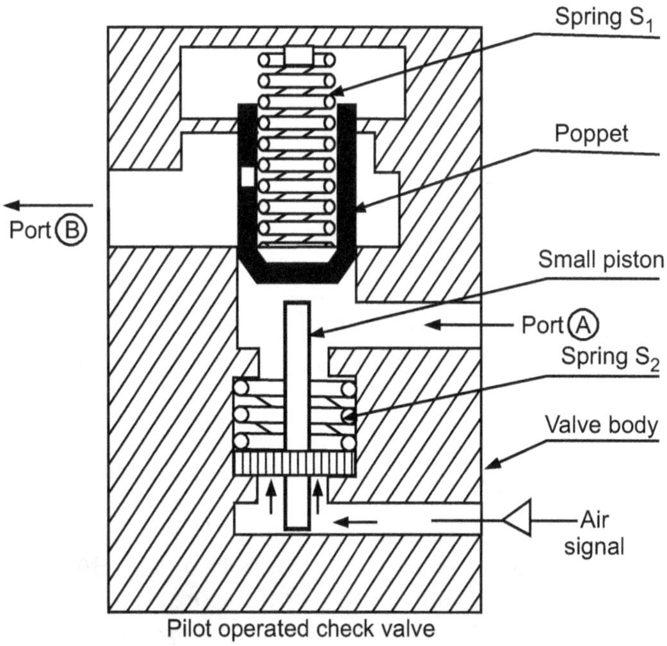

Fig. 4.33: Pilot Operated Check Valve

- It may be noted that, there cannot be air flow from (B) to (A).

Symbol of Pilot Operated Check Valve

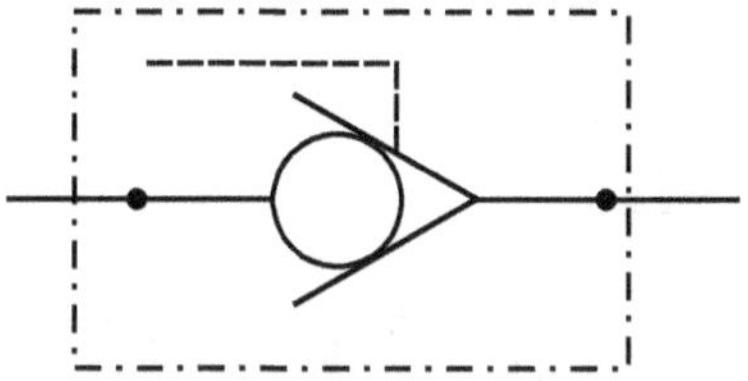

Fig. 4.34

Important:

- While drawing this symbol in pneumatic circuit we must draw the outside rectangle drawn with the help of centre line (— • — • —).

4.9.6 Double Check Valve (Shuttle Valve) (OR Gate)

- This valve is popularly known as Shuttle valve. (Name shuttle is used because, the valve element used in this valve is a ball and it moves in a valve in the same manner as that of Badmington shuttle). This valve is also known as **OR GATE**, because, it allows the air to flow from two alternative sources. Fig. 4.35 shows construction of Double check valve (shuttle valve) (OR GATE).
- **Construction:** In the valve Body there are 3 ports. A is consumer port i.e. connected to actuator. Port x and Port y are air supply ports. The valve element is free to move in the valve.
- **Working:** When air is supplied through port y, the ball will move towards left and will block port x. When air will be supplied through port x, the ball will move towards right and will block the port y; and air will flow from port x to port A. Thus, port A will receive air either from port x OR from port y. Hence it is called OR GATE.

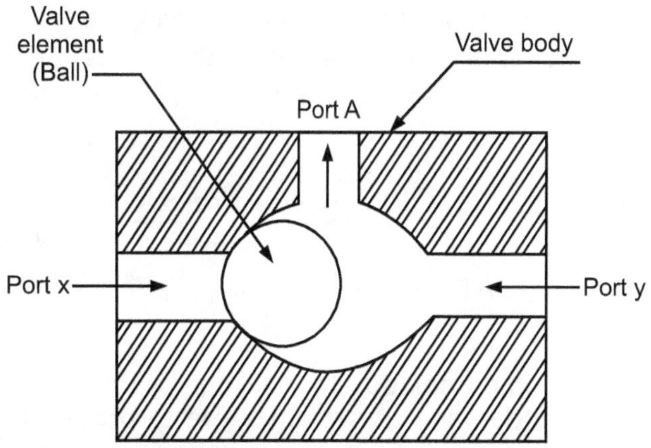

Fig. 4.35: Shuttle Valve

Symbol of Double Check Valve (Shuttle Valve)

- The symbol is designed on the basis of the construction of valve itself.

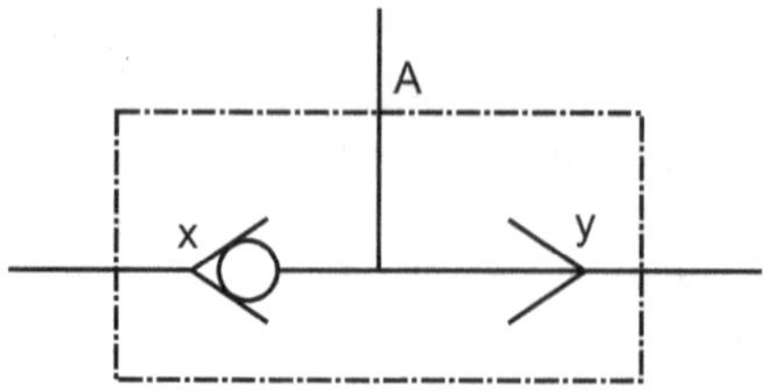

Fig. 4.36

Application of Double Check Valve

- The double check valve is used to actuate single acting cylinder.
- The typical circuit shows SA cylinder, the movement of which is possible through two sources. The circuit uses two DC valves. At present, valve (2) is in operation. The air is supplied to SA cylinder through port y and naturally port x is blocked. When piston moves from left to right (Advance of piston), it will return to its original position due to expansion of spring. And then value (1) is operated; so that Air will be supplied through port x.

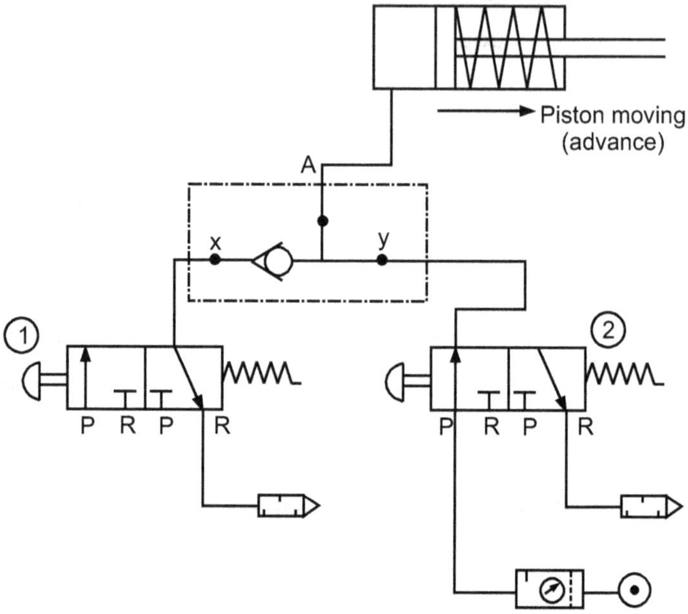

Fig. 4.37

4.9.7 Pressure Control Valves

- Refer following Fig. 4.38.
    - Here, the air travels from air receiver to consumer or load. The distance of pipe is long. Generally the air compressor is fitted outside the factor shed to avoid its noise. During the flow of air from receiver to load, there are considerable leakages which reduces the pressure.
    - It is called a 'pressure drop'. Hence, therefore, air pressure in the receiver is kept at higher level than the required by consumer or Load.
    - Suppose the pressure of air in receiver is say 8 bar and consumer or load requires say 5 bar then we have to reduce the pressure near that load. Here we need pressure control valve.

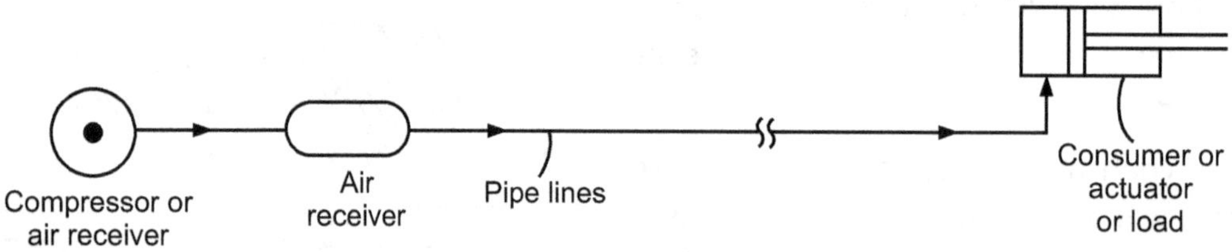

Fig. 4.38

- Sometimes due to blockage of load, the air starts accumulating in system, which may leads to damage of system. Here to protect the system we need pressure relief valve. This valve relieves, the accumulated dangerous air to atmosphere.

Important: Pressure control valve is actually a pressure reducing valve.

Now, we will see the various valves which take part in pressure control.

4.9.8 Pressure Relief Valve

- This valve is infact not a pressure control valve, but it is actually a safety valve. It operates when main pressure control system fails. It is a sort of stand-by arrangement.
- **Construction:** There is threaded joint between cap and body. By rotating the cap we can adjust the tension of spring. There are two (or one) exhaust ports and one inlet port.
- **Working:** The valve element is held closed by spring tension, according to required relief pressure. So normally there is no flow from inlet port to exhaust port.
- **Important:** Note that pressure relief valve is normally closed. When due to failure of main pressure control system, the air pressure starts increasing. When force due to air pressure exceeds the spring tension, the valve element lifts up and air starts flowing from inlet to exhaust. Once the inlet pressure falls below the safe pressure, the force of air reduces and spring expands the valve element will regain its original position so that the flow stops.
- **Location of Valve:** It is generally fitted near outlet of air receiver.

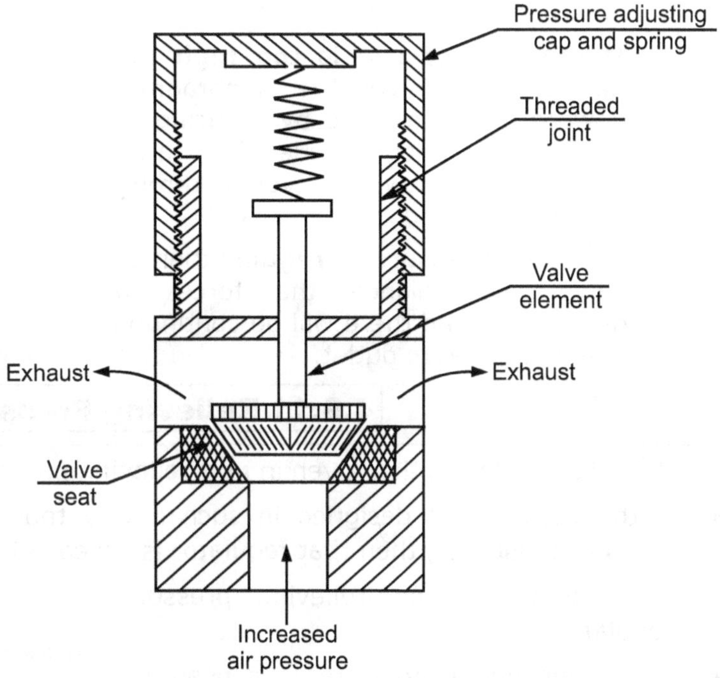

Fig. 4.39: Pressure Relief Valve

Symbol of Pressure Relief Valve

- The symbol shows exhaust port R as per construction. The arrow on spring shows that, the spring tension can be adjusted.

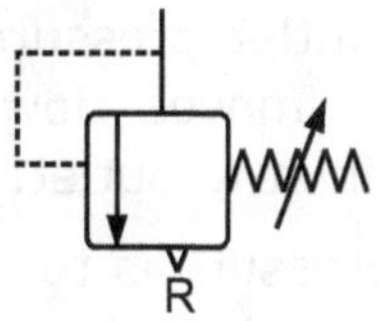

Fig. 4.40

4.9.9 Pressure Regulator

[W-09]

- Please refer Pressure Regulator in FRL unit.

4.9.10 Non-Relieving Pressure Regulator

- We are aware of the fact that, pressure regulator or pressure control valve is fitted near consumer or load (SA cylinder/DA cylinder/air motor/any other specially designed). We also know that, air pressure supplied to pressure regulator is of higher valve. Pressure regulator reduces this air pressure as required by consumer. To reduce the air pressure, some part of air must be vent out (exhausted to atmosphere). The venting of air is either through pressure regulator itself or through consumer.
- If the regulator is designed in such a manner, that air vents through consumer only (not through regulator), then that regulator **is called non-relieving pressure regulator**.
- Fig. 4.41 shows non-relieving pressure regulator.
- **Construction:** The diaphragm is fixed in the body. Outlet pressure is preset by pressure setting spring and adjusting screw. Valve element is attached to diaphragm.

- **Working:**
  - **If outlet pressure is too low** then spring will force down the flexible diaphragm and valve element will move downward because spring force is dominating. And air starts flowing from inlet to outlet to increase the outlet pressure to required valve.
  - **If outlet pressure is too high** then this pressure forces the diaphragm in upward direction there by reducing the air flow from inlet to outlet. Due to this reduction in flow, air vents through consumer or load.
  - **If outlet pressure is regular** then the valve will balance the forces on diaphragm, and there will be continuous flow from inlet to outlet.

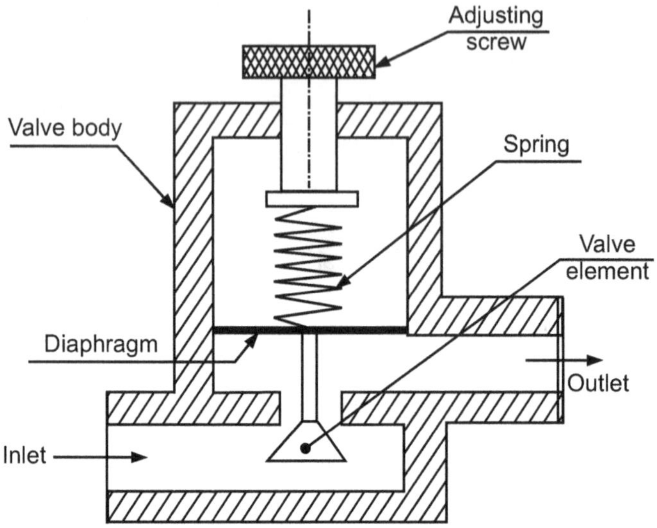

Fig. 4.41: Non-Relieving Pressure Regulator

4.9.11 Relieving Pressure Regulator

- Refer to the description given in non-relieving pressure regulator:
- If the regulator is designed in such a way that, the air vents through regulator only (not through consumer) then that regulator is called **relieving pressure regulator.**
- It is similar to non-relieving pressure regulator.
- Note that outlet pressure is sensed by diaphragm. The tension of diaphragm is adjusted by pressure adjusting spring.
- **Working:** If outlet pressure is too high, the diaphragm moves down closing the flow from the inlet to outlet.
- When outlet pressure is too low, then the inlet poppet valve is pushed open admitting more air to raise the pressure.
- If outlet pressure is regular then valve will balance admitting and venting small amount of air to keep load pressure at set valve.
- **Important:** Note that vent hole is shown on valve body i.e. air is venting out through valve only and not through consumer, in relieving pressure regulator.

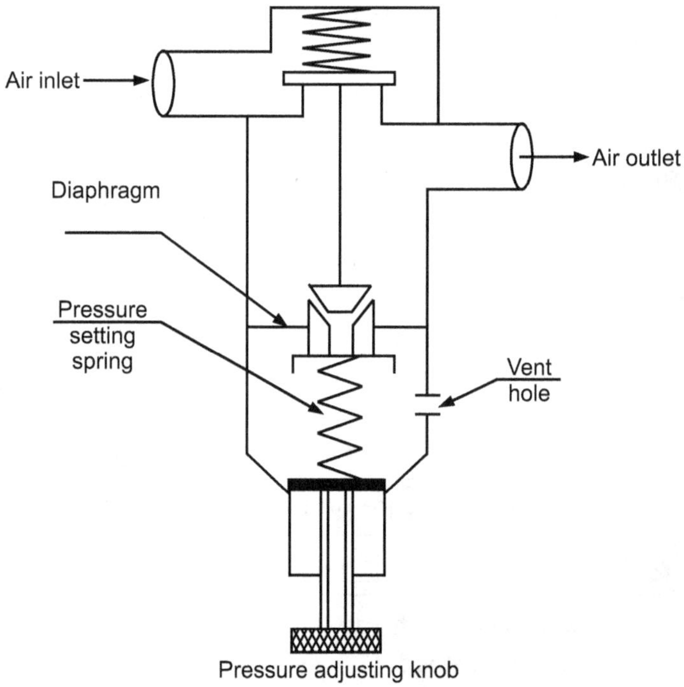

Fig. 4.42: Relieving Pressure Regulator

4.10 QUICK EXHAUST VALVE

- Quick-exhaust valves (QEVs - shortform) are ready-made piston valves or diaphragm valves, designed to quickly exhaust pneumatic cylinders.
- QEVs need a separate pilot valve for triggering. They can be very sensitive, triggering at a pressure drop of a few N/mm^2.
- Most QEVs are designed to be filled from the pilot side, and will typically work well using a 3-way valve as the pilot valve.

- Small QEVs are sometimes used as pilots for larger valves, often to allow the use of a blowgun as the trigger. Refer Figure depicted below. As pressure is applied through input port 1, the poppet shifts forward blocking the exhaust port 3 and allowing flow through output port 2.
- When pressure is removed from input port 1 back pressure at port 2 unseats the poppet and allows flow through exhaust port 3.

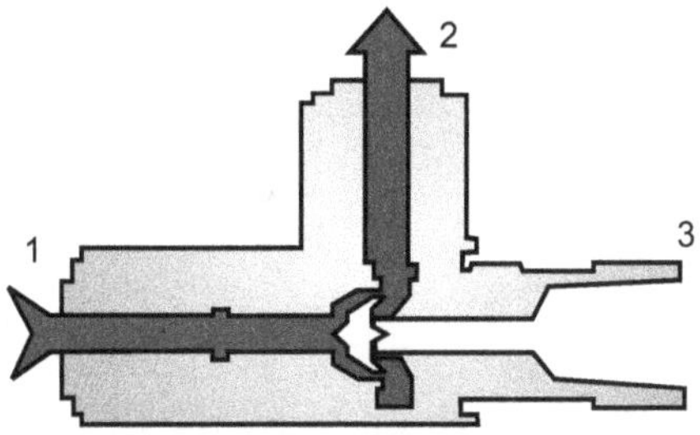

Fig. 4.43

4.11 FLUID OR OIL CONDUCTING ELEMENTS SUCH AS PIPELINES, HOSES, FITTINGS, SEALS, ACCUMULATORS (FLUID POWER PLUMBING)

- Connecting tubes or pipes are used to conduct hydraulic fluid or oil from one part of circuit to another is termed as fluid power plumbing.
- A typical hydraulic system is depicted below. The oil is taken by pump. Pump pressurises the oil. This oil actuates the actuator. The used oil is then returned to the oil tank. Hence hydraulic system is 'closed system'. The pressure lines and return lines shown in this block diagram are pipe lines.

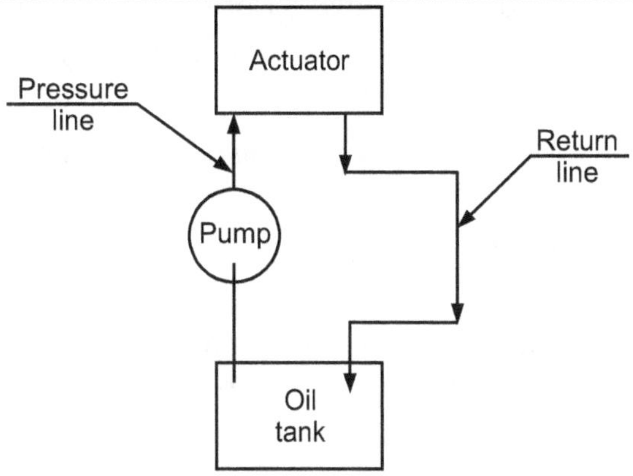

Fig. 4.44

4.11.1 Requirements of Good Fluid Power Plumbing [W-13]

(a) The pipes/tubes which carry the hydraulic fluid must have sufficient mechanical strength to support devices (pressure gauges/valves) mounted on it.
(b) The pipe/tube must withstand for the pressure of fluid it is conducting.
(c) Terminal points (union, flanges) must be provided at all junctions.
(d) Pipe must be kept clean and flushed regularly.
(e) The pipe must have correct required size.
(f) The material selection of pipe should be proper.
(g) Provision of a Air vents/bleeds must be made at proper locations.
(h) Do not use excessive number of elbows, bends because these create flow speed reduction.
(i) Use the pipe fittings only at required place.

4.11.2 Pipes/Tubes [W-13]

- Basically there are three types of pipes/tubes used in hydraulic systems.
  (i) Rigid pipes
  (ii) Semi-rigid pipes
  (iii) Flexible pipes
- Whether to select rigid/semi rigid/flexible depends on ;
  (a) Pressure oil it will carry.
  (b) Mechanical strength of pipe.
  (c) Type of fluid it will carry.
  (d) The location of the system.
    - Rigid pipe : Steel Seamless (Jointless) Pipes
    - Semi-rigid : Tubes (Copper/Aluminium/Steel)
    - Flexible : Nylon/Plastic/PVC

4.11.3 Materials for Pipes

[S-12, W-13]

(i) Steel: This material is most widely used pipe material in Hydraulic circuit. These are seamless pipes/tubes.

Tubes/Pipes are of annealed quantity (annealing is a heat treatment in which component is heated and cooled slowly). Mild steels or Soft Ductile Carbon steel (Ductility: Ability of material to be drawn into wires).

Stainless steel is also used for pipe/tube material but it is very costly. But are exclusively used in aircraft/marine fluid lines where pipes are exposed to varied atmosphere.

(ii) Copper: As we know copper is highly resistant to corrosion, so it can also be easily drawn into pipes/tubes. Copper tends to harden in high temperature. Hence used in medium pressures and low bore sizes. Some oils react with copper which limit the use of copper tubes. In old days hydraulic break systems of automobiles, copper tubes were used.

(iii) Aluminium: It is also highly resistant to corrosion and similar to copper, can be easily drawn into tubes. The additional 'profitable' property of aluminum is its light weight. Hence obviously these pipes are used in hydraulic systems used in aircraft. But similar to copper these pipes are costlier.

(iv) Zinc Coated Galvanised Pipes: These pipes are similar to house hold plumbing GI pipes. These are preferred for low pressure oil conducting.

4.11.4 Size of Pipe

[S-12, W-13]

- The pipes are specified in terms of its internal diameter (ID). Now we will look into actual specification. The outside diameter (OD) of pipe is fixed. The ID varies with its type. There are three types:

  (a) Standard (Schedule 40) [S-12]

  (b) Extra Strong (Schedule 80) [S-12]

  (c) Double Extra Strong (Schedule 160)

Let us take example of size $1\frac{1}{2}$" pipe

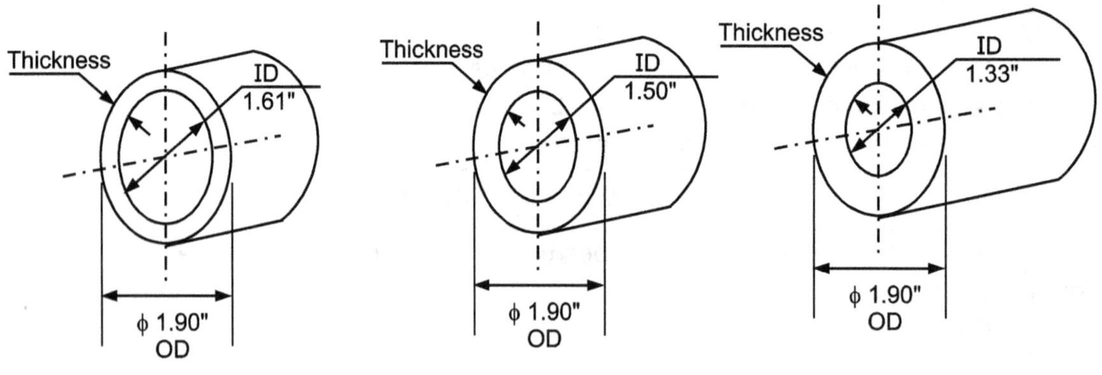

(a) Standard (Schedule 40) **(b) Extra Strong (Schedule 80)** **(c) Double Extra Strong (Schedule 160)**

Fig. 4.45

- We can easily understand that:
  - Pipe having size $1\frac{1}{2}$" will be having OD = 1.9" and the ID goes on decreasing as we go for stronger pipe. Thickness increases means ID reduces. Following are the sizes of different pipes.

(All dimensions are in Inch)

Size of pipe	OD	ID Std.	ID Extra Strong	ID Double Extra
$\frac{1}{2}$"	0.84	0.62	0.45	0.46
$\frac{3}{4}$"	1.05	0.82	0.74	0.61
1"	1.31	1.05	0.95	0.81
$1\frac{1}{2}$"	1.90	1.61	1.50	1.33

4.12 HOSES [W-10]

- In many hydraulic systems, the drive units are assemblies needed to move alongwith pipelines. In such cases, flexible pipes called 'hoses' are used.

General Construction of Hydraulic Hose

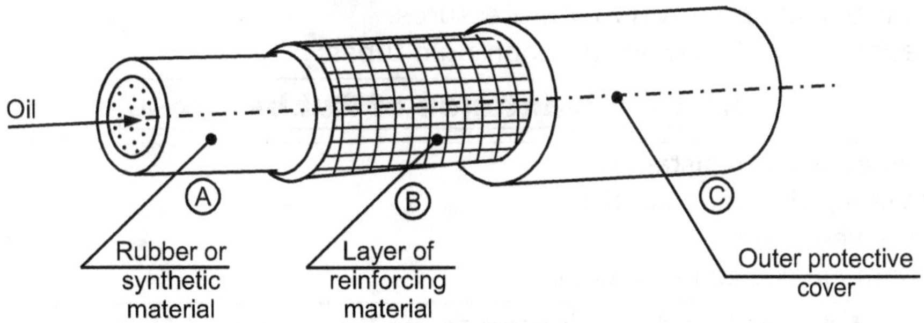

Fig. 4.46

- Hydraulic hose is a compound flexible pipe having three layers:

Layer A: This is the inner tube through which fluid/oil flows. Hence this layer comes in direct contact with pressurised hydraulic fluid. This layer is called 'hose material' layer. Following are the various materials used for this layer:

(a) Plastic

(b) Nylon

(c) Braided Nylon

(d) PVC

(e) Teflon

(f) Synthetic Elastomers

(g) Compounded Rubber/Natural Rubber

Layer B: This layer is called 'hose reinforcement'. This is the middle layer and it increases strength of inner layer. This layer provides structural strength to entire hose to withstand against hydraulic pressures of oil which is very high in hydraulic system. This layer (polylayer) is made up of following materials:

(a) Cotton

(b) Nylon

(c) Steel wires

(d) Synthetic yarn

(e) Rayon

Layer C: This layer is the outer layer and is called protective cover. The middle layer i.e. reinforcement layer is protected from corrosion, abrasion and other damages which can occur during accidents. Following materials are used for this layer:

(a) Neoprene

(b) Synthetic GRS rubber

(c) Cotton/Synthetic yarn.

- **Protective covers:** Many times Nylon sleeves, or plastic coil sleeves, steel coil etc. are also used over the hose. (Similar to protective plastic cover over domestic gas rubber pipe).
- **Colour of hose:** The outer layer colour of hose is either black or red. Generally black colour is preferred.

4.13 ADVANTAGES OF RUBBER HOSE [S-12]

- Generally rubber hoses are preferred all over the world. Rubber hose means layer (A) i.e. inner layer is of rubber. Following are the advantages of rubber hose:
  (i) Rubber hoses can be well equipped with quick connect – disconnect end fittings.
  (ii) Can be manufactured in long lengths.
  (iii) Capable of withstanding to very high pressures.
  (iv) They can absorb very heavy shocks than rigid tubes.

4.13.1 Disadvantages of Rubber Hose [S-12]

(i) Very poor in abrasion resistance.
(ii) Poor in resisting whether condition.
(iii) Initial cost is very high.
(iv) They can damage due to incompatible oil.

4.13.2 End Fittings or Coupling End of Hoses [W-12]

- The end fitting or coupling end is that part attached to the hose and other end is connected to the component of hydraulic circuit. Following are some end fittings:
- The types of end couplings are:
  (a) Male fixed coupling
  (b) Compression fittings
  (c) Quick coupling

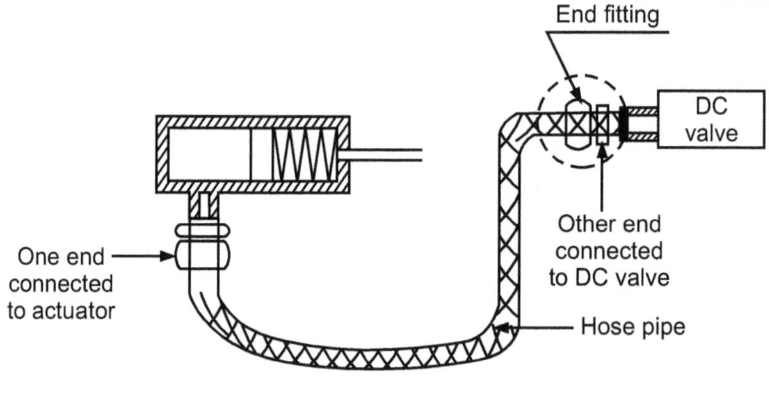

Fig. 4.47

(a) Male Fixed Coupling

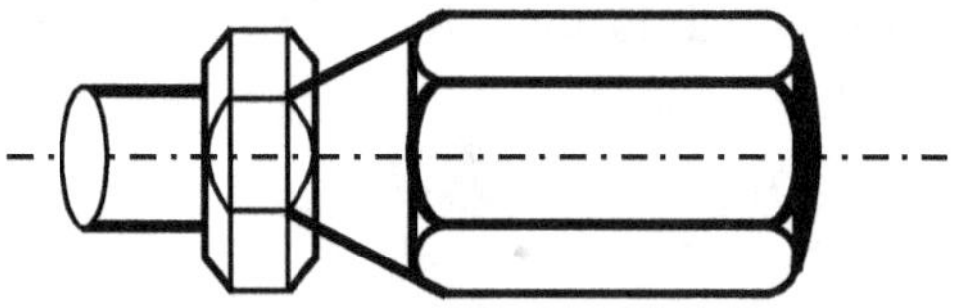

Fig. 4.48 (a)

(b) Compression fitting:

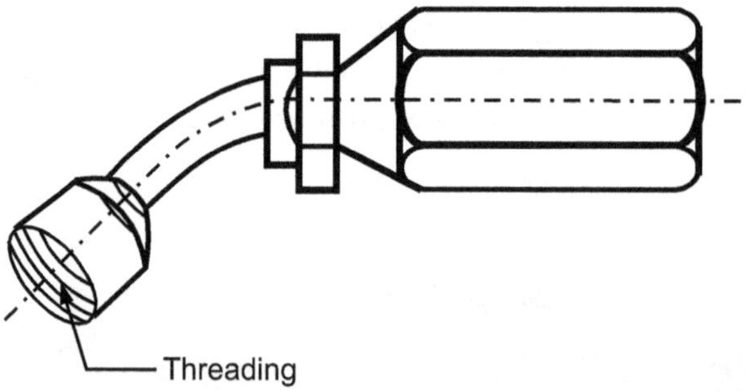

Fig. 4.48 (b)

Connected Disconnected

Fig. 4.48 (c)

- These end fittings are of steel and are crimped on rubber hose pipe by using crimping machine. (Crimping: Fixed with heavy pressure).

4.14 SEALS [W-09, S-10, S-14]

- Oil under pressure is moving in every hydraulic circuit. Since oil is a liquid it has the tendency to 'leak' through every gap/slot it find during movement. This leakage of oil results in (bad effects of leakage).
  - Loss of efficiency of hydraulic circuit.
  - During leakage, oil is coming out without doing any work hence there is loss of power.
  - Sometimes there is temperature rise due to leakage.
  - Hydraulic oil is costly and hence when it is leaked there is loss of money (monitory loss).
  - If leaked oil is dropped on 'hot' surfaces the there is possibility of fire. Hence there are fire hazards due to leakage.
  - When leaked oil falls on ground, it becomes slippery and there is possibility of accident.
- **Important:** Hydraulic circuit designer has to take lot of precautions at design stage itself to select proper sealing arrangement, for minimum leakage.
- **Seals:** The seal is an agent or element which prevents leakage of oil from hydraulic elements and protects the system from dust and dirt.

Functions of Seals are: [S-15]

(a) To stop leakage of oil.
(b) To maintain the pressure.
(c) To keep out contamination in the system.
(d) To enhance the working life of the system.
(e) To enhance the functional reliability of components over a longer period.

Classification of Hydraulic Seals: [S-15]

- These seals are classified on various basis. Let us see these basis:

4.14.1 Classification based on Applications

(a) Static seals: Static seals are those seals which are used to 'seal' two mating parts which are not moving or which are not having any relative motion between them. These seals are stationary. These are generally installed once and forgotten ! Example is sealing end covers of DA cylinder.

(b) Dynamic seals: Dynamic seals are those seals which are used to 'seal' two mating parts which are having relative motion between them. This means, seals are moving every now and then with component. For example, seals used between piston and cylinder. The motion encountered by dynamic seals may be either reciprocating or rotary or combination of both.

4.14.2 Classification based on Degree of Sealing (Extent of Sealing)

(a) Positive seals: Positive seals are those seals which do not allow any leakage (internal/external).

(b) Non-positive seals: Non-positive seals are those seals which permits a small amount of internal leakage.

4.14.3 Classification based on Shape of Seal

(a) 'O' Ring Seal: These are most common and simple seals with circular cross-section like 'O'. Hence called O-ring. The O-ring is used as static as well as dynamic seal.

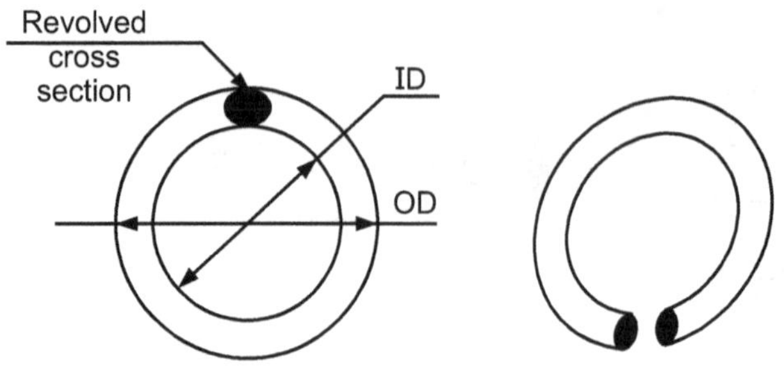

Fig. 4.49

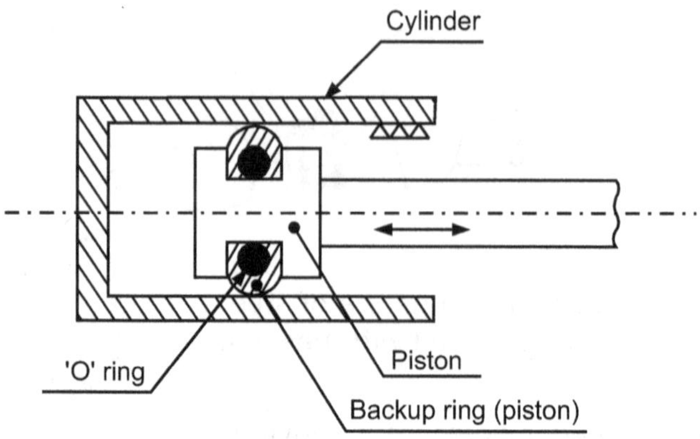

Fig. 4.50

The material used for O-ring is synthetic rubber and is specified by its ID/OD. The round cross-section of O-ring changes to oval shape under low pressure, hence O-rings are non-positive seals. O-rings are fitted with back ring.

The Fig. 4.50 depicts sealing of cylinder and piston by using 'O'-ring with backup ring.

(b) 'V' Ring Seal: 'V' ring seal is also called 'V' packing. It is commonly used in hydraulic systems. These seals are available in pieces to facilitate assembly and specific size. It has cross-section resembling to letter 'V'. There are male and female adapter at the extreme ends as shown. In between V-seal rings can be assembled.

These are made-up of fluorocarbon reinforced with cotton, asbestos or are made-up of neoprene reinforced with asbestos.

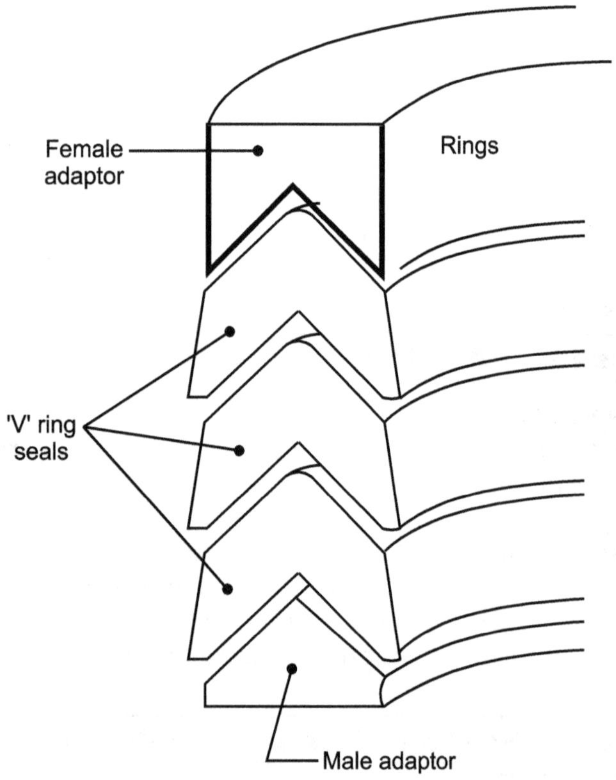

Fig. 4.51

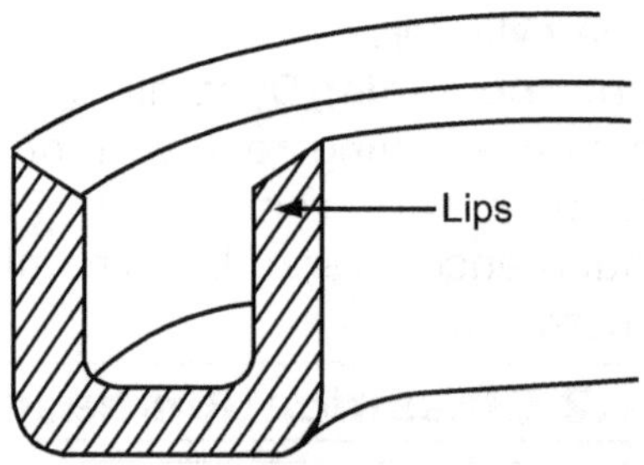

Fig. 4.52: Correct 'U' Shape Packing

(c) U-Packing Seal: U packing seal is most versatile in various applications and used as ID rod seal or OD piston seal. U packings are available in rectangular cross-section shape to suit piston seals.

- U-packing with long lips: Suitable for eccentric operations.
- U-packing with short lips: Suitable for pressure system.

U-packings are usually made-up of leather or fabric reinforced rubber. These seals are balanced seals and no staking like V-ring seal is required.

(d) **T-ring Seals:** These are very modern seals. These seals can be installed as 'O'-ring. But these seals are equally effective as static as well as dynamic seals.

T-ring seals can be used on rod as well as piston.

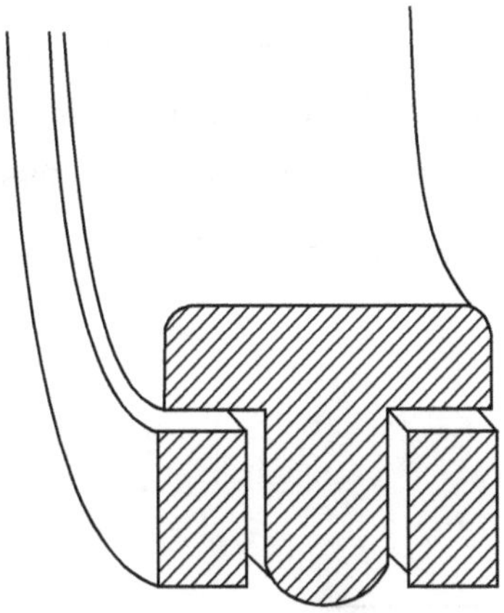

Fig. 4.53

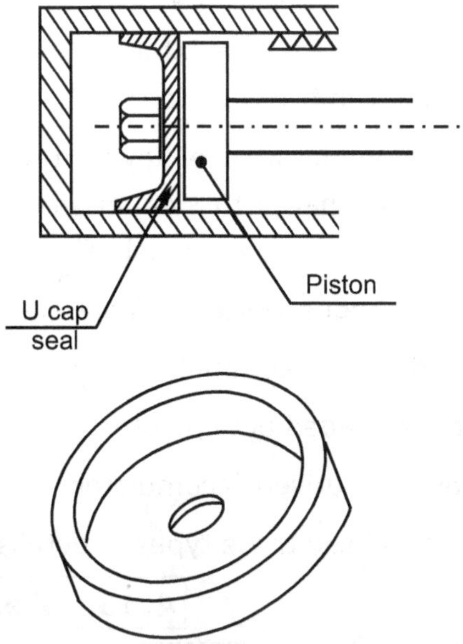

Fig. 4.54

(e) **Cupseals:** These seals are most commonly used with piston as shown in the Fig. 4.54. These are available in variety of geometric shapes to suit the applications. These seals can take higher pressures upto 700 bar. The material used for these seals is polyurethane or leather. For low pressure applications, neoprene is used. (See Fig. 4.54)

4.14.4 Selection Criteria for Seals [W-16]

- Before selecting proper seal the design engineer has to check following points for specific applications. The selection criteria for seals is as under:
  (a) Type of fluid/oil used in system. Whether it is compatible with oil used.
  (b) Maximum temperature of system in working condition.
  (c) Functional reliability expected.
  (d) Expected life of seal.
  (e) Cost of seal.
  (f) Working pressure of system.
  (g) Environmental conditions.

4.14.5 Various Reasons for Seal Failure [S-09, 16, W-11]

- Many a times, seals fail and heavy internal or external leakage starts. Following are the general causes of seal failure.
  (a) Incompatibility of seal material with oil. If we change the type of oil then there is possibility of chemical reactions between new oil and old seal.
  (b) Low speed of actuators is responsible for seal failure.
  (c) If seals are not correctly installed then there is possibility of failure of seal.
  (d) Unbalanced shaft, bend shaft, improper bearings, can cause vibrations on shaft and seal can fail.
  (e) High temperature of oil (medium) can burn the seals.
  (f) Old seals get hardened and fail.
  (g) If seal is excessively squeezed then it can fail.

4.15 ACCUMULATORS [W-10, 11, S-11, 14, 15, 16]

- Function of hydraulic accumulator: The main function of hydraulic accumulator is to store excess hydraulic energy and on demand make the stored energy available to the system.
- The function of accumulator is similar to the function of flywheel in the IC engine/steam engine or capacitor in the electric circuit.
- Since accumulators are having the ability to store excess energy and also having ability to release the energy to system when system is in bad need of energy, the hydraulic systems using accumulators are most efficient systems because there is very little energy loss.
- **Types of Accumulators:** There are three basic types of hydraulic accumulators:

  (a) Dead weight accumulator: Very old type.

  (b) Spring loaded accumulator.

  (c) Gas pressurised accumulator.
- Now we will see these types in details.

4.15.1 Dead Weight Accumulator [S-16]

- This accumulator consists of a sliding piston in a cylinder. The piston rod diameter is much bigger. The oil under pressure usually from pump enters into the cylinder through port P. The pressurised oil forces the piston upwards, until it reaches position B. Dead weight W kept on top of piston rod is in the form of concrete/steel/other heavy material.
- This weight is chosen to exert a predetermined pressure on the fluid entering the cylinder.
- When piston reaches at point B, the accumulator is said to be 'fully charged'. Now the oil in the cylinder is having 'pressure energy' due to dead weight W acting on it. The energy is stored by oil in cylinder.
- Now when the system in which this accumulator is connected, it demands hydraulic oil under pressure, then pressurised oil starts flowing out of port P. When oil starts going out the piston will move down. Due to dead weight pressure of oil will be maintained.

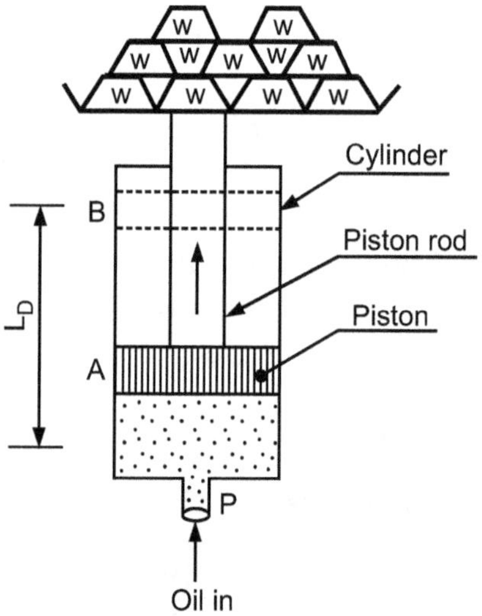

Fig. 4.55

Advantages of Dead Weight Accumulator:

- Pressure remains constant for full stroke due to dead weight.
- This accumulator can supply large amount of oil under pressure.

Disadvantages:

- These accumulators are bigger in size and occupy more space and are bulky.

4.15.2 Spring Loaded Accumulator

[S-16]

- This is a modified version of dead weight accumulator. In this accumulator, there is a spring loaded piston which moves up and down in cylinder. The oil under pressure usually from pump enters into the cylinder through port P. This oil forces the piston upwards causing the spring to compress. The top dead position of the piston shown by point 'B' in the Fig. 4.56 will be decided by **full compression length** of spring.
- When piston reaches point B by fully compressing the spring, the accumulator is said to be 'fully charged'. Now the oil in the cylinder will be having 'pressure energy' due to resilience of the spring.
- Now when system in which this accumulator is connected, if demands hydraulic oil under pressure, then spring starts expanding thereby pushing the piston downwards and pressurised oil will come out of the port 'P'. When spring will be fully expanded (reaches to its free length), the pressure of oil coming out of the accumulator will be almost minimum.

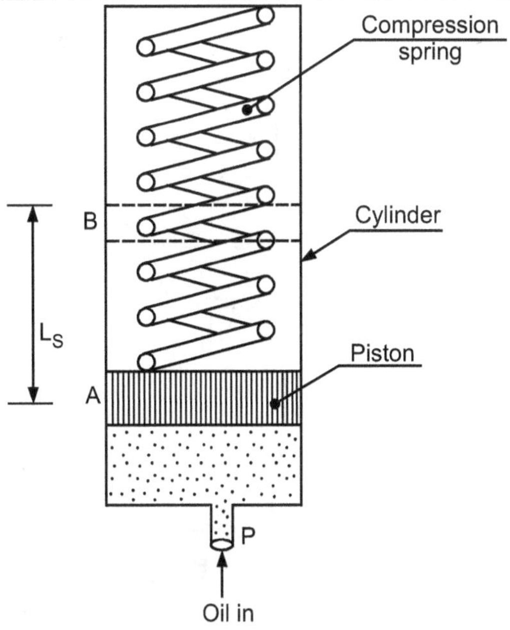

Fig. 4.56

Advantages of Spring Loaded Accumulator:

1. All moving parts are enclosed in the cylinder. No parts outside.
2. Compact design: If we use compound spring, the design will be more compact and handy.

Disadvantages:

1. When spring starts expanding it gives off the stored energy very quickly. And hence as spring starts expanding the pressure of oil below the piston goes on decreasing. So we cannot get uniform pressure in a stroke.
2. When spring is fully compressed, its length is called solid length. Due to this solid length, the stroke of piston becomes limited, as compared to dead weight accumulator.

4.15.3 Gas Pressurised Accumulator : (Bladder Type Accumulator/Gas Filled Bladder Accumulator)

[S-09, 16]

- This is further modified version of the accumulator what we have seen.
- **Principle of working:** In this accumulator a gas charged bag/bladder is fixed in a shell of accumulator. When pressurised oil enters into accumulator, the gas bag compresses. When system requires oil under pressure, the oil goes out and bladder expands.
- **Actual construction:** Fig. 4.57 (a) shows static position of accumulator. Oil comes in and also goes out through port (A). We can fill the gas through port (B) so that flexible bag will expand. To operate this accumulator first we have to fully precharge the gas into the gas bag. The position of the gas bag after fully precharging the gas (This gas is generally Nitrogen gas or any inert gas) is shown below. When we fill the gas in the bag through port (B), the gas bag fully expands. During this step oil is under normal pressure.

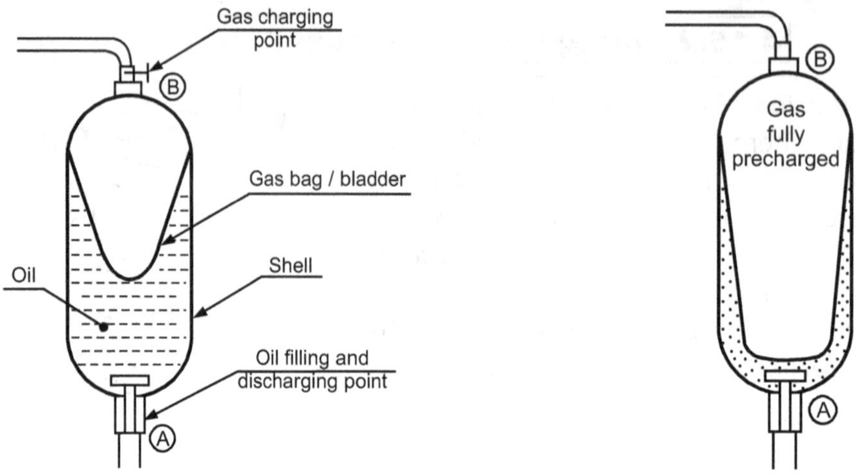

(a) (b)

Fig. 4.57

- Now the oil under pressure will start accumulating into the accumulator through port A. The pressure of oil is higher than the pressure of gas filled in the gas bag. Due to this pressure difference the oil will compress the gas bag. This is shown below. The compressed gas and bag are clearly seen. Now oil under pressure is stored in the shell. This is 'fully charged' position of this accumulator.

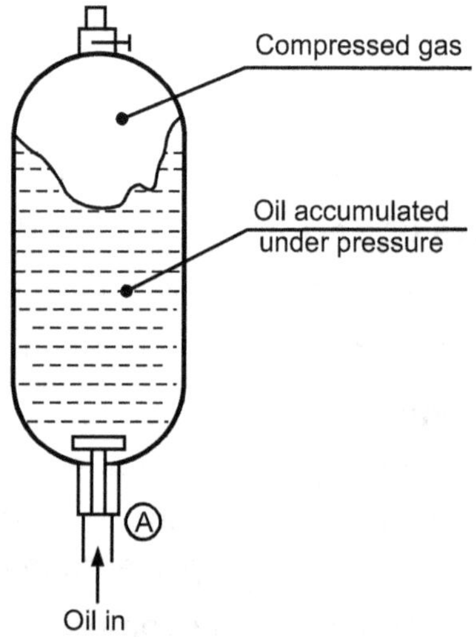

Fig. 4.58

- Now when system in which this accumulator is connected, if demands hydraulic oil under pressure, then oil will start flowing out through port A. As oil goes out, the pressure on gas bag will reduce slowly and the bag will expand. As bladder goes on expanding, the pressure of outgoing oil will go on reducing.

Advantages of Gas Pressurised Accumulator:

1. The accumulator is compact and light in weight.
2. Gas bag is of flexible material like rubber. Hence, bag gives quick response to minute changes in expansion and compression.
3. It is having very few functional parts and hence cheap.

Disadvantages:

1. Pressure of outgoing oil will not be constant. As gas bag (bladder) goes on expanding, the pressure of oil reduces.
2. Volume of oil stored in the accumulator is small.
3. We have to change the gas bag after specific period of service.
4. We cannot handle high temperature fluids in this accumulator.

Symbol of Accumulator:

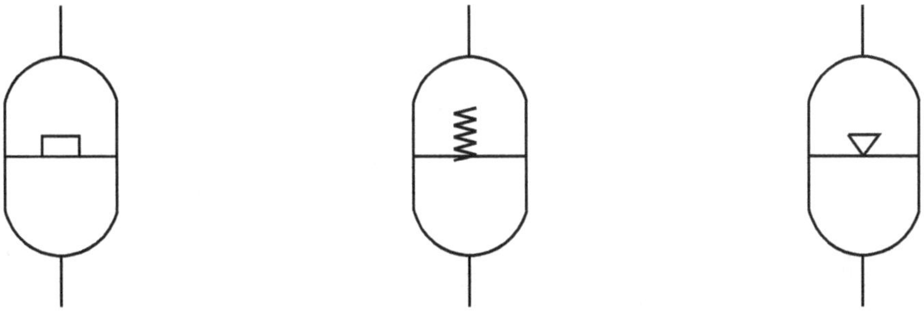

(a) Dead Weight Accumulator (b) Spring Loaded Accumulator (c) Gas Pressurized Accumulator

Fig. 4.59

4.16 ACCESSORIES OF HYDRAULIC SYSTEM

- We have already seen general layout of hydraulic system in Chapter 1. We will have a quick look at all components; of hydraulic system: [S-12]
  (a) Oil storage tank or Oil reservoir.
  (b) Oil conditioning elements such as oil filters, heat exchangers, heaters, strainer.
  (c) Pump to pressurise oil.
  (d) Fluid controlling elements such as direction control valves, flow control valves, pressure control valves, pressure relief valves, check valves etc.
  (e) Actuators – linear and rotary (cylinders and hydraulic motors).
  (f) Fluid or oil conducting elements such as pipe lines, hoses, fittings, seals, gaskets, accumulators.
  (g) Medium i.e. Hydraulic oil.
- In this chapter, we will study the various components listed above in detail along with their symbols.

4.17 OIL STORAGE TANK/OIL RESERVOIR [S-12]

- In any hydraulic system oil is pressurized by pump and is doing useful work by operating an actuator.
- Used oil is then returning for reuse. Hence any hydraulic system is called 'closed System' (In pneumatic system we are using air as a medium. Used air is vented to atmosphere. We are not reusing it. Hence it is called a 'open system'. Closed system creates the requirement of oil reservoir. This is closed box in which hydraulic oil is stored. Oil is going out through reservoir as well as returning to reservoir.

Construction:

- The above three orthographic projections of oil tank easily clarify the construction of same. Oil reservoirs are generally constructed from MILD STEEL plates with the side walls to encourage the heat transfer. The inside surface of tank is made very smooth and is treated with high quality rust preventive coating or paint.
  - **Oil level indicator** is a glass tube through which we can see the level of oil in oil tank.
  - **Inspection cover** is a removable plate for inspection of tank.
  - The return and suction areas are separated by baffle plate. This plate prevents dirt, sludge from return line to enter into suction line.
  - **Oil filling point** is a hole on cover plate. On this hole a mesh type filter is fitted. New oil is poured through this point.

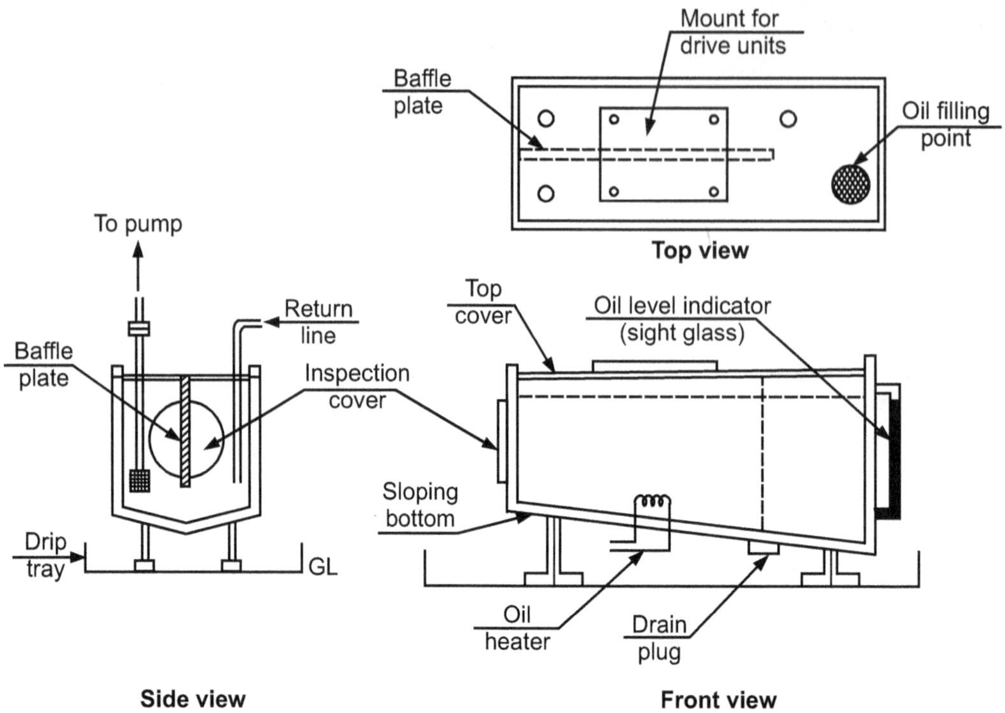

Fig. 4.60: Construction of Oil Reservoir

- **Electric oil heater** is at the bottom of tank. In cold weather when oil solidifies oil heater liquifies it.
- Drain plug is used for removing the old oil through reservoir during maintenance.
- Eventually there is oil spoilage around the tank (either from leakage or during oil filing). Hence to collect such undue spoilage MS/tin sheet drip trays are kept below the tank.
- Sloping bottom helps in collecting foreign particles/contaminants at the side wall of tank, so that its removal becomes easy.

Some important aspects to be remembered in respect of oil tank

1. We have to maintain oil level in oil tank. If oil level is too low, then pump may suck air with oil which may damage the system.
2. Normally an oil tank is designed to hold about 3 to 4 times the volume of fluid or oil taken by pump each minute.
3. For example, If pump takes 4 lits. of oil/min. then reservoir for this system must be large enough to hold 4 lit. × 4 times = 16 lit. of oil.
4. Oil tank should be drained periodically for cleaning.
5. Oil tank is also serving as a heat exchanger. The returning oil is 'hot'. To cool this oil, heat exchangers are used. The wall of tank plays major role in cooling the oil.

Symbol of Oil Reservoir/Oil Tank:

- The symbol of oil tank is similar to its construction (Refer Fig. 4.61 (a)).

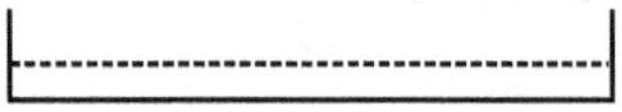

Fig. 4.61 (a): Symbol [S-12]

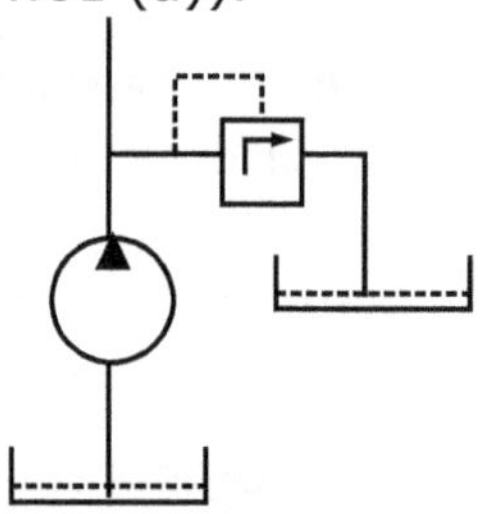

Fig. 4.61 (b)

- In many hydraulic circuit, the symbol of tank is shown separately. But it does not mean that there are many oil tanks. It is a suitable method of drawing a circuit. In the figure shown, there is only one oil tank (Refer Fig. 4.61 (b)).

4.18 OIL FILTERS [W-10, 16, S-11, 12]

- An absolutely clean fluid is essential for reliable operation of hydraulic system.
- Now sophisticated machines like CNC/VMC machines use high pressure oil which demands very very clean oil.
- To clean the oil is a main function of oil filters. Filters are cleaning stations used to arrest the unwanted materials/particles.
- **The hydraulic oil is having following general contaminants:**
  (a) Particles (micron size – metallic/fibrous/dirt)
  (b) Water
  (c) Chemical reaction products such as sludge
  (d) Rubber chips or particles of seals/hoses carried by flowing oil.
  (e) Other fluids.
- **Effect of contaminants:** Hydraulic systems are adversely affected by contamination. If we do not use filters, then all these contaminants will flow everywhere with oil, and will block or jam the moving parts. This jamming may result in damaging the entire system. Even dirt particles of 20 μ (1 μ = 1 micron = one millionth of meter) size can arrest or stop the entire hydraulic system.
- Filters can be classified as:
  (a) Full flow filters
  (b) Proportional flow filters

(a) Full flow filters: As shown in Fig. 4.62 oil comes in through port (A). Passes through filter element and goes out through port (B). In this filter all flow passes through filter hence is called full flow filter.

This is very efficient filter but the only drawback of this filter is there is a large pressure drop. When filter element of the filter is packed with the arrested particle then the pressure drop still increases.

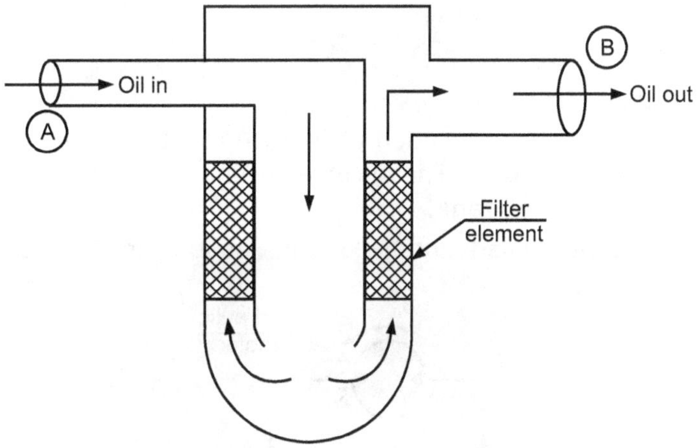

Fig. 4.62: Full Flow Filter

(b) Proportional flow filters: In this type main oil flow passes through venturi, which creates localized low pressure area inside the filter element. We can see that on outside of filter element there is high pressure oil.

Due to pressure difference created across filter element proportionate quantity of oil passes through filter element. In this filter the pressure drop across the filter is very low. Hence, is having wide applications.

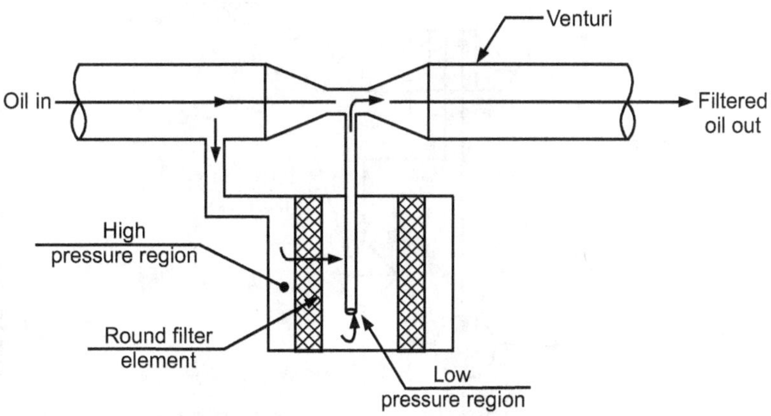

Fig. 4.63: Proportional Flow Filter

- **Filtration materials:** Filter material is wire mesh/screen or paper, cotton, cellulose. Following is the list of filtration materials or filter media used in hydraulic oil filters.
  (a) Felt
  (b) Paper
  (c) Fabric
  (d) Wire mesh
  (e) Wire cloth
  (f) Glass fibre
  (g) Paper disc
  (h) Phenolic resin
  (i) Sintered woven wire
  (j) Sintered PTFE
  (k) Membrane filter
  (l) Magnetic filter cloth

[S-12]

- **Symbol of filter:** The symbol of filter is as shown below: (Refer Fig. 4.64).

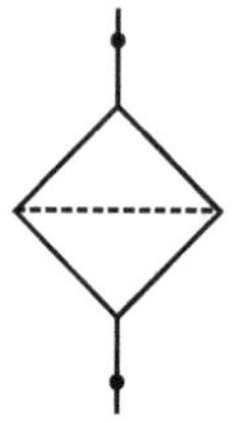

Fig. 4.64

- **Location of filter:** Filter can be located at three suitable locations in Hydraulic circuit.
  (a) Return line filter
  (b) Suction line filter
  (c) Pressure line filter

(a) Return line filter:

- This is common location of filter. As you can see that oil is supplying to motor through port (A). Motor is rotating. The oil is returning through port (B) via DC valve through filter F_1. The oil is collected in oil tank.
- The major disadvantage of this location is that, contaminants are removed only after leaving the system.

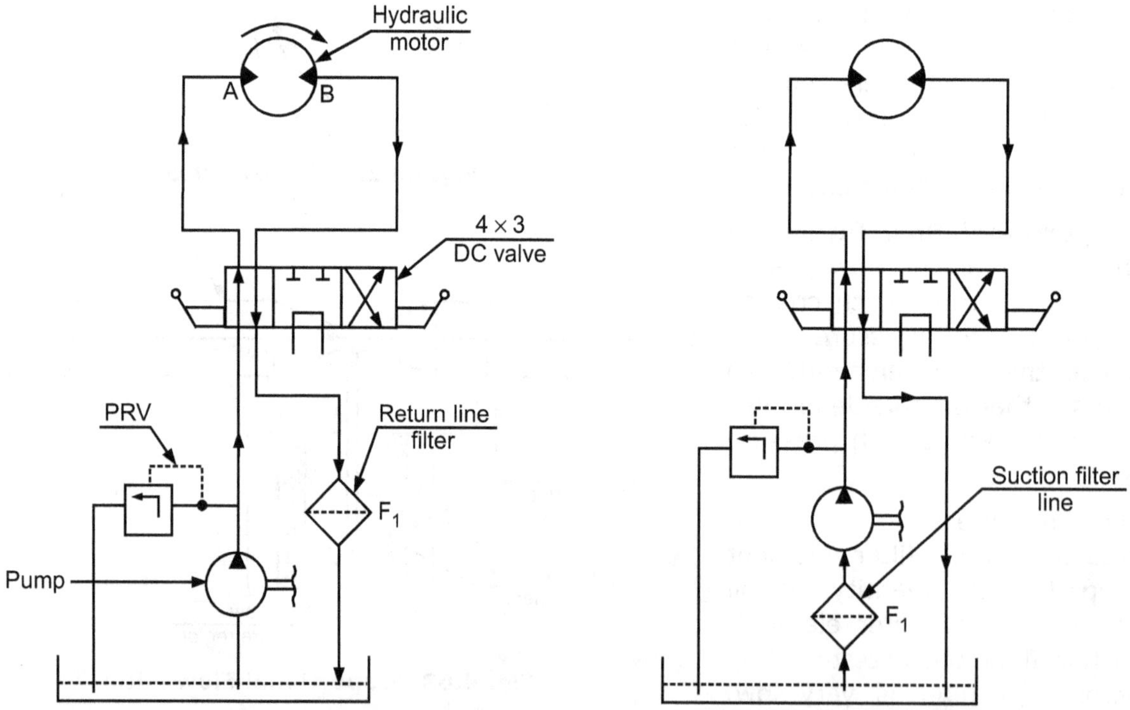

(a) Return Line Filter **(b) Pressure Line Filter**

Fig. 4.65

(b) Suction line filter:

- Now, in this location we have fixed the filter on suction line of pump. Hence oil will be filtered before entering into the pump. This prevents random entry of large and other contaminants (Refer Fig. 4.65 (c)).

(c) Pressure line filter:

- In this arrangement the filter is located after pump or on delivery side of pump. Full pressure is created by pump and this oil is entering into the filter. This arrangement prevents entry of contaminants generated in the pump into components like valves, actuators etc. Since filter is subjected to high pressure oil, its construction must be robust.

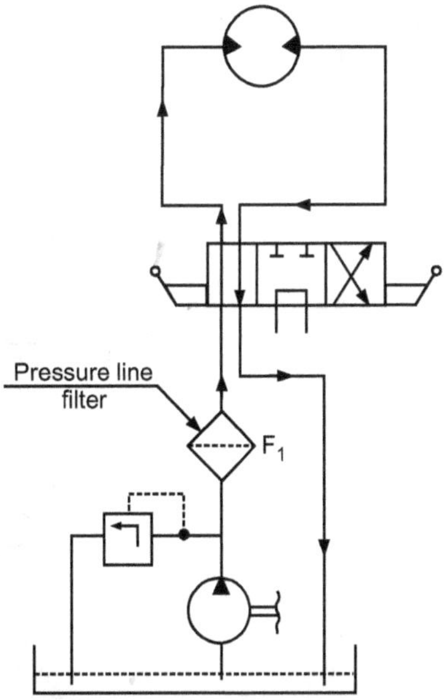

Fig. 4.65 (c)

4.19 HEAT EXCHANGERS

- Heat is generated in hydraulic system whenever oil flows from high pressure to low pressure.
- Like any other electromechanical system (electric motor), hydraulic system too gets affected by generation of heat which may sometime leads to overheating of the hydraulic oil.
- To reduce the temperature of oil to room temperature (or below 50°) proper heat exchangers are needed to fit in the system.
- **Reasons or Sources of HEAT in Hydraulic System:**
  - **(a) Pump:** The unutilized power of motor which drives the pump results in generation of heat.
  - **(b) Friction:** Loss due to friction in pipes, fittings, joints results in heat generation.
  - **(c) Blockages in filters, valves etc.**: Blockages create heat in the back portions.
  - **(d) Oil viscosity:** If viscosity of selected oil is not compatible to system then heat is generated.
  - **(e) Pressure relief valve:** If this safety valve is operated frequently when oil gets heated.
- **Ways of reducing the temperature of 'HOT' oil**
  - (i) In small hydraulic systems heat loss through walls of oil tank is sufficient to cool the oil.
  - (ii) In large systems the volume of oil in oil tank is large. If this oil is heated then we have to use additional cooling systems i.e. heat exchangers. 'Shell and Tube Heat Exchanger' is most commonly used heat exchanger in large hydraulic system.
    - o In this heat exchanger oil flows over the tubes, whereas water flows through the tubes. Heat is 'exchanged' and oil temperature goes down.
    - o **Important:** Oil flows in the opposite direction as that of water.

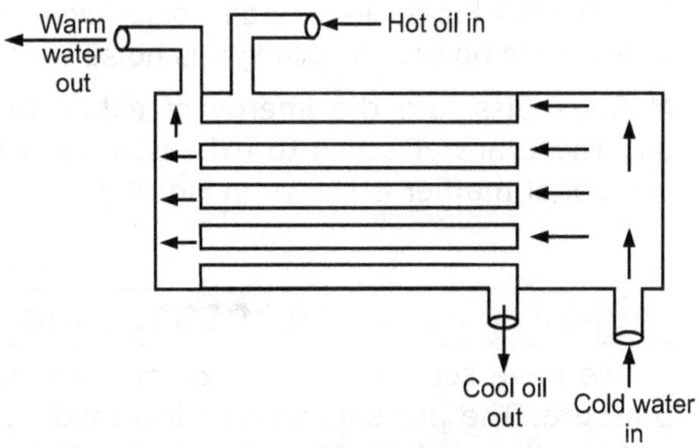

Fig. 4.66: Shell and Tube Heat Exchanger

(iii) Instead of using water as a cooling medium, air can also be used to cool the oil. This is called air cooler.

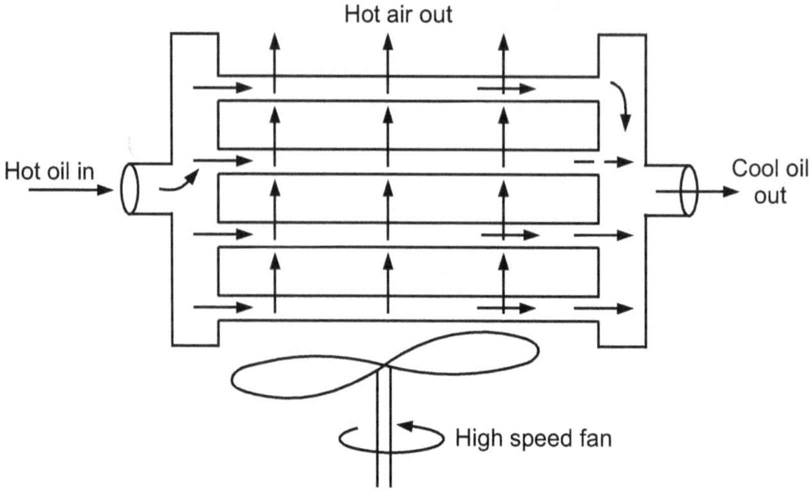

Fig. 4.67: Air Cooler

This heat exchanger uses high speed fan which moves the air on the tubes through which hot oil flows. It is a common type of heat exchanger.

Air cooler is noisy and occupies more space than water cooler.

- **Symbol of Heat Exchanger**

Fig. 4.68

4.20 HEATERS [W-16]

- When we are designing a hydraulic system which will work in cold whether, then oil solidifies due to temperatures below 5°C. In this case we have to use 'Thermostatically controlled oil heaters'. It heats the oil so that its viscosity increases and it can flow in the system smoothly.

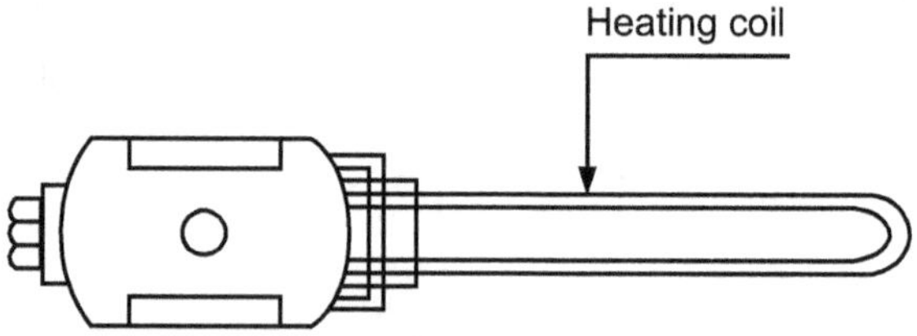

Fig. 4.69: Thermostatically Controlled Oil Heater

4.21 MUFFLERS OR SILENCERS

- Compressed air when moves or acts on component it makes high noise which disturbs human ear. It also increases overall noise level of industrial unit. The automobile engine also makes unbearable noise. To reduce this noise we have to use Silencer or Muffler.
- Mufflers dissipate the energy of exhausting air. These are attached to exhaust port. The symbol of muffler is shown in Fig. 4.70. [S-10]

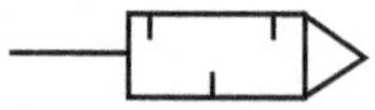

Fig. 4.70

4.22 ACCESSORIES OF PNEUMATIC SYSTEM

- As we have seen, the air is compressed by air compressor and is stored in air receiver under pressure. The pressurized air (flow medium) is transmitted to pneumatic circuits through well-laid pipeline system. If there is leakages in pipe line/joints/ fittings, then compressed air will be wasted and the efficiency of entire pneumatic system will drop to low level. So care should be taken while selecting and laying pipeline to achieve minimum leakage.

- The general pneumatic pipe line layout can be drawn in following manner.

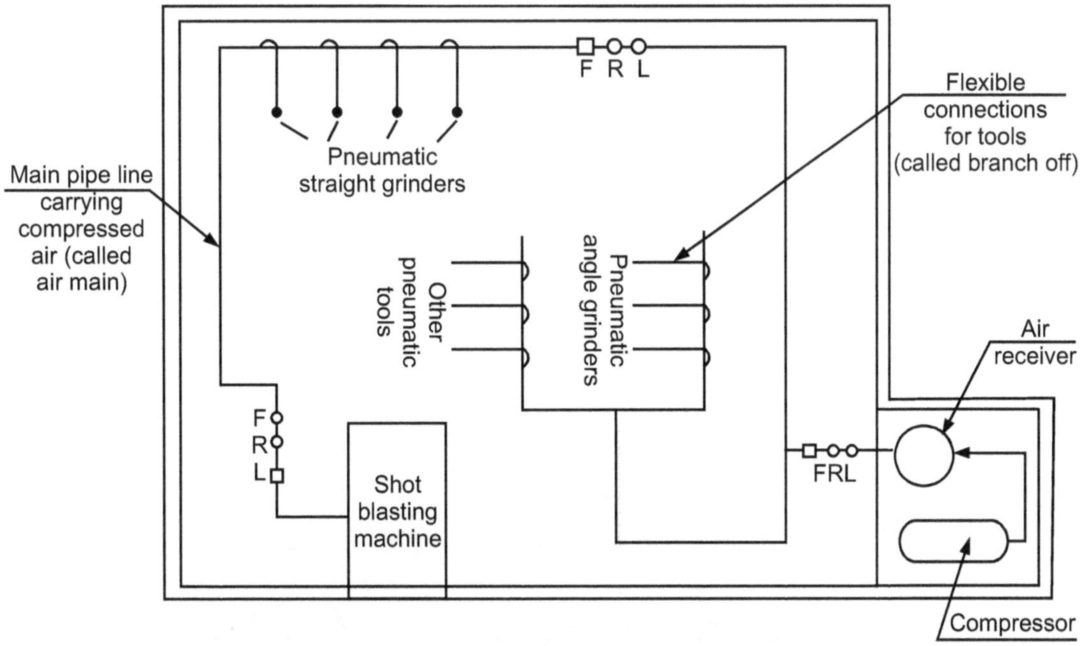

Fig. 4.71: The View of Fettling shop in Foundry

- Fig. 4.71 shows a typical layout of fettling shop in foundry using pneumatic tools and Shot Blasting Machine. Fettling ship is a department which cleans the castings in foundry. Shot Blasting machine is a machine which removes, the sand particles adhered to the castings during melt solidification process. In this machine metal shots (grit of metal) are bombarded on casting (by using compressed air) in a closed chamber of machine. Due to high force of shots, the adhered sand falls down and casting becomes clean.
- Angle Grinders and straight grinders are pneumatic tools which removes extra metal flow on parting line of casting.
- As you can see from the figure, compressor room is a separate room. This arrangement is done to avoid hearing of noise of compressor (especially if compressor is reciprocating type).

4.22.1 Pipeline of Pneumatic System [W-10]

- While laying compressed air pipeline, the pressure drop from Air receiver to consumption point should be kept as low as possible. It should be limited to 0.1 bar under standard conditions.

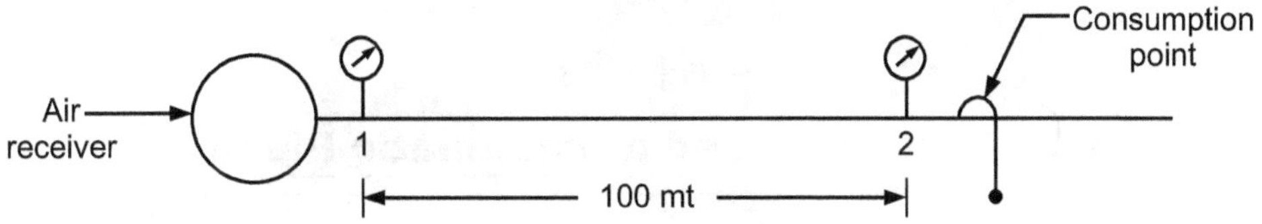

Fig. 4.72

- If at point (1) pressure gauge shows pressure of 7 bar then at point (2) pressure gauges should show pressure of 7 – 0.1 = 6.9 bar. This means that, there is pressure drop of 0.1 bar for distance of 100 meter pipe line.
- The pressure drop can be found out by using 'Nomogram'. Nomogram is a graphical method of finding pressure drop. It is a simple method and can be understandable to a layman.
- **Pipe materials:** The main line or 'Air Main' is the pipe line coming form air receiver. The branch pipe is a branch taken out from air main which is also known as 'Branch-off'. The selection of pipeline material is basically dependant on the pressure of air it will carry.

- Following pipe materials are used for various air pressures:

			Pressure
(a)	Stainless steel	....	2500 – 4500 Bar
(b)	Galvanized Iron (GI) pipes (used in water plumbing)	....	1000 Bar
(c)	Copper pipes	....	200 Bar
(d)	Aluminum pipes	....	125 Bar
(e)	Polyethelene	....	12 to 15 Bar
(f)	Nylon	....	7 to 10 Bar
(g)	Vinyl	....	8 to 10 Bar
(h)	Rubber	....	3 to 7 Bar

- In general 'Air main' pipelines are GI pipes and 'Branch-off' pipes used are Nylon, Polyethelene or copper pipes.
- **Tips of Good Piping System:** Following are the tips of standard and good piping system to reduce pressure drops and other maintenance problems.

  (a) If possible do not embed the main pipe in bricks. Use narrow open ducts.

  (b) The 'air mains' must be given slight slope in the direction of flow of air.

  (c) Do not turminate the air main near branching.

  (d) Use water traps and air dryers at all possible points.

  (e) Branching should be form top of main line shown as under:

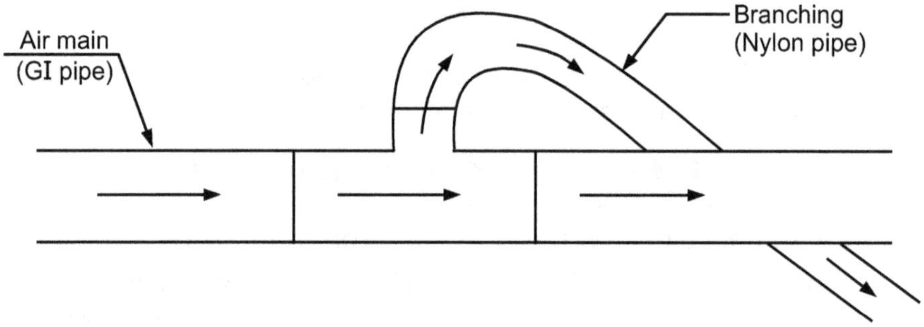

Fig. 4.73

4.22.2 Fittings used in Pneumatic Pipeline [S-10]

- When two pipes need to be joined we have to use pipe fittings. Welding or soldering is another alternative method of joining two pipes. This method makes the permanent joint. So also the air leakage is also almost zero. But permanent joints like welding/ soldering creates the effect of scaling due to heat. Hence in pneumatic systems permanent joints are not popular.
- Pipe fittings in pneumatic system can be classified in four types:

  (a) Bell mouth tube fitting (Also called flared fitting).

  (b) Compression type fitting (Also called ferrule fitting).

  (c) Screwed connections.

  (d) Quick connectors.

(a) Bell Mouth Tube Fitting (Also called Flared Fitting)

- In below figure pipe (A) is jointed to pipe (B) by using flared fitting. The end of pipe (B) is flared (Made in conical shape) by using flaring tool. Flaring Angle is 37° or 45°.
- To pipe (A) Male Connector end is fitted. And then nut is tightened on it by using packing sleeve.
- By loosening nut, we can disconnect the pipes.

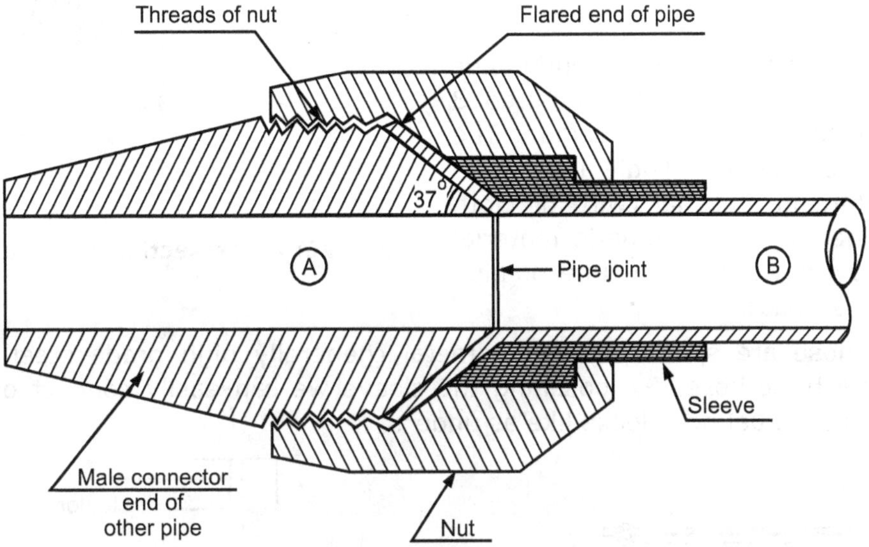

Fig. 4.74: Flared Fitting (Bell mouth tube fitting)

(b) Compression Type Fitting (Ferrule fitting)

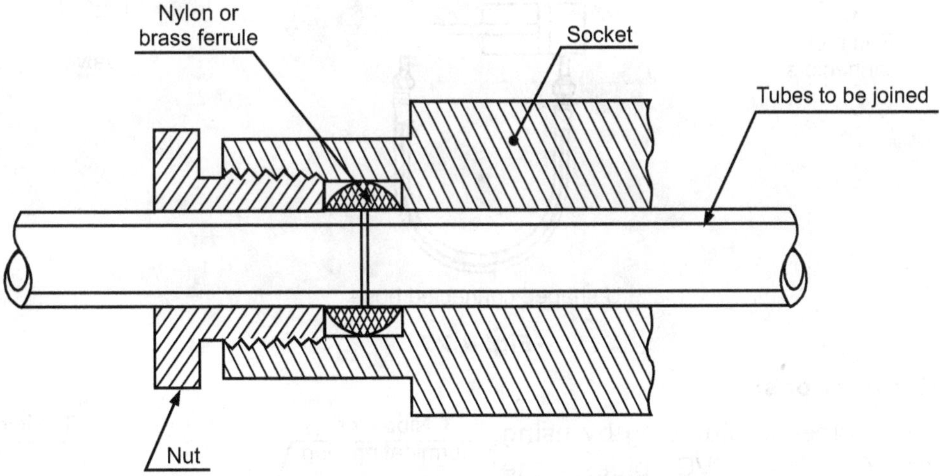

Fig. 4.75: Ferrule Fitting (Compression type fitting)

- To avoid flaring (expanding pipe into conical shape) flareless joints like ferrule fittings are used in pneumatics. In this joint nut and ferrule (Ferrule is a flexible member like rubber made-up of nylon or brass) are placed on tube and other tube which is to be joined is inserted, along with socket. Nut is tightened in socket threading. The joint is complete. The ferrule element will occupy entire space and will make perfect seal as well as perfect joint.
- By loosening the nut we can disconnect the joint.
- Wall thickness is a limiting factor in this joint. The joint creates compressive force on pipe near ferrule which must be comfortably taken by pipe wall. If wall is too thin then there is possibility of tearing of the pipe at joint.

4.22.3 Hoses used in Pneumatic Systems [W-11, 13, 16; S-15]

- We have seen that for higher pressures we have to use stronger non-flexible pipes like GI pipes or thick PVC pipes. But if our requirement is flexible pipe at high air pressures then 'Pneumatic Hoses' is best choice.
- Fig. 4.76 shows cross-section of 'pneumatic hose' pipe. It is specially manu-factured 3 layer pipe, generally black in colour. The outer layer is of plastic. Middle layer is a layer of Reinforcing material (wire or cloth braid) and inner layer is synthetic material layer. Inner layer is actually a tube or pipe through which air passes.

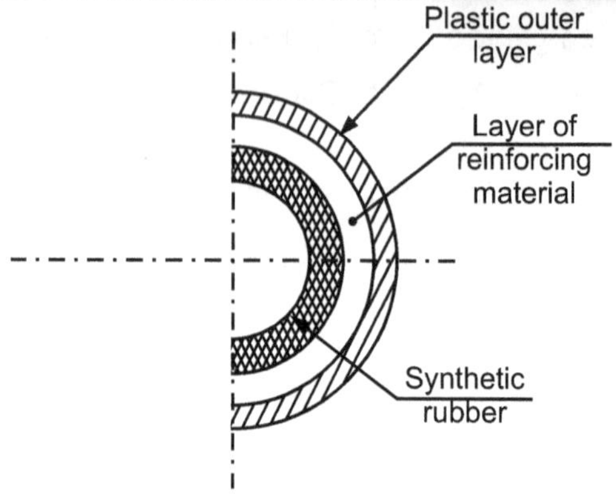

Fig. 4.76: Cross-section of Pneumatic Hose Pipe

- Two ends of hose are special. The end metal (generally high grade steel) connectors are crimped on the basic hose. By loosening the sleeve we can easily connect or disconnect the hose from circuit. Actual hose looks like as under:

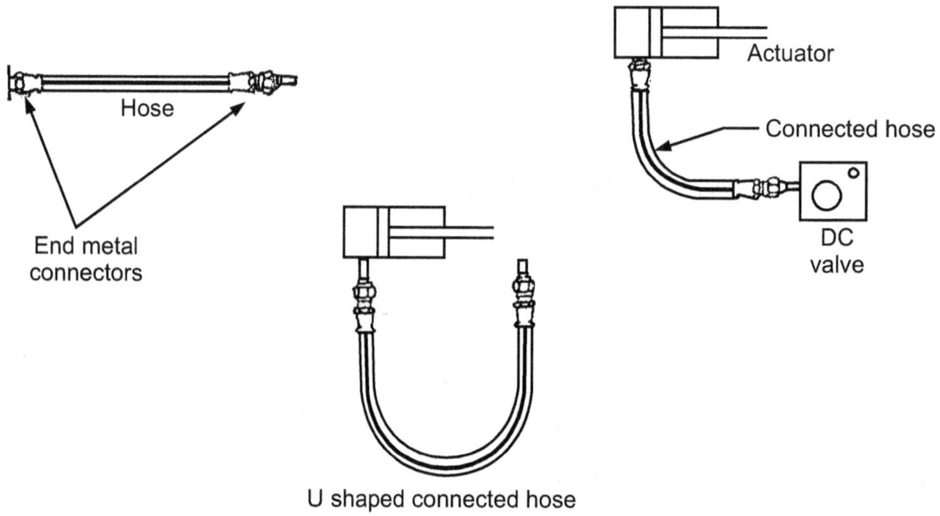

Fig. 4.77

(c) Screwed Connections:

- Similar to water pipeline plumbing by using GI pipe or 'C' class PVC pipes, the pneumatic pipelines can be fitted. We can comfortably use GI pipes/'C' class PVC pipes. Almost all factories using pneumatic power are using GI pipes. The screwed connections can be easily disconnected. We can use 'tee', 'unions', 'reducers', 'elbow', 'nipples' as fitting joints to take the pipe wherever we need. British standard pipe threads (BSP) are most favourable screw threads used in pneumatic piping.

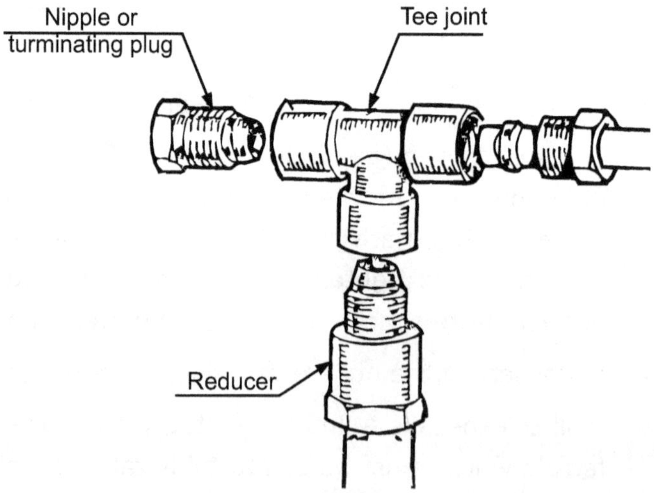

Fig. 4.78

(d) Quick Connectors:

- This type of fitting is used for plastic tubing and flexible hoses. These connectors are used where there is frequent connection and disconnection is required. It required perfect sealing arrangement to avoid leakage. Socket type high grade steel connector is crimped (crimping is an operation in which steel socket is fixed on end of hose pipe or flexible pipe by using very high pressures) at one end of pipe. Similarly pin type (Fernel connector) is crimped on the end of other pipe. By moving sleeve on socket we can disconnect the joint quickly.

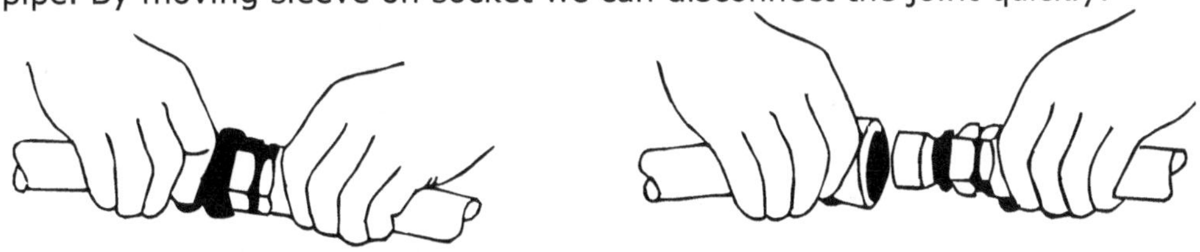

Connected Disconnected

Fig. 4.79

4.22.4 Seals

[S-12]

- In case of Linear Actuators, we are admitting pressurized air in the cylinder. This air finds every gap to leakout. The possible leakage regions are:
  (a) Piston and cylinder bore.
  (b) Between end of cylinder and cylinder end cover
  (c) Through other openings.
- To avoid leakage of air we have to use seals.
- The life of seals depends on friction as well as speed of piston.
- In pneumatic systems following seals are used.
  (a)'O' ring seal: 'O'ring is most commonly used static seal. It is moulded synthetic ring, having round cross-section. 'O' ring gives very effective sealing at very high pressures (Refer Fig. 4.80 (a)).

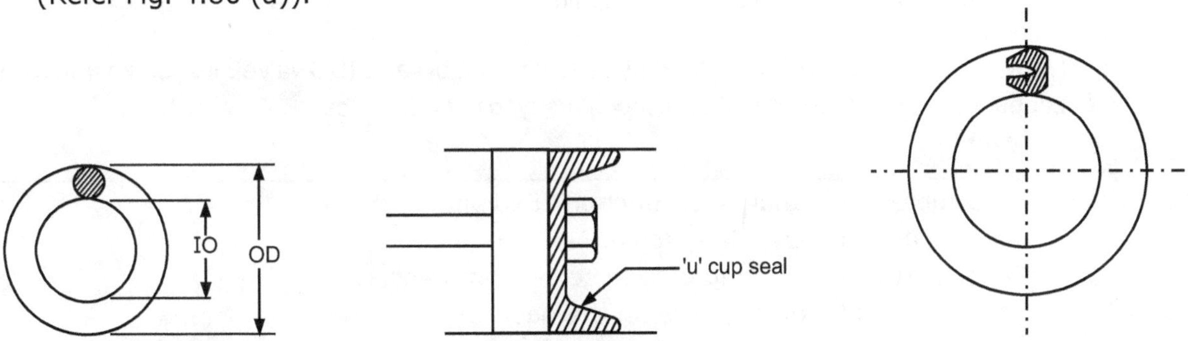

(a) Oring Seal **(b) 'u' Cap Seal** **(c) u/v Ring Seal**

Fig. 4.80

(b) 'U' cup seal: It is used in SA cylinder and is clamped in front of piston with bolt to achieve perfect sealing (Refer Fig. 4.80 (b)).

(c) 'U' or 'V' Ring Seals: These are similar to 'O'rings but the cross-section is 'U' or 'V' shaped. These are effective at high pressures. The air pressure forces the two lips of 'U' apart which creates perfect sealing. Common elastomeric materials such as Teflon, Synthetic nitrile rubber or neoprene are used for seals (Refer Fig. 4.80 (c)).

Important Points

- Compressor is a heart of pneumatic system. It gives energy to air.
- In general compressors are classified as:
  (i) Positive displacement (rotary or reciprocating) compressor
  (ii) Dynamic or turbo compressor.
- Reciprocating air compressor and screw air compressor are widely used in almost all type of industries to obtain compressed air.
- Reciprocating air compressor works similar to IC Engine, the only difference is that, the crank is rotated by electric motor or engine or turbine.
- Single acting and double acting reciprocating air compressor are its types.

- Mono screw (single screw) and twin screw (two screws) are 2 types of screw air compressor.
- Twin screw air compressor (positive displacement rotary compressor) is widely used air compressor.
- In twin screw air compressor two meshing screws are rotating. The air is trapped between meshing screw threads. The trapped air is carried along the discharge port which goes to air receiver.
- In rotary vane or sliding vane compressor can deliver air pressure up to 3 bar.
- Vane compressor is positive displacement compressor.
- Construction of vane pump is similar to hydraulic vane pump.
- Vane freely slides in the in the slot of rotor. When rotor rotates the vanes fly out in slot due to centrifugal force.
- Lobe compressor (roots blower) is also positive displacement compressor.
- Lobe compressor can create pressure up to 2 bar. But the delivery of air is very high.
- Compression ratio = Volume of inlet air / Volume of outlet air.
- Centrifugal type compressor and axial type compressor are dynamic or turbo compressors.
- Centrifugal compressor is known as blower. Its impeller rotates at very high speeds.
- Axial compressors are used for heavy discharge of air.
- Pressure requirement/volume requirement/space availability are 3 basic selection criterias for compressor.
- Air receiver stores compressed air. It is MS fabricated cylinder. It is pressure vessel.
- Air contains water particles called moisture. When air is humid then moisture percentage is higher.
- This moisture from air must be separated otherwise it will carry to the components with air and the components will rust.
- Water trap is used in circuit to separate the moisture. Hygroscopic chemicals absorbs moisture hence used in circuit to remove moisture. Refrigeration dryers are very common in industry.
- FRL unit is must in pneumatic circuit.
- Filter is first element in FRL unit and is having function of arresting the foreign particles.
- Regulator is second element in FRL unit and is having function of regulating air pressure.
- Lubricator is third element in FRL unit. In lubricator fine oil particles are mixed with air which lubricates the components of pneumatic circuits.
- Direction of flow of compressed air is controlled by direction control valve (DC valve). it acts like traffic police.
- Seat type (poppet type) and spool type (sliding spool) are types of DC valves.

Practice Questions

1. What is air compressor? What is its function? Explain
2. How reciprocating compressor is specified?
3. Explain the construction, working of twin screw compressor.
4. Give classification of air compressors on the basis of:
   (i) Stages, (ii) Drive of compressor, (iii) Cooling, (iv) Action
5. Write the names of four types of positive displacement compressor.
6. Write construction and working of Doubled acting reciprocating compressor with neat sketch.
7. What is intercooler? What is its function explain with sketch?
8. Draw sketch of vane compressor and label the parts.
9. What is compression ratio if inlet air pressure in 2 bar and outlet air pressure is 8 bar?
10. When you will use AXIAL TYPE AIR COMPRESSOR? Explain
11. Write any three selection criteria's of Air compressor.
12. What is the function of air receiver? Explain and draw its symbol.
13. What will happen if moisture from air is not removed?
14. What are Hygroscopic Chemical Dryers? How they remove moisture? Explain
15. Write with neat sketch, the working of water trap.
16. What is FRL unit? Explain its function.
17. Draw separate and combined symbol of FRL unit.
18. On what principle Lubricator works? Explain with sketch.
19. Write any four tips for good piping in workshop using pneumatic system.

20. Write any four materials of which pneumatic pipe is made.
21. What Ferrul Fitting? Explain
22. Draw constructional details of pneumatic hose. Why hose is required in pneumatic circuits?
23. What are the various functions of seals? Write types of seals used.
24. Write any two materials of which pneumatic pipe is made.
25. What is FRL? State its importance.
26. What is function of accumulator?
27. Compare forward stroke of DA cylinder and return stroke of DA cylinder with figure.
28. Write short note on dynamic or turbo compressor with figure.
29. What is the function of Air receiver? Explain.
30. Draw neat constructional figure of variable flow control valve with built-in check valve.

■■■

Chapter 5

OIL HYDRAULIC CIRCUITS

Weightage of Marks = 10, Teaching Hours = 06

Syllabus

5.1 Control of Single Acting, Double Acting Hydraulic Cylinders, Motors

5.2 Speed Control Meter-in Meter-out Bleed Off Circuit

5.3 Regenerative, Synchronizing, Counter Balance, Sequencing Circuits, Two Pump Unloading Circuits

5.4 Hydraulic Circuit for Milling Machine, Grinding Machine, Shaping Machine

About This Chapter

At the end of this chapter students will be able to:

- Sketch and label simple oil hydraulic circuits
- Describe working of given oil hydraulic circuits
- Develop oil hydraulic circuit for different applications
- Select appropriate components required for given simple hydraulic circuits

5.1 INTRODUCTION

- We have already seen the basic facts relating to hydraulic circuit elements, their construction and working.
- Now let us see how hydraulic system is constructed by using all these elements to obtain useful mechanical work.
- Hydraulic circuit is a symbolic representation of system which operates on push button/pilot signal/electromechanical circuits and thereby achieving flow of pressurized oil to operate the motion of actuator.
- Hydraulic circuit is a combination of various hydraulic elements arranged in a systematic manner to perform specific task.
- **Elements of Hydraulic Circuit:** **[S-10]**
  (a) Oil tank or oil reservoir
  (b) Oil filter
  (c) Oil pressurizing pump
  (d) Direction control valve/Flow control valve and other valves
  (e) Actuators (DA Cylinder/SA Cylinder/Motor)
  (f) Pipelines
- **We will see following hydraulic circuits:**
  1. Control of Single Acting Cylinder (SA Cylinder)
  2. Control of Double Acting Cylinder (DA Cylinder)
  3. Speed Control of DA Cylinder by using:
     (a) Meter-in-Circuit
     (b) Meter-out-Circuit
     (c) Bleed-off-Circuit

4. Regenerative Circuit
5. Speed control of Bi-directional Hydraulic Motor
6. Synchronizing Circuit
7. Sequencing Circuit
8. Hydraulic Circuit for Shaping Machine
9. Hydraulic Circuit for Milling Machine
10. Hydraulic Circuit for Grinding Machine

5.1.1 Circuit No. 1 : Control of Single Acting [SA] Cylinder

[S-12]

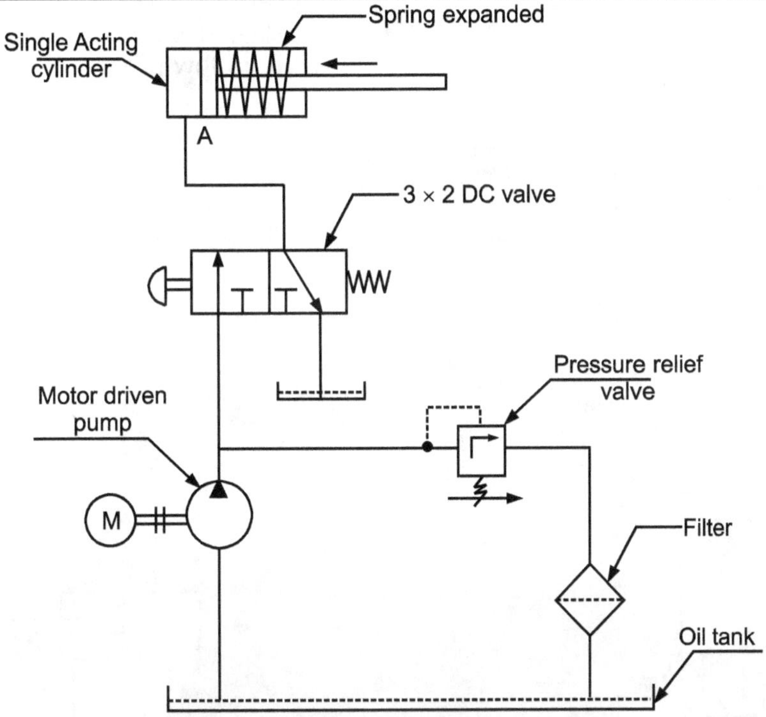

Fig. 5.1 (a): Non-Actuated Position

- The circuit shows all components of hydraulic system to operate single acting cylinder called SA cylinder. It is linear actuator.

(a) Non-actuated Position: Push button of DC valve is not operated. That means pump is not supplying oil to SA cylinder through port A. The spring has expanded and oil present on piston side (came in during earlier actuated stroke) is returning to oil reservoir via DC valve.

(b) Actuated Position: Now push button is operated and flow of pressurised oil will start via DC valve to SA cylinder. Oil will enter through port A and will push the piston of SA cylinder towards right, thereby compressing the spring. When push button is released, flow through port A will stop and spring will expand.

- This expansion will return the oil to oil tank as mentioned in non-actuated position.

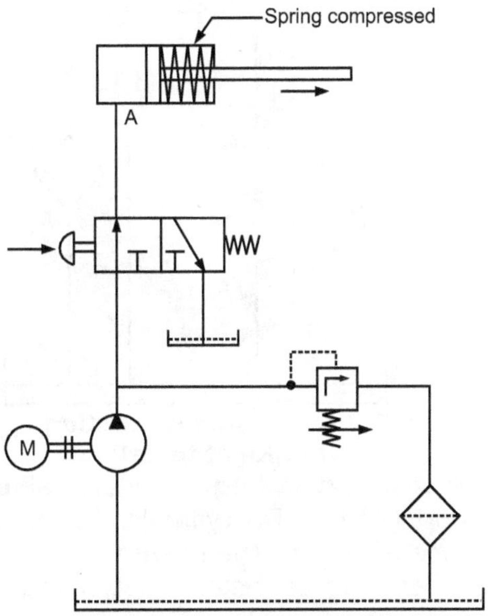

Fig. 5.1 (b): Actuated Position

5.1.2 Circuit No. 2 : Control of Double Acting Cylinder (DA Cylinder)

[S-10, 12]

The circuit in Fig. 5.2 (a) and (b) shows how to control reciprocating movement of DA cylinder (Note that in DA cylinder, there is no spring on piston rod).

- There is 4 × 3 solenoid operated DC valve.
- Pressure relief valve will maintain the system pressure.
- All return lines from system are brought back to reservoir.
- We have used return line filter to clean the returning low pressure oil.

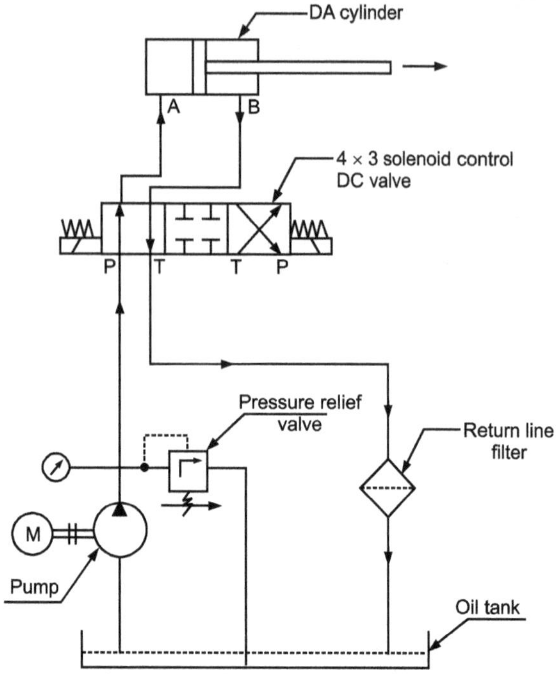

Fig. 5.2 (a): Piston Extending (Moving Left to Right)

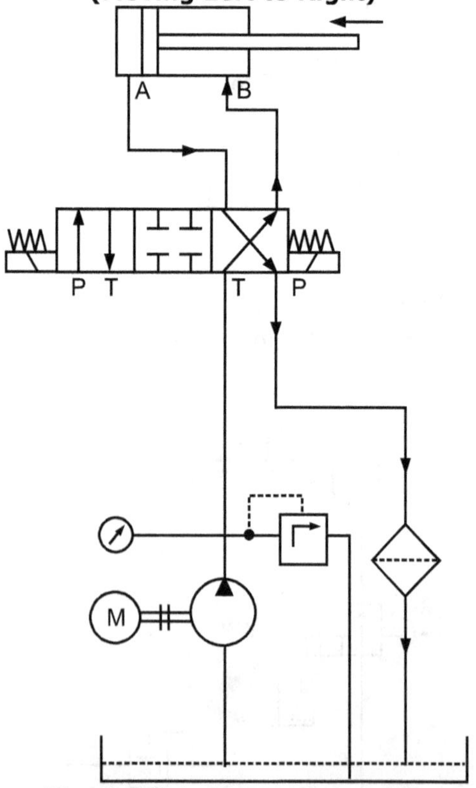

Fig. 5.2 (b): Piston Retracting (Moving Right to Left)

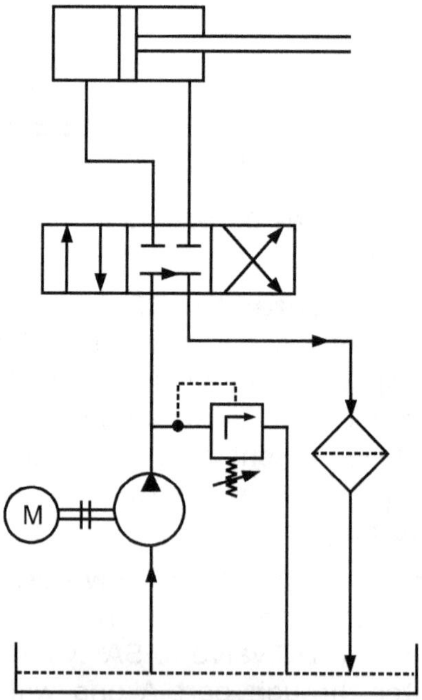

Fig. 5.2 (c): No Movement of Piston (Hold Position)

(a) Piston Extending: Solenoid valve operates at set value and pressurized oil will enter from port (A) into DA cylinder. Piston will move from left to right. Since oil is admitted on front face of piston, the movement is called 'Piston Extending'. During this movement oil on the other side of piston (piston rod side) will automatically return to oil tank via DC valve through port (T).

(b) Piston Retracting: Now again solenoid valve will operate to its second setting. Now pressurized oil will be supplied through port (B) via DC valve. Due to pressure of oil, piston will move from right to left. The oil on the other side of piston will return to oil tank via DC valve through port (P).

(c) Intermediate Position (No movement of piston): In this position, centre port of DC valve operates. It is clear from Fig. 5.2 (c) that, pressurized oil is not going to DA cylinder. It is returning to tank. This is called 'Hold' position. It keeps the actuator on 'Hold'. There is no movement of piston on either side.

5.1.3 Circuit No. 3 : Speed Control of DA Cylinder-Meter-In-Circuit

[S-10, 12, 15, 16; W-10]

- Double acting cylinder (DA cylinder) is cylinder-piston pair. Piston reciprocates in the cylinder. It is having linear motion (meter per second). In many applications, we have to control this speed. Meter-in-Circuit is one of the speed control circuits of DA cylinder.
- **Meter-in-Circuit:** We must know that to control the speed, we have to control the flow of pressurized oil (lit/sec.) entering into the DA cylinder.
- In meter-in-circuit, the flow control valve is placed in primary line (also called pressure line), directly before the load. The arrangement is depicted in circuit diagrams shown in Fig. 5.3 (a) and (b).

[S-12, W-16]

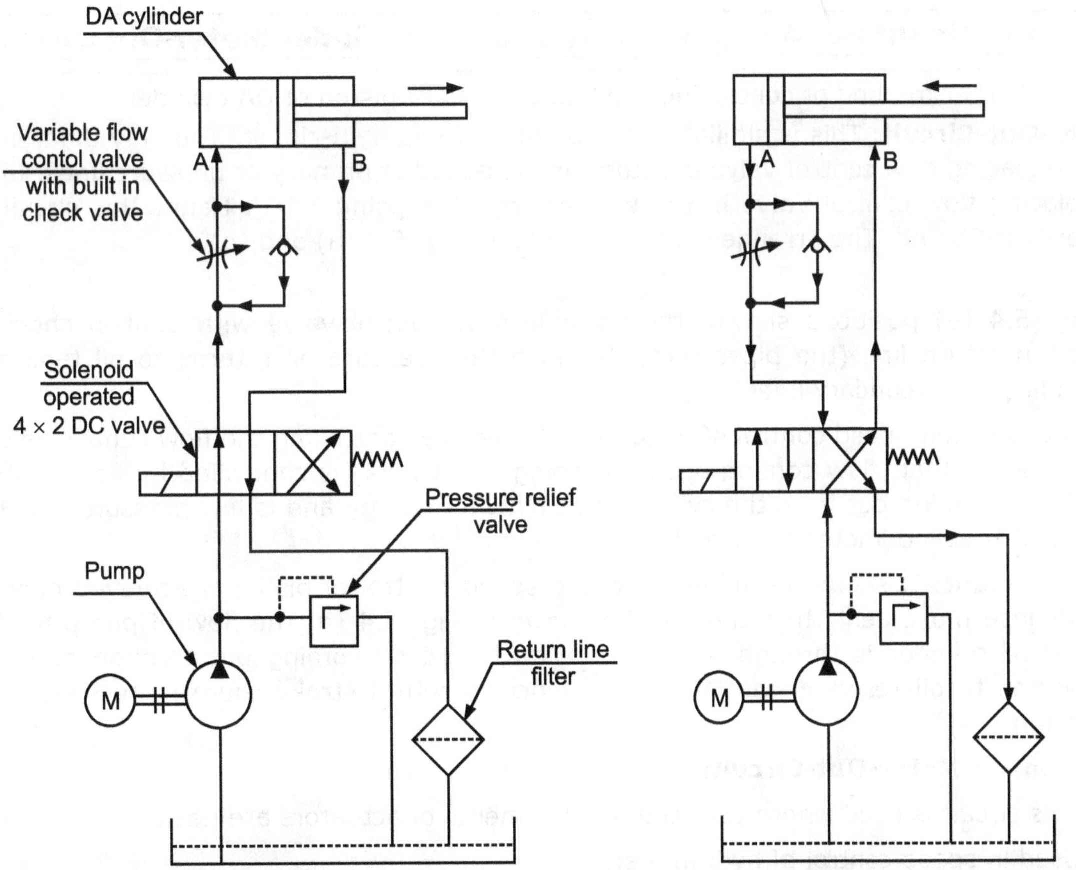

(a) Speed Control When Piston Advanced **(b) No Speed Control When Piston Retracts or Returns**

Fig. 5.3: Meter-In-Circuit

- In Fig. 5.3 (a) position shown, the variable flow control valve with built-in check valve is placed in primary line or pressure line (primary line or pressure line is a pipeline which carries pressurized fluid). Pump controls many devices and its output is higher than needed by any individual actuator. So flow restrictor or flow control valve is placed to set the flow of each actuator. Hence this circuit is known as 'meter-in-circuit'.
- **Important:** It should be noted that, the speed control of the piston (shown in the circuit) is achieved only in piston advance movement. We cannot control the speed of piston in return stroke. In return stroke, the returning oil comes out through port (A) and goes to oil tank via check valve and DC valve. Carefully look at the arrows shown in Fig. 5.3 (b).

Applications of Meter-in-Circuit:

(a) This circuit is used where finer speed control is required.

(b) When load is needed to move then this circuit is used.

(c) This circuit is suitable for resisting load actuated by the cylinder.

Advantages of Meter-in-Circuit:

(a) Gives best results when used in higher pressure systems.

(b) Relatively small friction to face by piston which increases life of piston.

(c) Suitable for very low piston rod speeds.

Disadvantages of Meter-in-Circuit:

(a) Not suitable for low pressure systems.

(b) We are placing flow control valve in pressure line. In this valve throttling of fluid takes place which heats the oil and hot oil is admitted into cylinder.

(c) There is pressure on one side of piston every time hence, load actuated by piston rod is not held firmly in the position.

5.1.4 Circuit No. 4 : Speed Control of DA Cylinder Meter-Out-Circuit [W-10]

- This is another method of controlling the linear speed of piston of DA cylinder.
- **Meter-Out-Circuit:** This is similar circuit as that of meter-in-circuit. The only difference is that we are placing flow control valve in return lines instead of primary or pressure lines. In short we are placing flow control valve in the way where oil is going 'OUT'. Hence this circuit is called 'Meter-Out-Circuit'. The arrangement is depicted in Fig. 5.4 (a) and (b).

[S-12]

- In Fig. 5.4 (a) positions shown, the variable flow control valve with built-in check valve is placed in return line (the pipes through which the pressure oil returns to oil tank are called return lines or secondary lines).
  - In this circuit speed control of piston is achieved by controlling the flow control valve in such a manner that, flow coming out (and going to oil tank) is controlled by flow control valve. The oil coming out from the cylinder has used its energy and is low pressure oil. Hence this circuit is called 'meter-out-circuit'.
  - **Important:** It should be noted that, the speed control of piston is achieved only in piston advance movement. In return stroke shown in Fig. 5.4 (b) the flow of pump to the piston end of cylinder is through flow control valve and oil coming out through port A directly returns to oil reservoir via DC valve. Hence in return stroke there is no speed control of piston.

Applications of Meter-Out-Circuit:

(a) This circuit is used where very stable movements of actuators are needed.

(b) Used in speed control of hydrometer.

(c) This circuit is best suited for drilling/boring or reaming operation. On these machines if clamping of job is loose, the tool drags the workpiece.

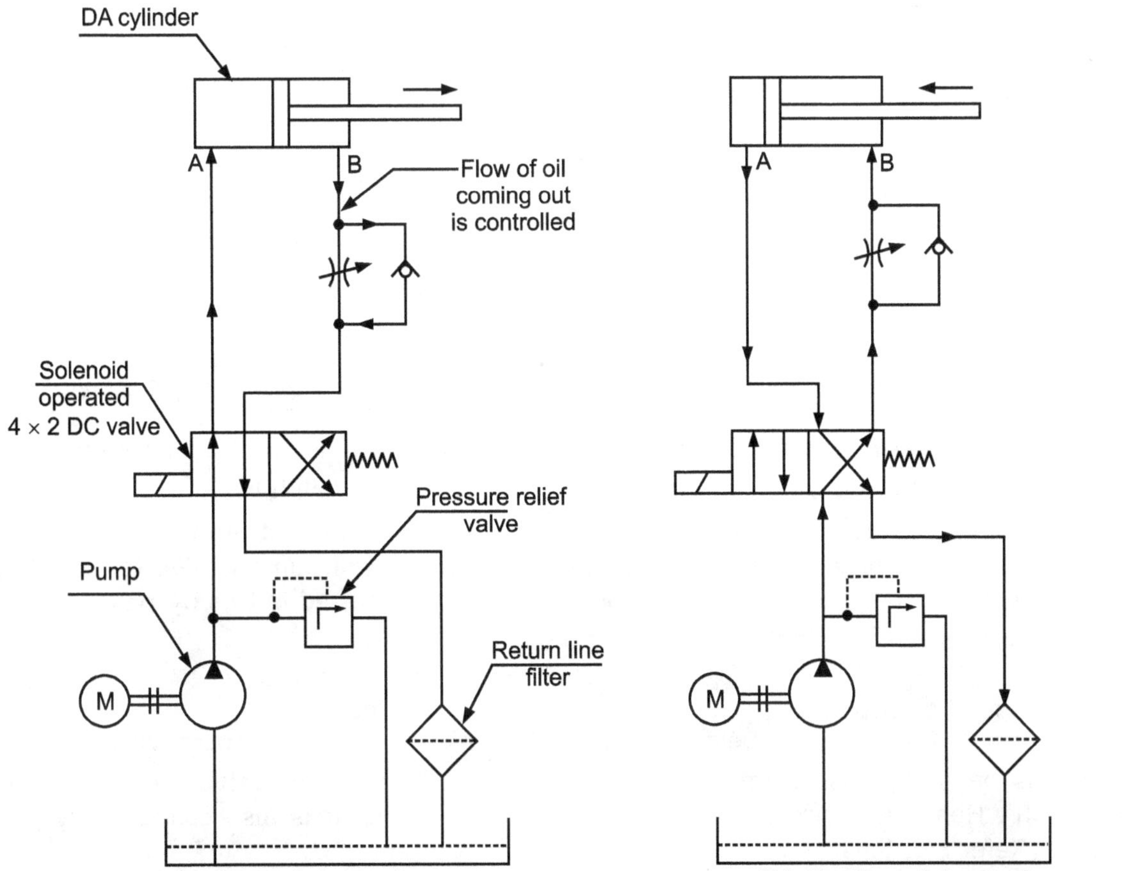

(a) Speed Control When Piston Advanced

(b) No Speed Control When Piston Retracts or Returns

Fig. 5.4: Meter-Out-Circuit

Advantages of Meter-Out-Circuit:

(a) The actuator movement is more stable.

(b) Heat generated due to throttling at flow control valve is given to the oil tank.

(c) Provides positive speed control of the cylinder.

Disadvantages of Meter-Out-Circuit:

(a) Since on both sides of piston there is pressure, there is possibility of higher friction.

(b) Pump works against the maximum pressure.

Compare Meter-In-Circuit and Meter Out Circuit **[W-11, 16]**

Meter-In-Circuit	Meter-Out-Circuit
(a) 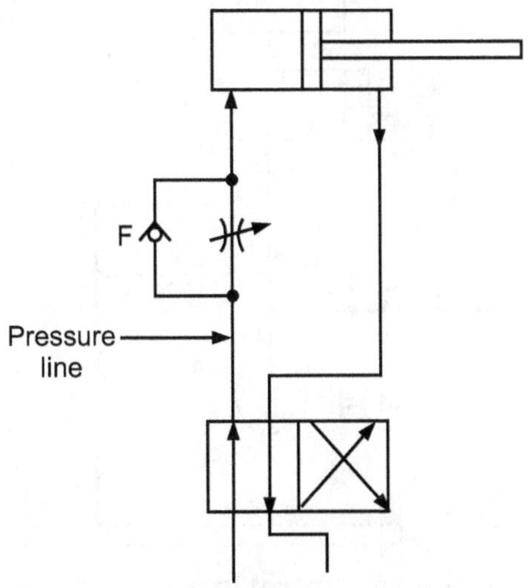 **Fig. 5.5 (a)**	(a) 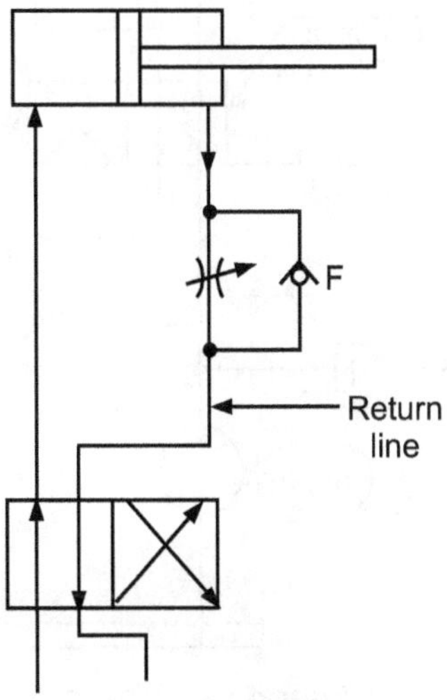 **Fig. 5.5 (b)**

contd. ...

(b) This is speed control circuit.	(b) This is also speed control circuit.
(c) Flow control valve 'F' is placed in pressure line.	(c) Flow control valve 'F' is placed in return line.
(d) Speed of piston can be controlled in Advance Stroke only.	(d) Speed of piston can be controlled in Advanced stroke only.
(e) This circuit is used where finer controls are required.	(e) This circuit is best suited for drilling/ boring where tool drags the workpiece.

5.1.5 Circuit No. 5 : Speed Control of DA Cylinder – Bleed-Off-Circuit [S-15]

- This is still another method of controlling the linear speed of piston of DA cylinder.
- This is slightly different circuit than 'meter-in' and 'meter-out' circuit.
- In this circuit neither the 'inflow' to the actuator nor the 'out flow' from the actuator is controlled. Instead of this, the pressurized fluid coming out of pump is diverted and bypassed to the oil reservoir. In this circuit the flow of control valve is placed in this by-pass line.
- Due to by-pass arrangement, the speed of piston is given by,

The speed of piston $\Rightarrow$ Difference between pump delivery flow and flow being by-passed to tank through flow control valve

- The oil is being by-passed immediately after pump. Another English word for by-pass flow is 'Bleed-off'. Hence this circuit is known as bleed-off circuit. It is also called as by-pass control circuit. (Refer Fig. 5.6)
- **Important:** In this circuit, we can control the speed of position in advance stroke as well as return stroke. [W-16]

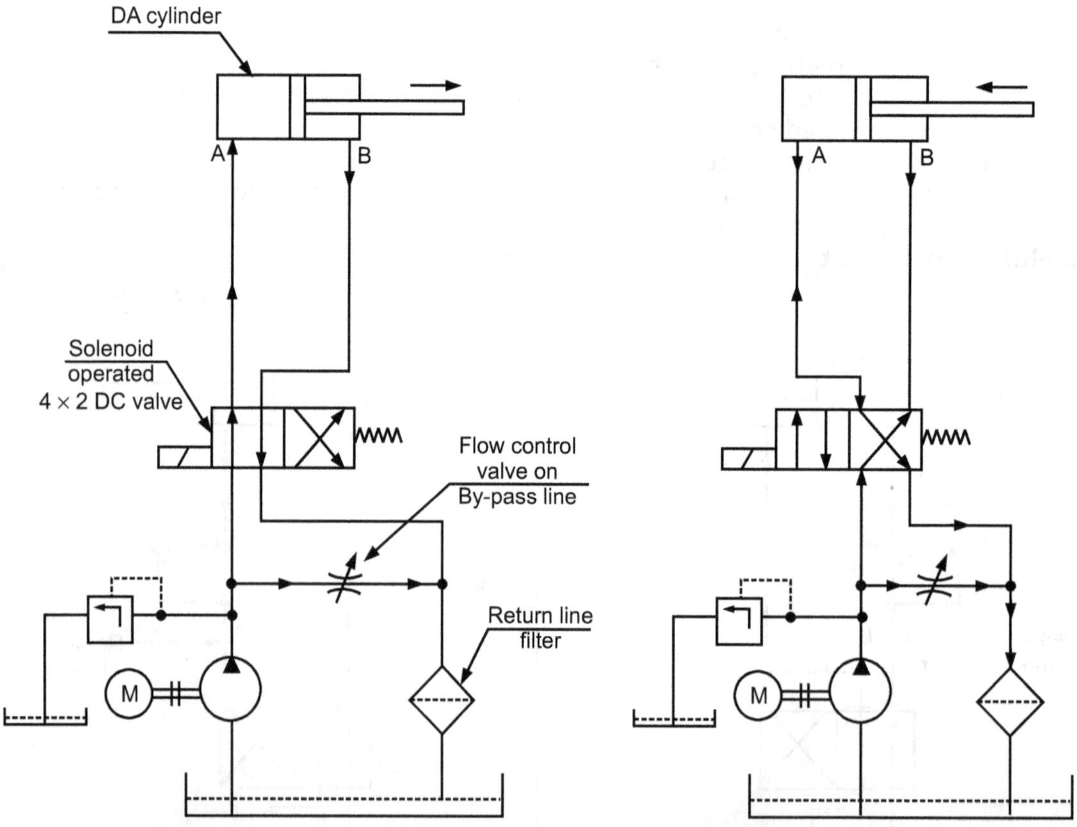

(a) Speed Control in Advance Stroke **(b) Speed Control in Return Stroke**

Fig. 5.6: Bleed-Off-Circuit

Applications of Bleed-off-Circuit:

(a) This circuit is suitable when pressure is constant.

(b) This circuit is used where precise speed control is not required.

(c) This circuit is used for control of broach in broaching machine. It is also used in hydraulic shaping machine.

Advantages of Bleed-off-Circuit:

(a) In this circuit, there is no excess flow going through pressure relief valve and system is efficient.

(b) Heat generated due to throttling is fed to tank.

Disadvantages of Bleed-off-Circuit:

(a) The circuit does not allow positive speed control.

(b) Speed adjustment is upto average value.

(c) This circuit is not suitable for accumulator circuit.

(d) If flow rate of pump fluctuates, the speed of actuator fluctuates.

5.1.6 Circuit No. 6 : Unloading Circuits

Hydraulic Pump Unloading

- All hydraulic systems, at least those larger than fractional horsepower size, should be designed in such a way that the pumps(s) will be allowed to idle, or run in an unloaded condition during periods in the machine operation when the output actuators, cylinders or motor, are not working. If the pump is allowed to discharge across the system relief valve, under pressure, while the actuators are not working, the normal full system horsepower is being changed from fluid power energy into heat energy which is carried back into the oil reservoir by the returning oil. Thermometer placed at the inlet and outlet of the relief valve would show an increases of oil temperature as the oil passes through the valve.

Problems if Circuit is not Unloaded:

- The penalties for failing to unload the pump during non-working periods are:
  1. High consumption of input power from the electric power line or high fuel consumption of an engine.
  2. Life expectancy is reduced on all components which continue to run without being unloaded. The hydraulic pump, for example, is designed for (so many) hours of operation at full rating. A common design figure for many gear pumps is 3000 hours. In theory, at least, a gear pump could be worn out by operating it for 3000 hours across a relief valve at full pressure without the pump ever performing a minutes work.
  3. Overheating of the oil in the system is another penalty. Overheating chemically degrades the oil making it unfit for further service. It shortens the life on all soft seals in the entire system and increases the rate of wear in many components.

Why Unloading is required?

- In studying these systems remember that power flowing through the pipes in a hydraulic system consists of two components: (a) pressure and (b) flow.
- There can be no power in the pipes unless both pressure and flow are present. This works the other way around, too. The power level can be reduced to zero (pump unloaded) by reducing either the pressure or the flow to zero. It is not necessary to reduce both to zero at the same time. Some of the circuits unload the pump by reducing the pressure to zero. Other circuits maintain the pressure at full value but reduce the flow to zero.
- Hence by reducing pressure to zero or flow to zero unloading can be done.

Ways of Unloading the Pump

(a) Dump Valve

The simplest way to unload a pump is to by-pass the oil directly to tank by opening a 2-way manual valve. However, this method has limited use as the operator may forget to unload the pump when pressure is not required.

(b) Using 2-Way Solenoid Valve

A better way is to use a 2-way solenoid valve which is automatically operated with a limit switch. The switch may be placed so it is actuated by the machine at the end of a cycle and breaks current to the solenoid. For this type of circuit the solenoid valve should be a 2-way normally open type. (See Fig. 5.7)

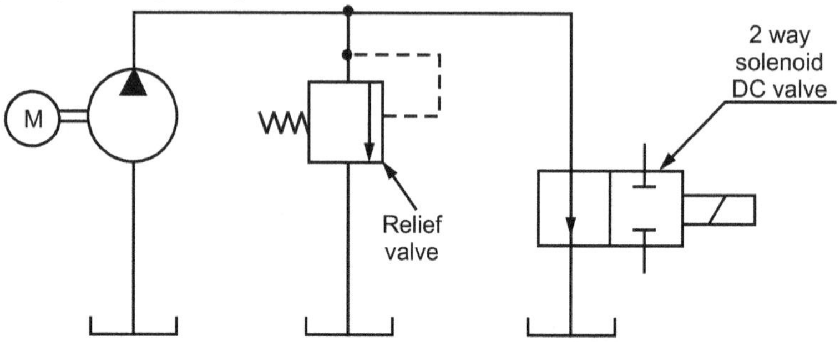

Fig. 5.7

(c) Pilot Control Valve

This is essentially the same circuit as the dump valve except that a small (1/4 inch) unloading valve is capable of unloading a very large oil flow by "dumping" the pilot or RC connection of a large pilot-operated relief valve. The relief valve is sized to handle the volume of flow. The "dump" valve may be a manual 2-way type, or a 2-way normally open solenoid valve of miniature size. (shown in above Figure).

(d) Tandem Center 4-Way Valves

This system may be applied equally well to either manual or solenoid 4-way valves, and for any number of such valves connected in series. However, the pressure losses may become high when more than 2 or 3 valves are used, and a pump capable of supplying sufficient pressure must be used to supply the losses.

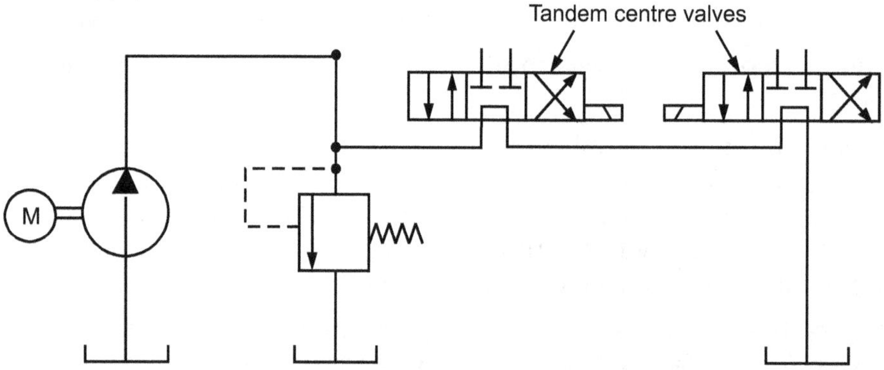

Fig. 5.8

Caution: Some valves are not capable of handling full system pressure on their tank ports, and these should not be used in tandem circuits of more pump pressure than their tank port rating. Check manufacturers rating.

(e) Low Pressure Unloading Circuit

This circuit may be used for counterbalancing a press ram or cylinder against downward drift due to gravity. The main circuit relief valve, V1, set for the maximum required working pressure of the system. This valve is sized to handle the full pump flow. An auxiliary relief valve, V2, works on the vent connection of V1, to reduce system pressure to a very low value when the control valve, V3, is shifted to retract the cylinder. V2 may be of miniature size, and because its outlet is connected to the blind end cylinder port instead of to tank, it is effective in reducing the setting of the main relief on the return stroke, and has no effect on the system pressure while the cylinder is extending.

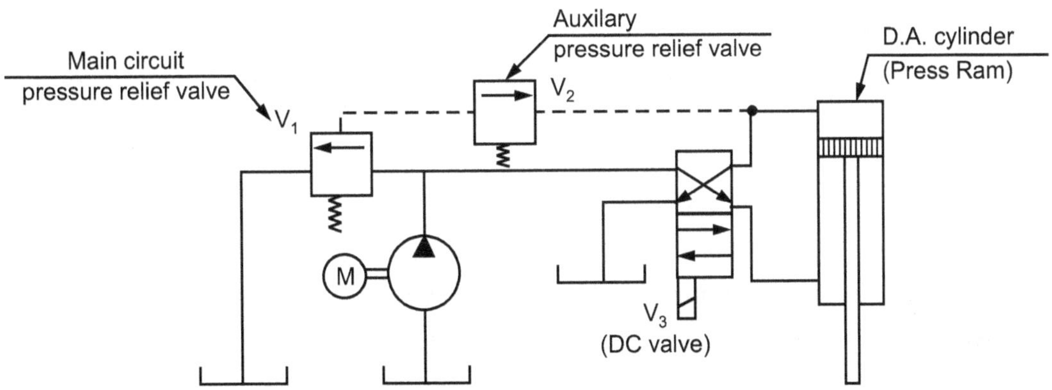

Fig. 5.9

(f) Cam Valve Unloading Circuit

A cam may be placed on the cylinder rod or a part of the machine located so it will depress the roller on a cam valve when the cylinder reaches the desired unloading position. The cam valve is a 2-way normally closed type, or a 4-way type, or unused ports that are plugged. When the directional valve, V2, is shifted to start the cylinder forward, the pump is automatically loaded to full pressure while the cylinder rod end is now vented to tank.

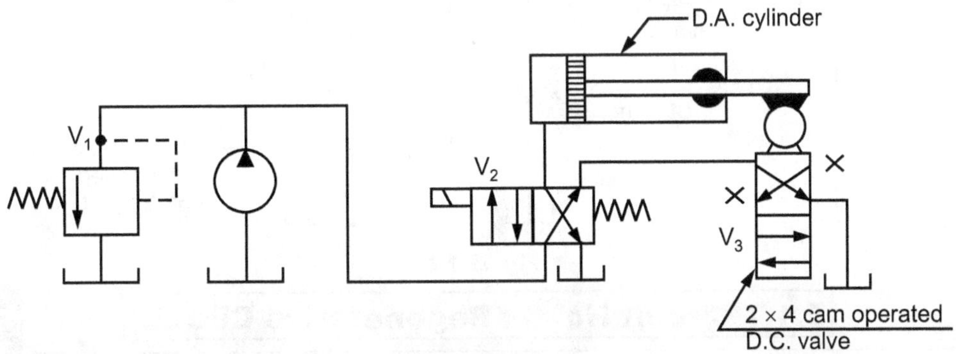

Fig. 5.10

(g) High Pressure Forward, Low Pressure Return Circuit

The system relief valve, V1, is set for the desired maximum pressure for the forward stroke, and the auxiliary relief valve, V2, is set for the minimum pressure which will retract the unloaded cylinder. When the cylinder retracts and bottoms out, pump flow discharges across V2. For circuits in which the cylinder is virtually at no load while retracting, this method will usually unload to a pressure low enough to minimize overheating of the oil.

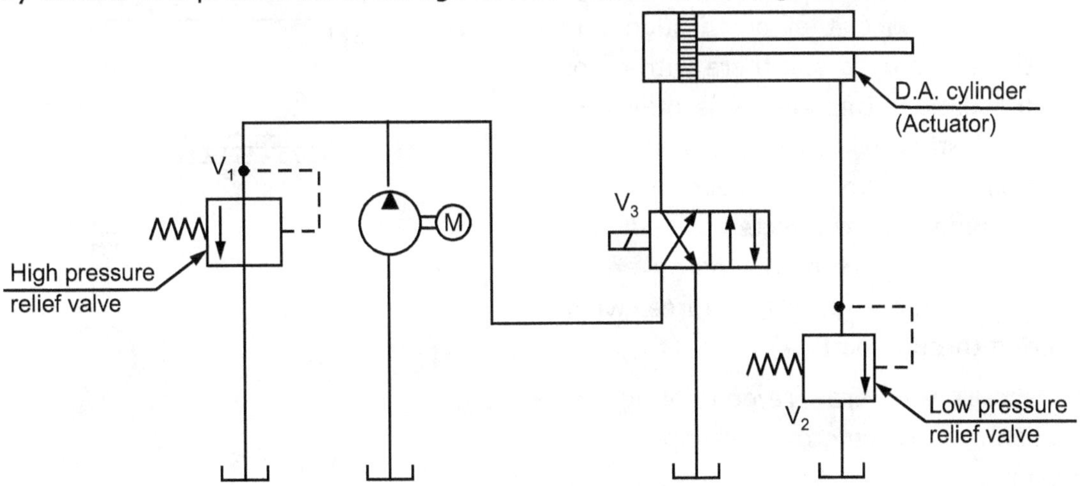

Fig. 5.11

(h) Pressure Compensated Pump Unloading Circuit (Stopping the Oil Flow)

Other circuits shown unload a pump by allowing free oil flow at very low or no pressure. Another method of unloading is to stop the oil flow. This is done by using a 4-way directional control valve which has a closed pressure port in center position, thus reducing the oil flow and the power level to zero. A pressure compensated type of pump is required in this type of unloading.

The compensator is usually a built-in part of the pump. It receives a pressure signal from the pump pressure line. When the system pressure builds up to a level set on the compensator adjustment, the compensator automatically reduces the pump displacement to zero, thus unloading the pump and the driving motor.

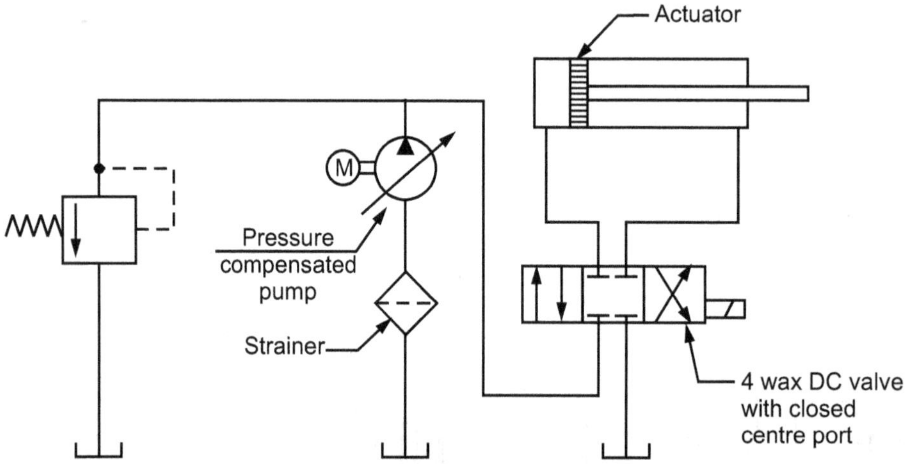

Fig. 5.12

5.1.7 Circuit No. 6 : Regenerative Circuit

[S-10, W-10, 16]

- The concept of re-generative circuit can be explained in the following manner: Consider the double acting cylinder. Pressurized oil from pump is admitting in cylinder cavity through port (A). Due to pressure force piston is moving from right to left. During this movement, the oil present on piston rod side of piston, starts coming out through port (B). This oil will return to the oil reservoir via DC valve.

  It is clear from Fig. 5.13 that, returning oil will enter in pressure pipe through pipe 'P'. During exit of oil through port (B), some energy is still there with oil on piston rod side. This energy is otherwise wasted if this oil directly goes to oil reservoir. To avoid wastage of this energy, pipe 'P' is connected so that, the pressurized oil gets more energy and it will create more pressure force while entering through port (A).

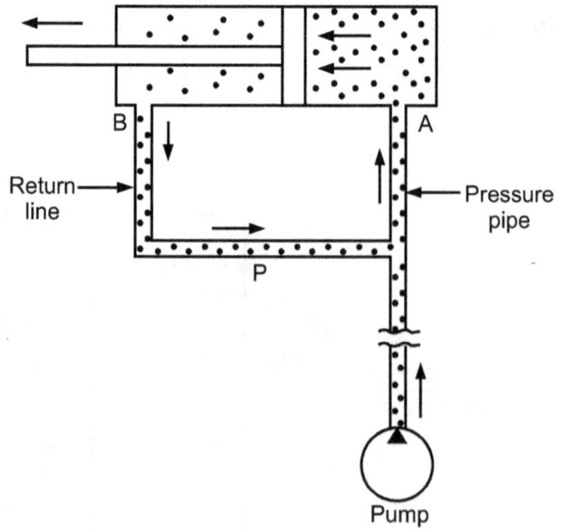

Fig. 5.13

  - In short we are re-generating or recovering the energy available with returning oil by using regeneration technique.

- **Regenerative Circuit:** A typical Regenerative circuit is depicted in Fig. 5.14. We are using 4 × 3 solenoid operated DC valve having regenerative central port. When DC valve is in central position then only we can achieve regeneration. When DC valve is in other two positions, there is no regeneration.
- **Important:** Regenerative action is possible only when oil is entering from piston side.

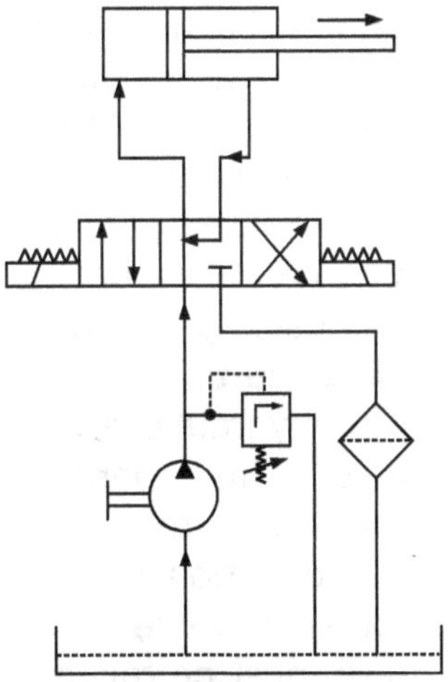

Fig. 5.14: Regenerative Circuit

5.1.8 Circuit No. 9 : Counter balance Valve and Circuits

[W-16

Counterbalance Valve:

- Counterbalance valves are used in hydraulic systems working with overriding (running away) or suspended load. They are designed to create backpressure at the return line of the actuator to prevent losing control over the load.
- **Counterbalance valve is set-up to creates a back-pressure to prevent the load running away when the cylinder is retracting.**
- They are modulating valves which allow free flow into the actuator and then block the reverse flow until they feel a pilot pressure inversely proportional to the load in the pilot line. They employ a check valve for the free flow direction, a relief valve to control the flow in the reverse direction and a pilot piston which lowers the relief valve setting.
- They can lock loads in a leak free mode and they are well suited for many clamping applications or to prevent negative loads from falling down in case of hose failure. For load lowering, they improve motion control in most systems because they compel the directional control valve to always meter positive pressure, also under overrunning load conditions.
- **For best safety, they should be fitted close to the actuator, either flange mounted or connected through metallic pipe.**
- Functions of counterbalance valve:
  1. Free upstream flow through the check valve for load lifting.
  2. Locking of reverse downstream flow when the directional valve is not operated, or the pump is stopped.
  3. Load lowering metered by the piston of the relief valve opened and controlled by the pilot pressure of the "lowering" flow delivered upstream by the directional control valve, also when the load tends to overrun.
  4. Pressure relief (with open centre directional spools) for pressure surges in the actuator caused by the inertia of the load, by oil expansion due to heating, or by external forces.

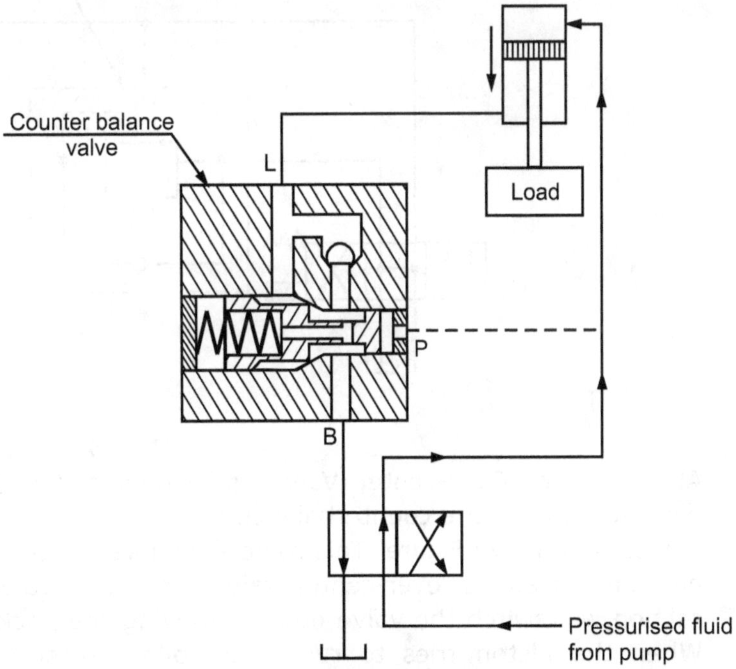

Fig. 5.15

- To understand this valve, see the circuit above (Fig. 5.15) which shows the construction details of valve.

- If a directional valve is shifted into position that lowers the load, then the fluid from the rod chamber of the cylinder can exit only if pressure at port P (pilot pressure) and port L (load pressure) create enough force to overcome the spring force and open the valve.
- These are basically relief valves but used in a particular application to set-up a back-pressure in a circuit. They are frequently employed to 'counterbalance' a load as shown in the circuit in Fig. Here the valve creates a back-pressure to prevent the load running away when the cylinder is retracting. The usual pressure setting is 1.3 times the load-induced pressure.

Counterbalance Circuit

- Following is the hydraulic circuit which incorporates, counterbalance valve. (See Fig. 5.16)
- A check valve is incorporated in the circuit in above figure to allow free-flow in the reverse direction (i.e. to bypass the counterbalance valve when raising the load). Care must be taken if using a conventional relief valve for this application as at some stage of operation the tank port will be subject to maximum circuit pressure. This is not permissible with many relief valves. The counterbalance valve shown has an integral check valve. A separate drain connection from the spring chamber is unnecessary because the pressure section of the valve is inoperative when the T port is pressurized (flow is through the check valve). When it is counterbalancing the back pressure at T should be kept to a minimum.

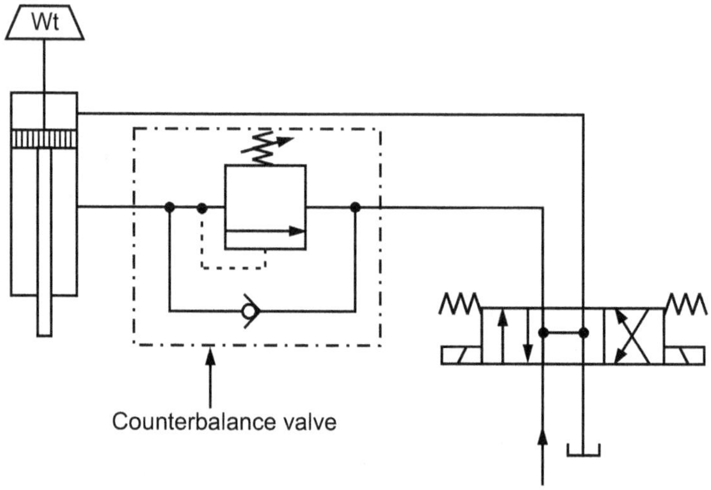

Fig. 5.16

Piloted Counterbalance Valve Circuit (also known as Over Center Valve/Brake Valve Circuit)

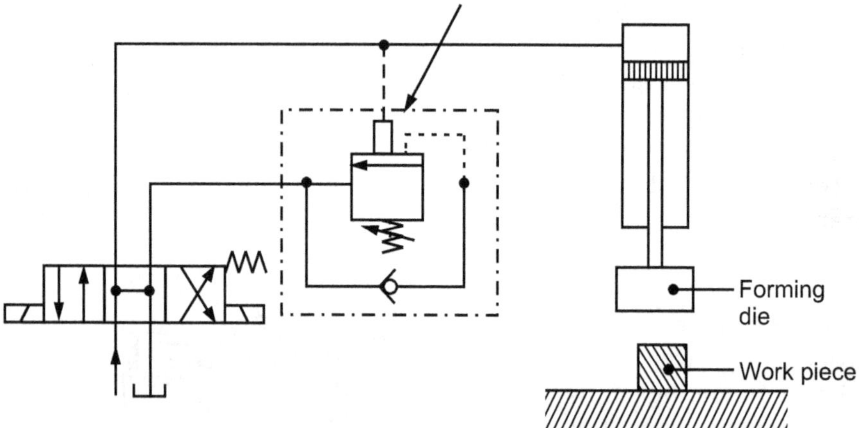

Fig. 5.17

- Above Figure Over-center Valve (pilot-operated counterbalance valve or brake valve). The disadvantage of the counterbalance valve can be overcome by using remote pilot operation as depicted in above Figure. This remotely piloted counterbalance valve shown diagrammatically is also known as an 'over-center valve' or 'brake valve'. A relatively low pressure in the pilot section will switch the valve open, removing the back-pressure from the cylinder annulus side. When the piston tries to run away, pilot pressure is lost and the counterbalance section switches back into circuit. During the forming part of the press operation, the valve is piloted open removing the back pressure and all the pressure on the full bore side is then available for pressing.

Winch Circuit using Piloted Counterbalance Valve (Over Center Valve)

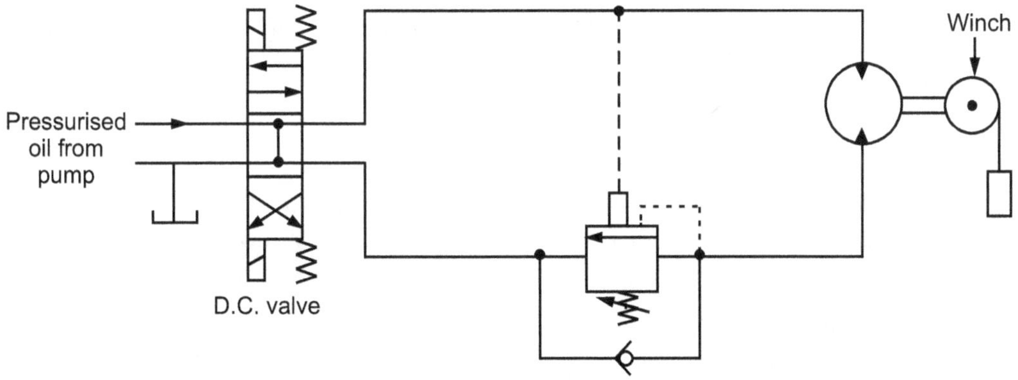

Fig. 5.18

- The over-center valve is frequently used in motor circuits (hydrostatic transmissions) as a brake valve. In Fig. 5.18, the circuit shows a simple winch driven by a hydraulic motor; the over-center valve will:

  (a) Hold the load in the neutral position.

  (b) Prevent over-run during lowering.
- The ratio between the pilot pressure to the direct pressure necessary to open the valve is generally from 2:1 to 10:1 according to the application.

5.1.9 Circuit No. 7 : Speed Control of Bi-directional Hydraulic Motor

- Hydraulic motors are now playing an important role in industrial applications.
- They convert hydraulic power from the hydraulic lines into reversible, variable speed and rotary mechanical power.
- Bi-directional hydraulic motor runs in clockwise as well as anticlockwise direction depending upon the position of DC valve.
- The speed of motor (RPM) can be controlled by two separate flow control valves.
- The circuit (in three position) is shown below in Fig. 5.19.

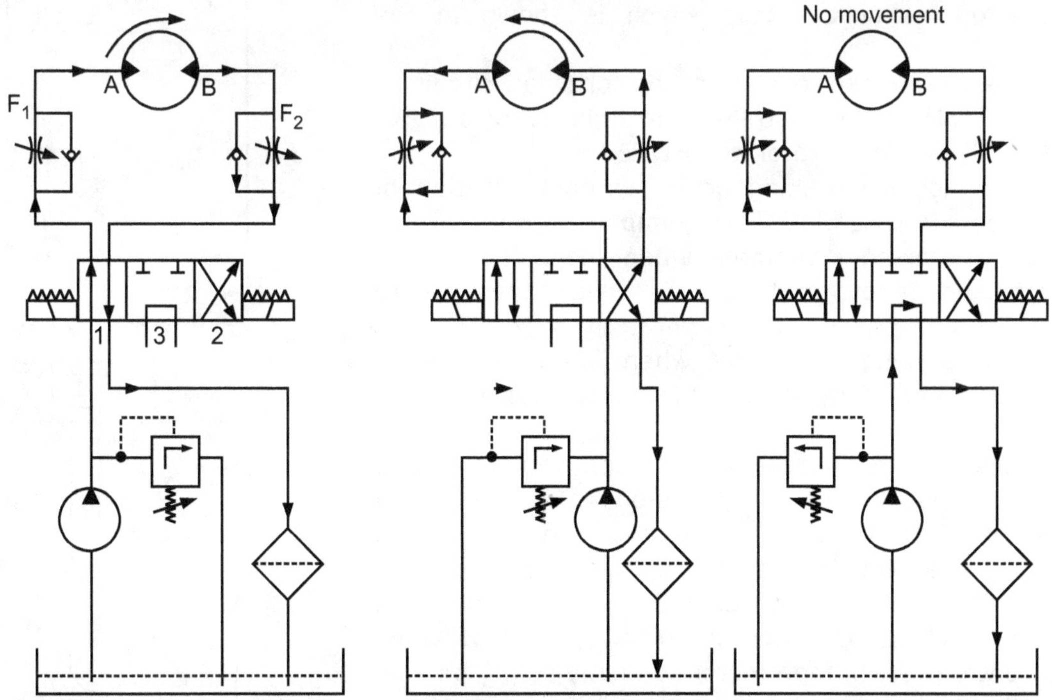

(a) Position 1: Motor Rotates in Clockwise Direction **(b) Position 2: Motor Rotates in Anticlockwise Direction** **(c) Position 3: No Rotation of Motor**

Fig. 5.19: Speed Control of Bi-directional Motor

- The circuit shown in Fig. 5.19 uses:
  - (i) 2 variable flow control valves with built-in check valve. (F_1 and F_2).
  - (ii) 4 × 3 DC valve solenoid operated having centre closed position.
  - (iii) Return line filter.
- **Position 1:** 1st position of DC valve is in operation. Pressurized oil enters into motor through port (A) and motor rotates clockwise. The used oil present in the motor goes back to oil tank through port (B) via check valve and DC valve. (carefully look at the arrow shown for oil movement). The speed of motor can be controlled by using flow control valve F_1.
- **Position 2:** 2nd position of DC valve is in operation. Pressurized oil enters into motor through port (B) and motor rotates anticlockwise. The used oil present in the motor goes back to oil tank through port (A) via check valve and DC valve. The speed is controlled by F_2.
- **Position 3:** 3rd position i.e. central position of DC valve is in operation. One can easily observe that pressurized oil goes to DC valve centre port but it goes to return line and collects in oil reservoir via return line filter. Oil is not going to motor. Hence in this position motor stops, no rotation of motor. This condition is called 'Blocked-off' condition.

5.1.10 Circuit No. 7 (A) : Braking and Replenishing of Hydraulic Motor

- **Drawback of circuit no. 7:** In the circuit of speed control of hydraulic motor by using 4 × 3 DC valve (Circuit No. 7) discussed earlier, we cannot run the motor at higher speeds.

  If motor runs at higher speed and if by mistake DC valve takes centre position (blocked-off position or no movement position), then motor suddenly stops. This sudden stoppage of motor creates serious shock in circuit and there is possibility of damage in hydraulic system.

  Hence, if the motor runs at higher speed, then we have to introduce safety device in the circuit to avoid shock.

  This modified circuit is called 'braking and replenishing circuit of hydraulic motor' which is shown in the Fig. 5.20.
- One can easily make out that the original circuit is modified by introducing Two Pressure Relief Valves (PRV-1 and PRV-2) between motor and F1/F2.
- In order to achieve braking action without creation of serious shock, PRV1 and PRV2 play important role.
- PRV1 is for braking movement when motor is running clockwise and if by mistake DC valve takes centre position.
- PRV2 is for braking movement when motor is running anticlockwise and if by mistake DC valve takes centre position.
- It is easily understandable that, when motor rotates (clockwise/anticlockwise) and if it stops suddenly, it will continue to rotate due to its inertia and load inertia. This rotation creates pumping action which results in sudden pressure rise in Line 1 or Line 2. Suppose motor is rotating in clockwise direction and if by mistake DC valve attains centre position then motor suddenly stops, but will continue to rotate due to inertia and pressure in line 1 will increase. Now PRV-1 will act and will reduce the pressure increased by by-passing the oil and protecting the system.

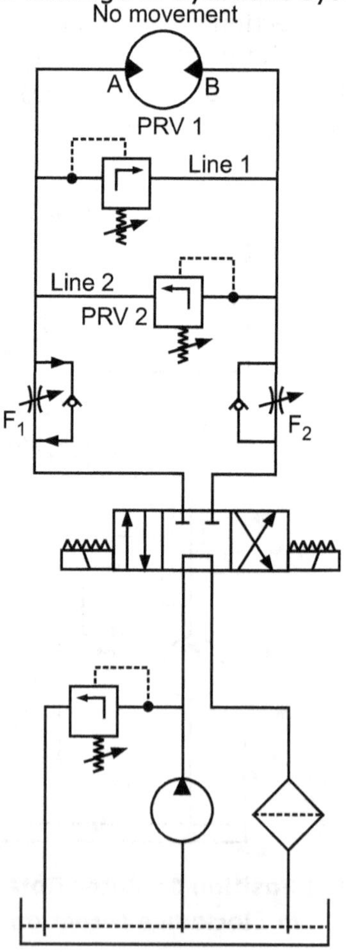

Fig. 5.20

5.1.11 Circuit No. 8: Synchronizing Circuit When Linear Actuators are Mechanically Coupled

- Synchronizing circuits are the circuits in which motions of two or more actuators are simultaneous (or synchronized).
- Fig. 5.21 shows synchronization of two DA cylinders when they are mechanically coupled.
- Cylinder D1 and D2 are mechanically coupled at piston rod end.
- DC valve attains position 1. Hence pressurized oil will be supplied to D_1 and D_2 through inlet ports A_1 and A_2 respectively.
- Since two cylinders are mechanically coupled they will simultaneously move 'up' and thus we are achieving synchronizing movement. The oil present in cylinder on piston rod side will return to oil tank through ports B_1 and B_2 via DC valve and return line filter.

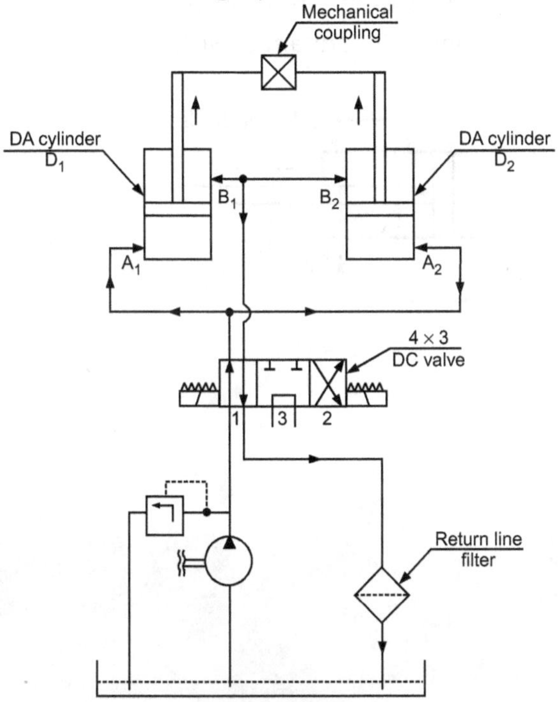

Fig. 5.21

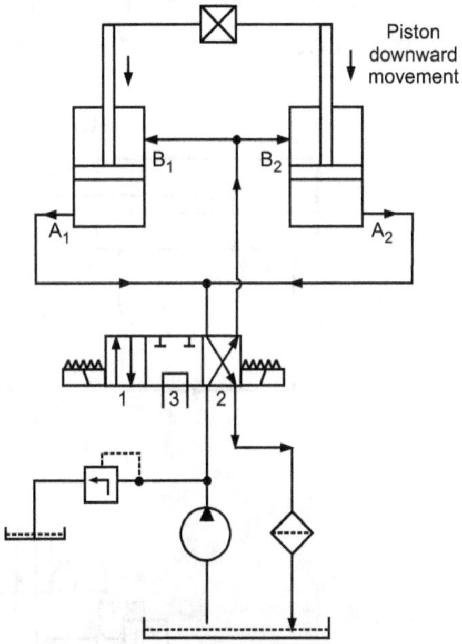

Fig. 5.22

- The other two positions are shown below:
- Now DC valve attains position 2 and oil will be supplied through ports B_1 and B_2. So that both pistons will simultaneously come down. Oil present on other side of piston will come out through ports A_1 and A_2 and will return to tank.
- When DC valve attains centre position (position 3) then, no oil will be supplied to port A_1/A_2/B_1/B_2.
- Oil from pump will be simply by-passed and will return to oil tank.
- In the position shown below in Fig. 5.23 there is no movement of any actuator. It is 'Hold' position.

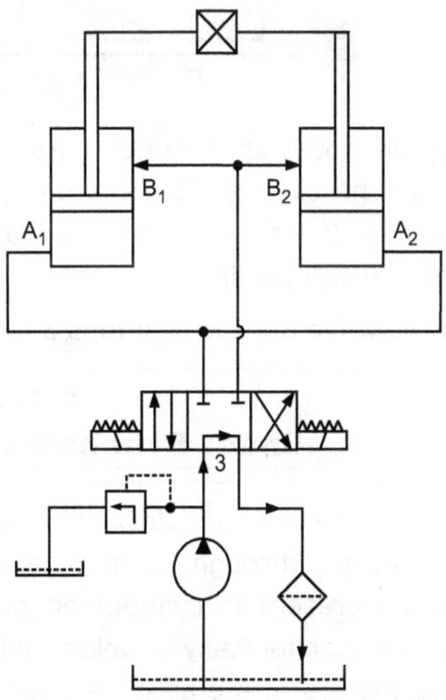

Fig. 5.23 (Hold position)

5.1.12 Circuit No. 8 (A) : Synchronizing Circuit using Flow Dividers

- The circuit can be used to operate two DA cylinders whose motion is to be synchronized.
- This circuit uses two flow dividing valves (F_1 and F_2).
- In the circuit shown above, the two DA cylinders are not mechanically coupled.
- Flow divider valves F_1 and F_2 are placed as shown.
- When pressurized fluid comes at F_1 and F_2 junctions, the flow of oil is equally divided in the branch and equal amount of oil enters into the cylinders through ports A_1 and A_2 ($Q_1 = Q_2$).
- If both pistons and cylinders are having same dimensions then, piston of both the cylinders will advance with same speed. If dimensions are different then same quantity of pressurized oil will act on different area of piston and naturally both pistons will advance with different linear speed.

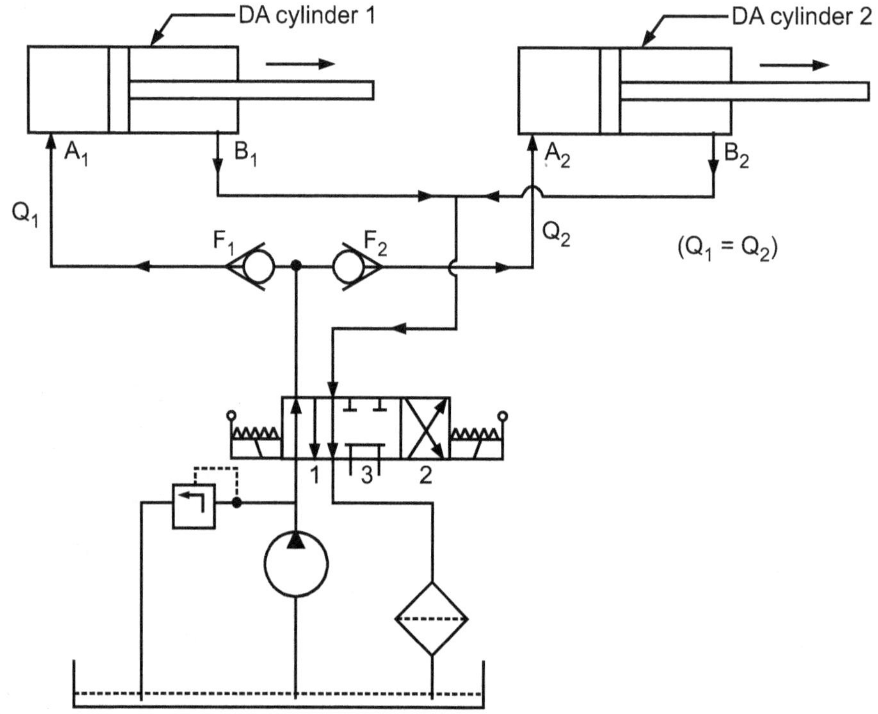

Fig. 5.24

- It may be noted that, synchronization of both pistons is possible only in advance stroke (as shown in the circuit). There is no synchronization in return stroke (i.e. when DC valve attains position 2). If we want synchronization in return stroke also, then we have to use additional flow divider in that line.
- When DC valve attains positions 3 (i.e. centre position) then there is no movement of pistons.

5.1.13 Circuit No. 8 (B): Cylinder Synchronizing with Linked Hydraulic Motors

- The circuit shows two hydraulic motors (M_1 and M_2) coupled mechanically. Here, the pressurized oil is entering through ports B_1 and B_2 i.e. pistons are retracting. During the movement of piston, oil present in cylinder on piston side will come out through ports A_1 and A_2 and will rotate two mechanically coupled hydraulic motors M_1 and M_2. The mechanical coupling ensures the two motors rotate at same speed. Hence, it must be noted that, two DA cylinders must be of same dimension.

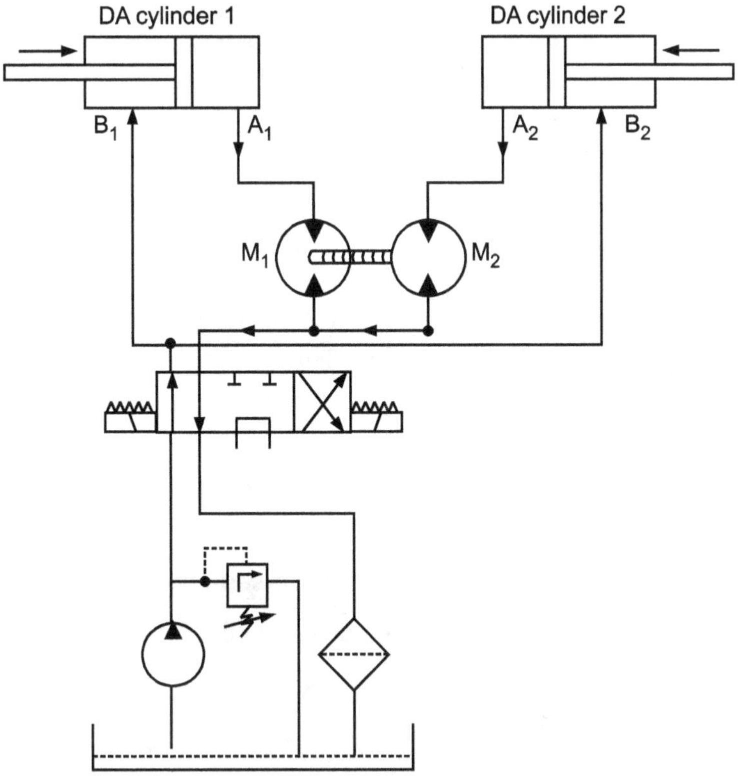

Fig. 5.25

5.1.14 Circuit No. 9 : Sequencing Circuit

[W – 09, S-10]

- Sequencing circuit is extensively used in industrial hydraulic systems. It is especially used in mass production.
- Sequence valve is 'close relative' of pressure relief valve; but sequence valve is used in hydraulic systems to cause various operations in sequential order i.e. one after the another.
- Let us take an example of drilling a hole in a workpiece.

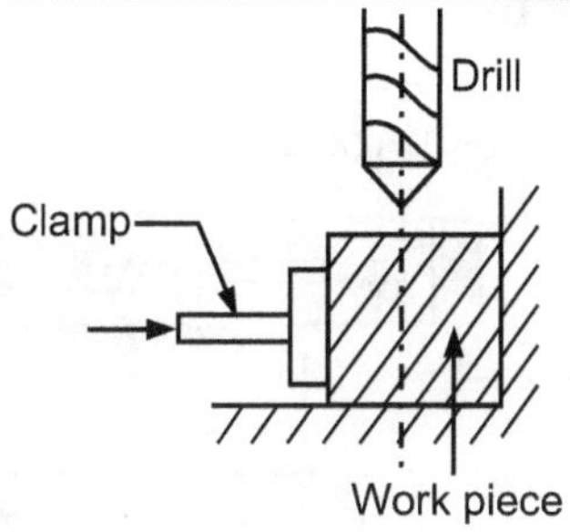

Fig. 5.26

  - Before drilling we have to clamp the workpiece firmly and then we have to carry out drilling.
    (a) Step 1: Clamping of workpiece
    (b) Step 2: Drilling
    (c) Step 3: Removing drill from drilled hole and declamping.
  - When an industrial unit goes for mass production (say 20,000 jobs/month) then it has to mechanize the above three steps. For mechanization, hydraulic system is the best solution.
  - There are two types of sequencing circuits:
    (a) Pressure Dependant Sequencing: In this circuit, second action starts when the first action is complete.
    For example, the above example, drilling starts only when clamping is complete.
    (b) Travel Dependant Sequencing: In this circuit second action starts when first action is partly complete. This means the actions overlap but in sequential order.
  - **Sequencing can be obtained by using:**
    (a) Sequencing valve.
    (b) Roller actuated DC valve with cam.
    (c) Time delay valve.
  - Sequencing can be used for obtaining actions like bar feeding, ejecting, lifting, pushing, clamping, in predetermined order.

5.1.15 Pressure Dependant Sequencing Circuit

- The circuit is used for drilling a hole in a workpiece. The sequence operation is:
  - (a) Clamping of workpiece
  - (b) Drilling
  - (c) Declamping and drill taken out from hole.
- The DC valve takes centre position (No. 3). No oil is supplied to either of cylinders C_1 or C_2. Now undrilled workpiece is kept on the fixture seat. The oil from pump is going to oil tank via DC valve and return line filter. So there is no movement of cylinder C_1 or C_2. No clamping (Fig. 5.27).
- Now worker will operate foot lever of DC valve and valve takes position 1. (Refer Fig. 5.28)
- Now, oil will start supplying directly to C_2 and through sequence valve to C_1.
- When oil enters through port A_2 of cylinder C_2, piston will advance and immediately clamps the workpiece.

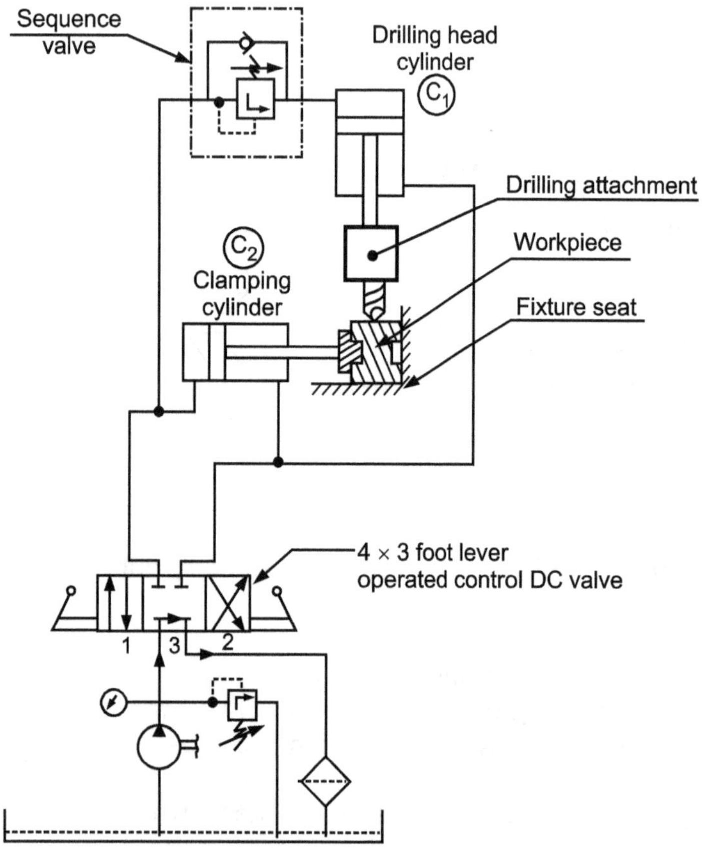

Fig. 5.27: No Movement of any Cylinder

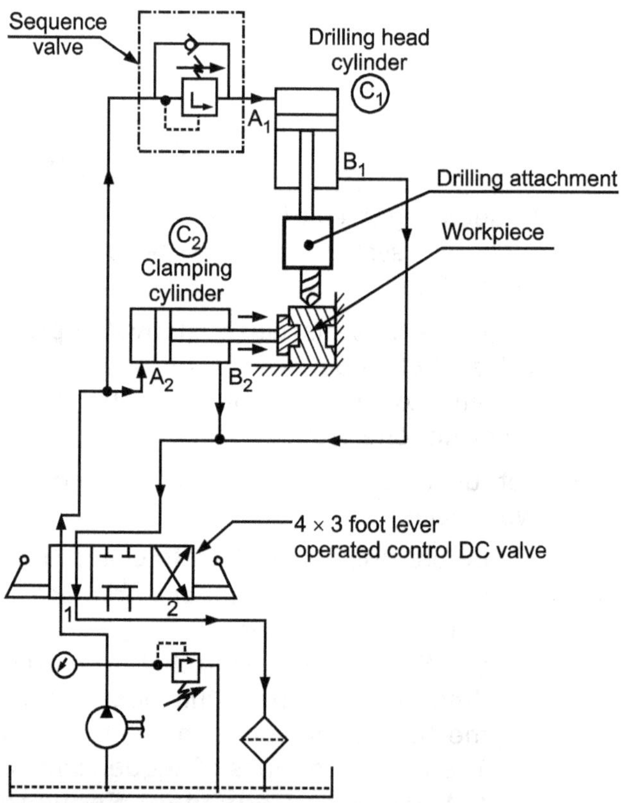

Fig. 5.28: Clamping Complete and Drilling Started

- At the same time oil flows towards port A_1 of cylinder C_1; but through sequence valve. Some higher pressure is set at pressure relief valve of sequence valve. When the pressure of flowing liquid reaches this set value, the sequence valve opens and oil enters through port A_1 into cylinder C_1. Due to this, position advances and comes down, so that drilling starts.
- When operator again operates foot lever of DC valve, it takes position 2. Both the pistons retract, workpiece declamps and drill comes out of the drilled hole.

5.1.16 Travel Dependant Sequencing Circuit using Cam

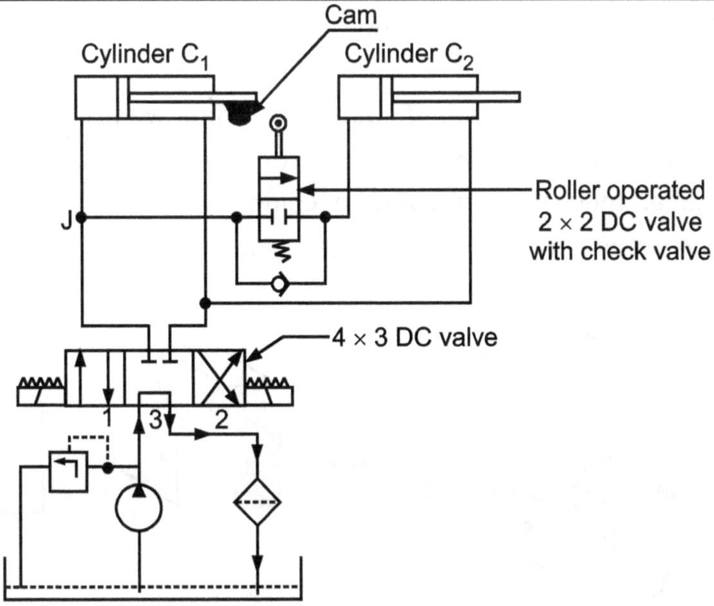

Fig. 5.29: No Cylinder Movement

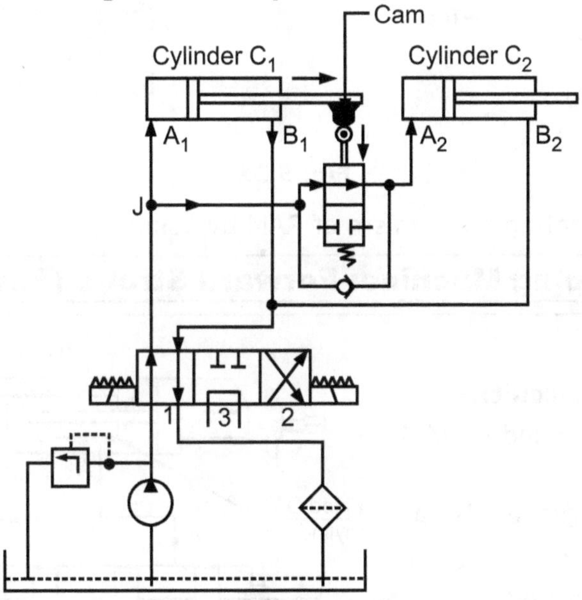

Fig. 5.30

- The circuit shown is travel dependent sequencing circuit using roller and cam pair. In this circuit, cylinders C_1 and C_2 will be operated in sequence. But movement of piston of C_1 and C_2 will overlap for some distance. Position shown in the circuit is no movement position. (Refer Fig. 5.29)
- Now DC valve takes position 1 (Refer Fig. 5.30). Then oil flow starts towards junction 'J'. At this junction flow divides. Some oil enters into cylinder C_1 through port A_1, piston moves towards right. During this movement cam presses down roller operated 2 × 2 DC valve. Then flow of port A_2 of cylinder C_2 starts and then the piston of this cylinder moves towards right. Thus movement of piston of C_2 will start only after cam operates 2 × 2 DC valve.

5.1.17 Hydraulic Circuit for Shaping Machine [W-11, S-16]

- **What is Shaping Operation?:** Shaping machine uses single point cutting tool. This tool makes reciprocating motion on workpiece. In half reciprocating stroke the tool cuts the metal and in second half, the tool returns to the original position.

- The tool head of shaping machine is shown in following Fig. 5.31. Fig. 5.32 also shows types of surfaces it is best able to produce.
- It may be noted that, shaping machine is not a production machine. It is widely used in tool rooms of machine shops.
- To generate reciprocating motion of ram of shaping machine we are having two alternatives:
  (a) By using 4 bar mechanism (cylinder, piston, connecting rod and crank) driven by motor. This machine is called mechanical shaping machine.
  (b) By using hydraulic DA cylinder. This machine is called hydraulic shaping machine.

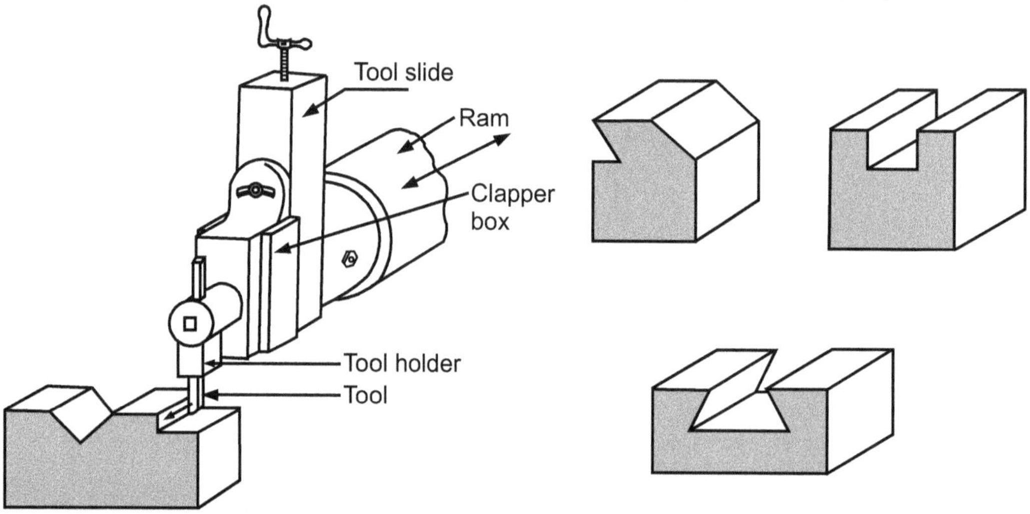

Fig. 5.31

- So we will study the operating mechanism of RAM by using hydraulic circuit.

5.1.18 Hydraulic Shaping Machine: Forward Stroke (Power Stroke) [S-09, W-10]

(a) Forward stroke:

- **Elements Used in the Circuit:**
  (a) Fixed DA cylinder – (Fixed in RAM of machine)
  (b) Special spool type DC valve. At the end of spool shaft stroke adjusting lever is attached.
  (c) Other elements like pump, pressure relief valve, return line filters (2) and oil reservoir.
- **Working:** As shown in Fig. 5.32, pump is supplying oil to DA cylinder through DC valve and through port B. Hence piston will move from right to left with force and this is a cutting stroke. At the end of stroke 'stroke changing lever' will touch the adjustable stopper S_1.

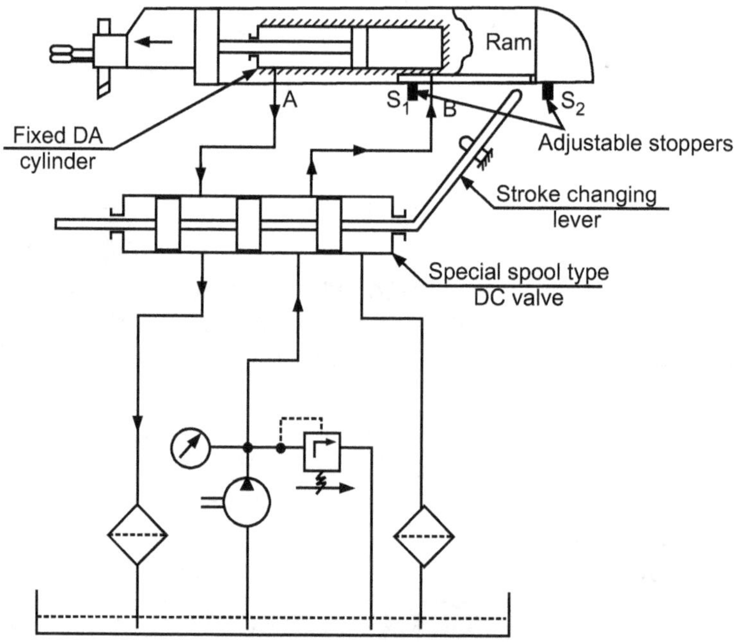

Fig. 5.32: Forward Stroke

(b) Backward Stroke (No metal cutting only tool returns) (Refer Fig. 5.33).

- **Working:** When lever touches the stopper S_1 then spool shifts to right and flow directions in DC valve change as shown. Now oil is entering through port A of DA cylinder and piston will move from left to right. **During this stroke, tool post slightly lifts and hing and no question of metal cutting.** Again the lever is stopped by stopper S_2 and the cycle repeats.

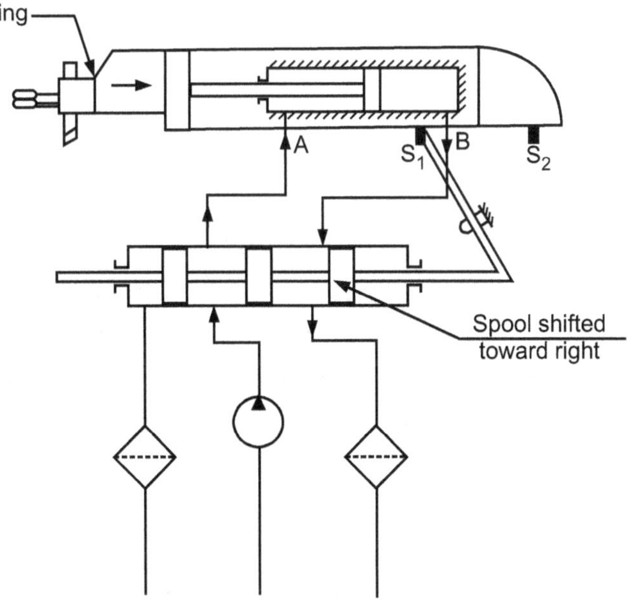

Fig. 5.33: Return Stroke

5.1.19 Hydraulic Circuit for Milling Machine [W-09, S-12]

What is Milling Operation?

- Milling is a process of removing metal by feeding the workpiece past a rotating 'multipoint tool cutter'. In milling the rate of metal removal is very rapid and hence is one of the important machines in mass production circuit. Various milling operations are given below:
- **Types of Milling Machines are:**
  (a) Horizontal milling machine.
  (b) Vertical milling machine.
  (c) Universal milling machine (Both horizontal and vertical).

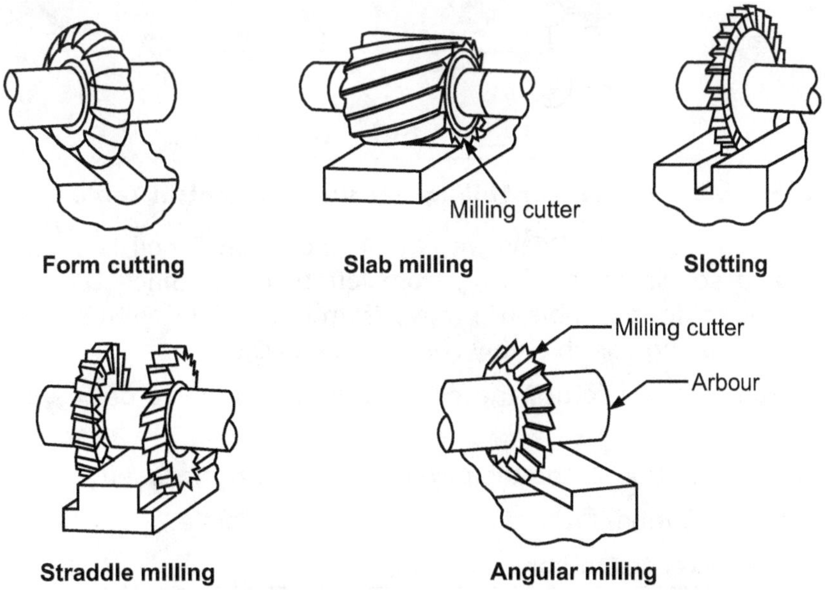

Fig. 5.34

- The cutter rotates on arbour of milling machine.
- The cutting feed is given to table of milling machine so that it reciprocates for metal cutting.

- This table movement is controlled by a hydraulic circuit.
- Note that the reciprocating movement of milling machine is comparatively slower than shaping machine. So also different feeds are required for milling different types of work.
- The Hydraulic circuit for table movement is shown in the Fig. 5.35.
- Elements used in circuit:

  (a) Fixed DA cylinder.

  (b) Flow control valve with built-in check valve (2 Nos).

  (c) 4 × 3 DC valve.

  (d) Main pump with pressure relief valve.

  (e) Pressure booster pump with pressure relief valve.

  (f) Return line filter.

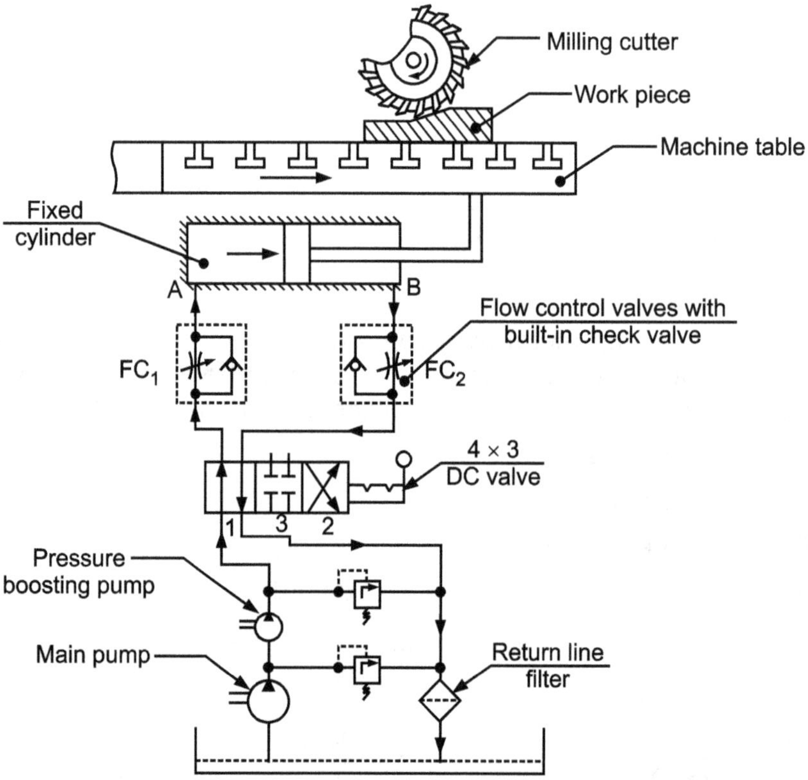

Fig. 5.35: Hydraulic Circuit For Milling Machine (To Control Table Movement)

- **Working:** As shown in Fig. 5.35, DC valve is in 1st position. So oil is supplying to DA cylinder through port 'A', and so piston is moving from left to right. Since DA cylinder movement is connected to machine table, the table will move from left to right with set feed (mm/min.). The feed is adjusted by adjusting variable flow control valve FC_1.

  The oil on piston rod side will return the oil tank via check valve of FC_2, DC valve and return line filter.
- When DC valve attains position 2, the table will move from right to left.
- When DC valve attains position 3, there is no movement of table.
- Booster pump further increases the pressure of oil.

5.2 SURFACE GRINDING MACHINE WORKING

- Surface Grinding Machine uses a rotating abrasive wheel to remove the material from the surface of the workpiece so as to create a flat surface with high surface finish. The grinding wheel revolves on a spindle and the workpiece is mounted on a reciprocating table.

- Surface grinding is the most common of the grinding operations. It is a finishing process that uses a rotating abrasive wheel to smooth the flat surface of metallic or non-metallic materials to give them a more refined look by removing the oxide layer and impurities on work piece surfaces. This will also attain a desired surface for a functional purpose.
- The surface grinder is composed of an abrasive wheel, a work holding device known as a chuck, and a reciprocating or rotary table. The chuck holds the material in place while it is being working on. It can do this one of the two ways: ferromagnetic pieces are held in place by a magnetic chuck, while non-ferromagnetic and non-metallic pieces are held in place by vacuum or mechanical means. A machine vise (made from ferromagnetic steel or cast iron) placed on the magnetic chuck can be used to hold non-ferromagnetic workpieces if only a magnetic chuck is available.
- Factors to be consider in surface grinding are the material of the grinding wheel and the material of the piece being worked on.
- Typical workpiece materials include cast iron and mild steel. These two materials do not tend to clog the grinding wheel while being processed. Other materials are aluminium, stainless steel, brass and some plastics. When grinding at high temperatures, the material tends to become weakened and is more inclined to corrode. This can also result in a loss of magnetism in materials where this is applicable.

5.3 SURFACE GRINDING MACHINE WITH HYDRAULIC CIRCUIT

- In surface grinder machine, the workpiece is clamped on the machine table. The machine table is made is made to reciprocate. That is the workpiece is made to reciprocate against a rotating multipoint cutting tool (grinding wheel). Thus the cutting tool cuts the metal.
- Continuous reciprocation of ram can be obtained either by using solenoid DCV or by using pilot-operated DCV. You need limit switches if solenoid DCV is chosen and limit valves if pilot-operated DCV is chosen.
- In this example, pilot operated DCV and limit valves are used.

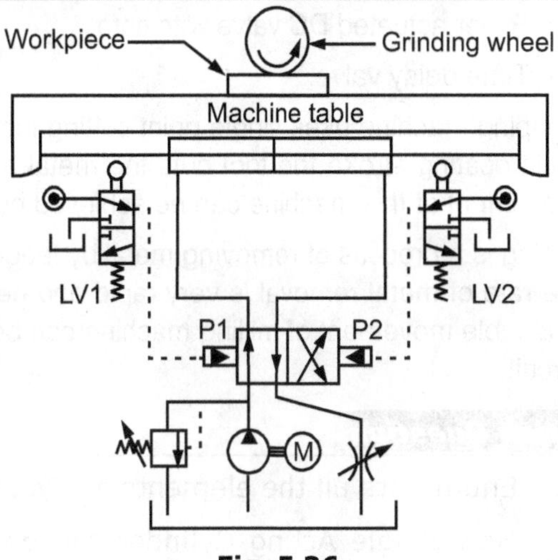

Fig. 5.36

- 4/2 double-pilot DCV is used with two limit valves on either end of the piston rod. The piston rod, while reciprocating, actuates these limit valves, which intern actuates 4/2 DCV. Thus automatic reciprocation of ram is achieved.
- In first position of 4/2 DCV, oil flows from P - A and B - T. Hence the piston extends.
- In second position of 4/2 DCV, oil flows from P - B and A - T. Hence the piston retracts.
- Flow control valve is placed in return line to the reservoir, hence it is meter out circuit with same speed in both strokes.

Important Points

- For control of **SA** Cylinder we have to use 3 × 2 DC Valve in circuit.
- For control of **DA** Cylinder we have to use 4 × 3 DC Valve in circuit.
- When 4 × 3 valve attains centre position then the movement of piston in DA cylinder stops.
- Meter-in-circuit/meter-out-circuit/bleed-off circuit are the circuits used to control speed of DA Cylinder.
- In meter-in-circuit variable flow control valve is placed in primary line (pressure line).

- Meter-in-circuit gives best results when used in high pressure system. However it is not suitable for low pressure systems. We can achieve finer speed control with this circuit.
- In meter-out-circuit flow control valve is placed in return line.
- Meter-out-circuit provides positive speed control of cylinder. However since on both sides of the piston there is pressure, there is possibility of higher friction.
- In bleed-off circuit the pressurized oil coming out of pump is diverted (bleed-off) and is taken to oil tank.
- In bleed-off circuit there is no excess flow of oil through pressure relief valve. However this circuit does not allow positive speed control.
- In regenerative circuit we are regenerating or recovering the energy available with returning oil using regenerative technique.
- Regenerative action is possible only when oil is entering from piston side.
- Bi-directional motor can rotate in clockwise or anti-clockwise direction depending upon position of DC valve.
- The speed of bi-directional motor can be controlled by two separate flow control valves.
- Sequencing circuit is extensively used in industrial hydraulic systems used especially in mass production.
- Sequence valve is 'close relative' of pressure relief valve; but sequence valve is used in hydraulic systems to cause various operations in sequential order i.e. one after the another.
- Sequencing can be obtained by using:
  (a) Sequencing valve.
  (b) Roller actuated DC valve with cam.
  (c) Time delay valve.
- Shaping machine uses single point cutting tool. This tool makes reciprocating motion on the workpiece. In half reciprocating stroke the tool cuts the metal and in 2nd half tool returns to original position. The reciprocating movement of this machine can be achieved by using the hydraulic circuit.
- Milling is a process of removing metal by feeding the workpiece past a rotating 'multipoint tool cutter'. In milling the rate of metal removal is very rapid and hence is one of the important machine in mass production circuit. The table movement of milling machine can be achieved by using DA cylinder and two flow control valve in the circuit.

Practice Questions

1. Enumerate all the elements of Hydraulic Circuit.
2. How Single Acting Cylinder can be controlled by using 3 × 2 DC Valve? Draw circuit and explain.
3. How speed of DA cylinder is controlled? Explain and name the 3 circuits used to control the speed.
4. Compare Meter-in-Circuit and Meter-out-Circuit.
5. Draw bleed-off circuit and label the circuit.
6. Explain the principle of regenerative circuit. What is the main advantage of this circuit?
7. How speed of hydraulic motor is controlled? Explain it with suitable circuit diagram.
8. Why braking and replenishing of hydraulic motor is required? Explain.
9. Write short note on synchronization circuit.
10. What is pressure dependent sequencing circuit? Explain the concept with circuit diagram.
11. Draw Hydraulic circuit for Shaping Machine and explain its working
12. Draw Hydraulic circuit for Milling Machine and explain its working.
13. Draw Hydraulic circuit for Grinding Machine and explain its working.

14. State which of the following circuit is more effective for clamping of component during drilling operation (i) Meter-in (ii) Meter-out. Justify your selection.
15. Explain with neat sketch the working and application of meter-in circuit.
16. State any two applications of sequencing circuit.
17. Sketch and explain 'meter-in hydraulic circuit' to control the speed of extension of DAC Explain why meter-in circuit is not preferred for over-running loads.
18. Explain the principle of regenerative circuit.
19. Why unloading circuits are required? Explain.
20. Draw neat figure of Travel dependant sequencing circuits with cam.
21. What is Counterbalance valve? Explain.

■■■

Chapter 6

PNEUMATIC CIRCUITS

Weightage of Marks = 10, Teaching Hours = 06

Syllabus

6.1 Direct/Indirect Control of Single and Double Acting Air Cylinders, Motors

6.2 Speed Control Circuit for Cylinders and Motors

6.3 Sequencing Circuits, Logic AND/OR Circuit, Time Delay Circuits, Piston Continuous Back and Forth required for Low Cost Automation

6.4 Hydro-pneumatic Applications

About This Chapter

At the end of this chapter students will be able to:

- Sketch and label simple pneumatic circuits
- Describe working of pneumatic circuits
- Develop pneumatic circuit for specified applications
- Select appropriate components required for given simple pneumatic circuits

6.1 INTRODUCTION

- Any circuit diagram is symbolic representation of system which operates on push buttons or pilot signals and the actuator operates so that we can obtain useful work.
- Pneumatic circuit is combination of various pneumatic elements arranged in a systematic manner to perform certain task.
- **Elements of Pneumatic Circuits**
  (a) Source of compressed air – Compressor working on electrical power/diesel engine/steam turbine.
  (b) Air storage element – Air tank.
  (c) Air Servicing unit – FRL unit.
  (d) Flow control and direction control valves and other valves.
  (e) Actuators.

We will study following Pneumatic Circuits:

1. Control of single acting cylinder.
2. Control of double acting cylinder.
3. Pilot control of single acting cylinder.
4. Pilot control of double acting cylinder.
5. Actuation of pneumatic motor (Air motor)
6. Speed control of double acting cylinder.
7. Speed control of bi-directional air motor.

8. Sequencing circuit of 2 double acting cylinders.
9. Sequencing circuit with DA cylinder, SA cylinder and roller operated DC valve.

6.1.1 Circuit No. 1 : Control of Single Acting [SA] Cylinder

- The circuit shows [Fig. 6.1 (a)] all components of pneumatic system to actuate single acting (SA) cylinder. The position shown in figure is non-actuated position. The circuit works as: compressor supplies air; which stores in receiver. FRL unit services the air and supplies air to DC valve. But air does not go to SA cylinder because the air supply is connected to R port. Hence SA cylinder is not actuating.

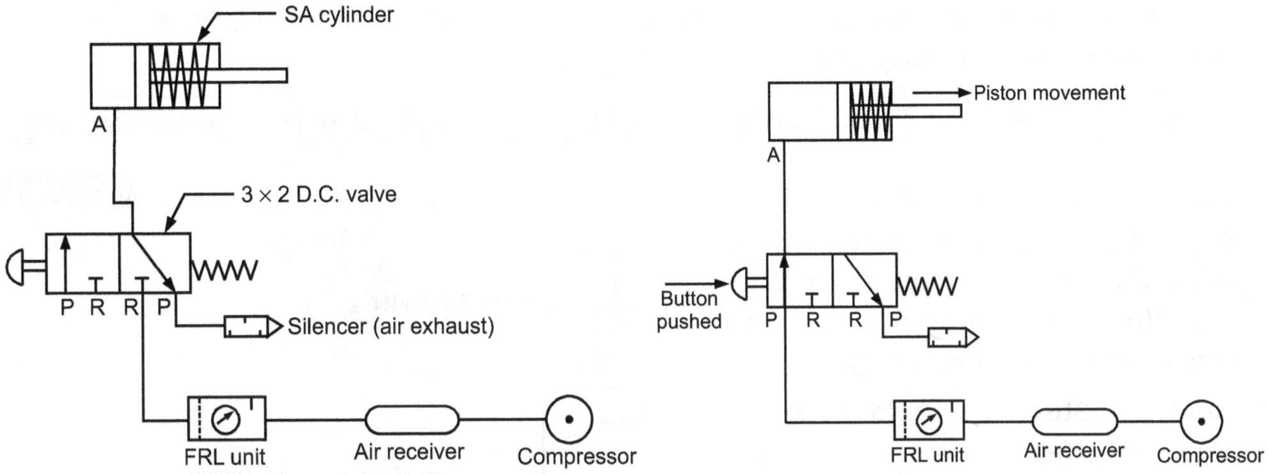

(a) Non-Actuated Position **(b) Actuated Position**

Fig. 6.1

- Now to actuate the SA cylinder, we have to push the button of 3 × 2 DC valve. When button is pushed, the air will be admitted in SA cylinder, through port A and compressed air will expand inside the cylinder. This will result in pushing of piston of SA cylinder from left to right. When air supply will stop, then spring will expand and (as shown in Fig. 6.1 (a)) the piston will regain its original position. During this movement air will be exhausted through DC valve.

6.1.2 Circuit No. 2 : Control of Double Acting [DA] Cylinder

[S-16]

- Now we will see the circuit showing control of DA cylinder using 4 × 2 Push button operated DC valve.

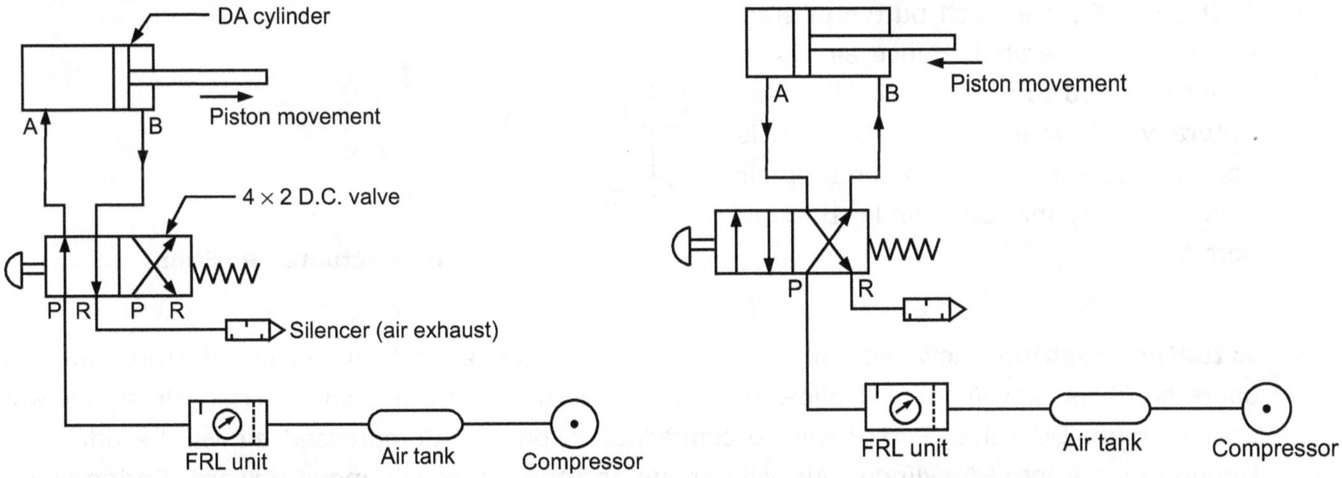

(a) Piston Moving From Left to Right **(b) Piston Moving From Right to Left**

Fig. 6.2

- In the position shown port P (Pressure Port) is connected to port A (Consumer Port) of DA cylinder. The compressed air will be admitted in DA cylinder and it will expand. Due to this expansion, piston of DA cylinder will move from left to right. (See Fig. 6.2 (a))
- During this movement, the air present on piston rod side of piston will come out through port B and will be exhausted through DC valve and Silencer.
- Now push button will be operated. This action, connects port P to port B of DA cylinder and port R to port A. Now compressed air will be admitted in DA cylinder through port B and the air will expand. This expansion results into movement of piston from Right to Left. During this movement, the air present on other side of piston will come out from port A and will be exhausted through DC valve and silencer through port R. (See Fig. 6.2 (b))

6.1.3 Circuit No. 3 : Pilot Control Single Acting Cylinder (Impulse Circuit)

[W-11, W-13]

- Now let us see the control of SA cylinder using two numbers 3 × 2 DC valve (one for pilot signal and other form operation of SA cylinder).
- **Solution:** Students should note that pilot operated means, the main direction control valve of SA circuit will be operated by using compressed air (instead of push button).
- You can easily make out the difference in circuit (1) and circuit (3). In fact both circuits are shown to operate SA cylinder. But the above shown circuit is slightly different. The main 3 × 2 DC valve actuates with air signal (force of compressed air).
- In the Fig. 6.3 the push button of start valve is not operated. Hence air signal is not supplied to main 3 × 2 DC valve. Naturally this main 3 × 2 DC valve is not working and so, the compressed air is not entering into SA cylinder through port A.

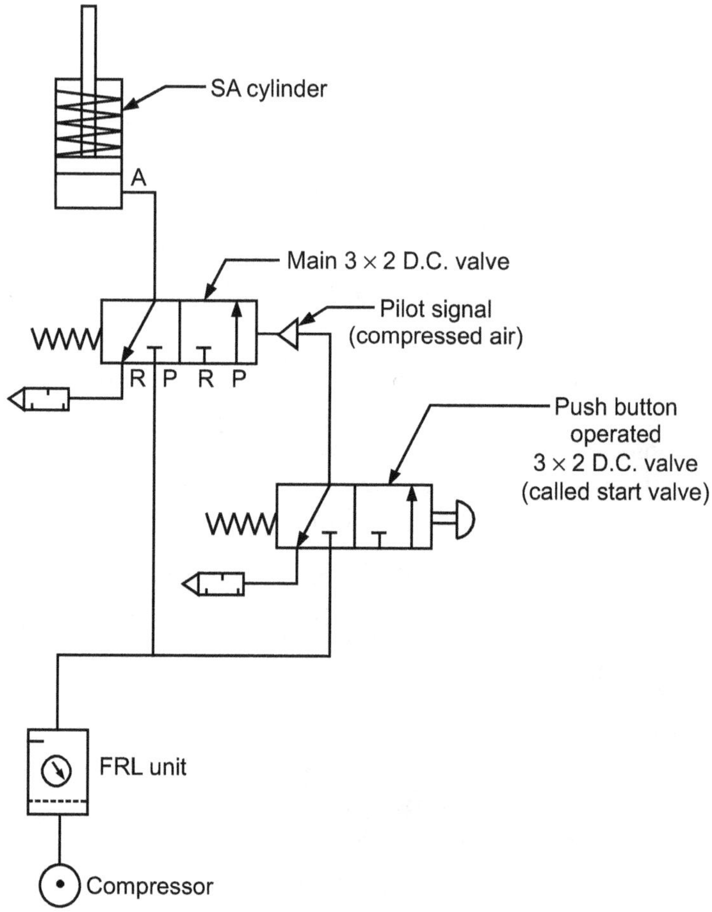

Fig. 6.3: Non-actuated Position

- **Actuated Position:** Now we will see the actuated position: Push button of start valve is operated. This movement will allow the air signal to flow from point B. This air signal will actuate main DC valve. Port P will be connected to port A. Compressed air will be admitted through port A into SA cylinder. Air will expand and the piston will move upward. Spring of SA cylinder will compress. When push button of start valve will be released, the air signal will stop and the supply of air through port 'A' will be cut-off. Now the spring will expand and piston will come its original position.

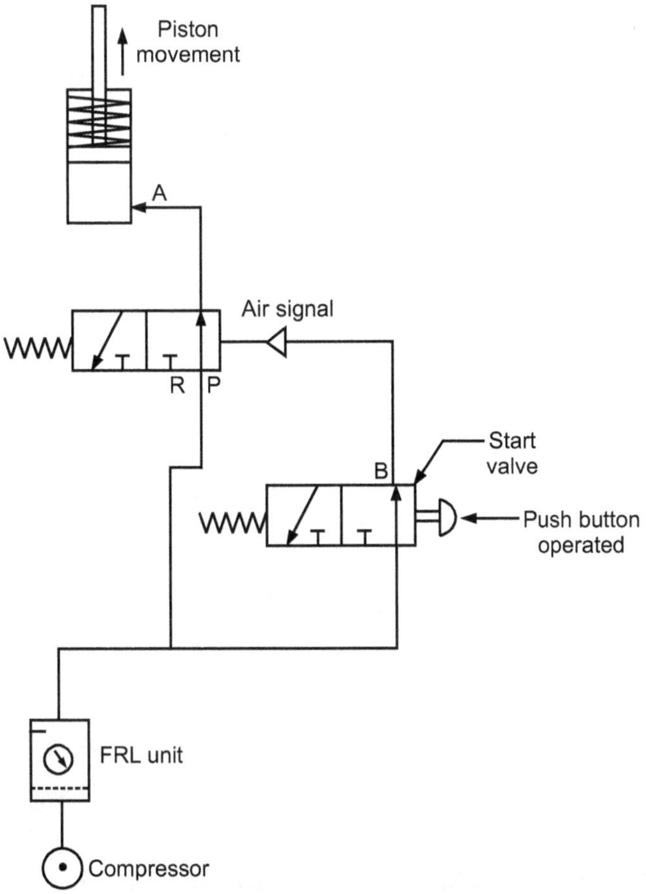

Fig. 6.4: Actuated Position

6.1.4 Circuit No. 4 : Pilot Operated Double Acting Cylinder (Impulse Circuit)

[W-10, 11, 16; S-16]

- Similar to pilot operated SA cylinder, DA cylinder can also be operated by pilot signal. This circuit operates with 4 × 2 DC valve pilot operated and 3 × 2 DC valve called start valve. This valve is push button operated.
- The (Fig. 6.5) the operation of DA cylinder with Air signal. The push button of start valve is not operated. Hence no air signal to main DC valve. DC valve is in the position shown, in which pressure port 'P' is connected to port 'A' of DA cylinder and piston is moving upward.

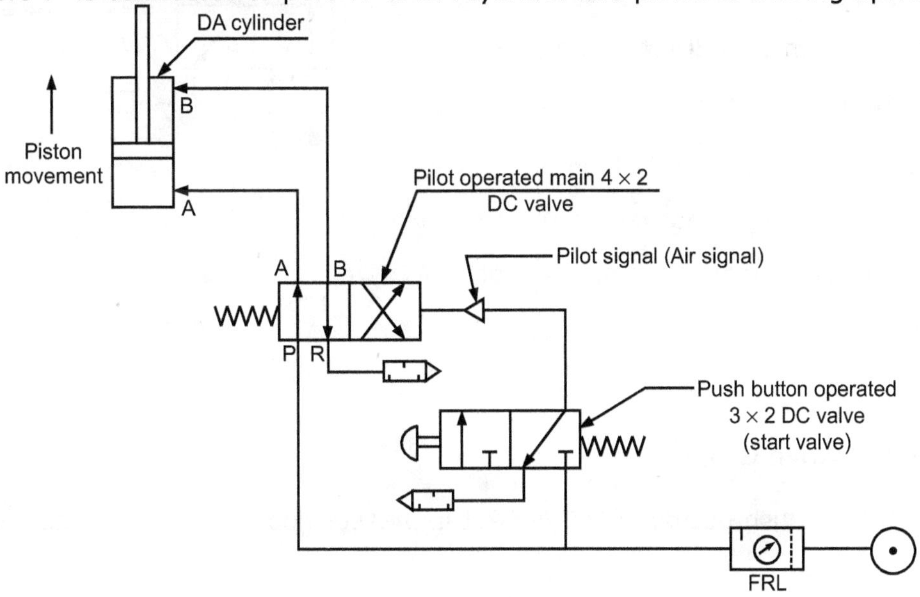

Fig. 6.5

- Now the push button of start valve is operated. The air signal will be supplied to main 4 × 2 DC valve. Now port P will be connected to Port B and piston starts moving down because air is admitted through port B in DA cylinder. (See Fig. 6.6)

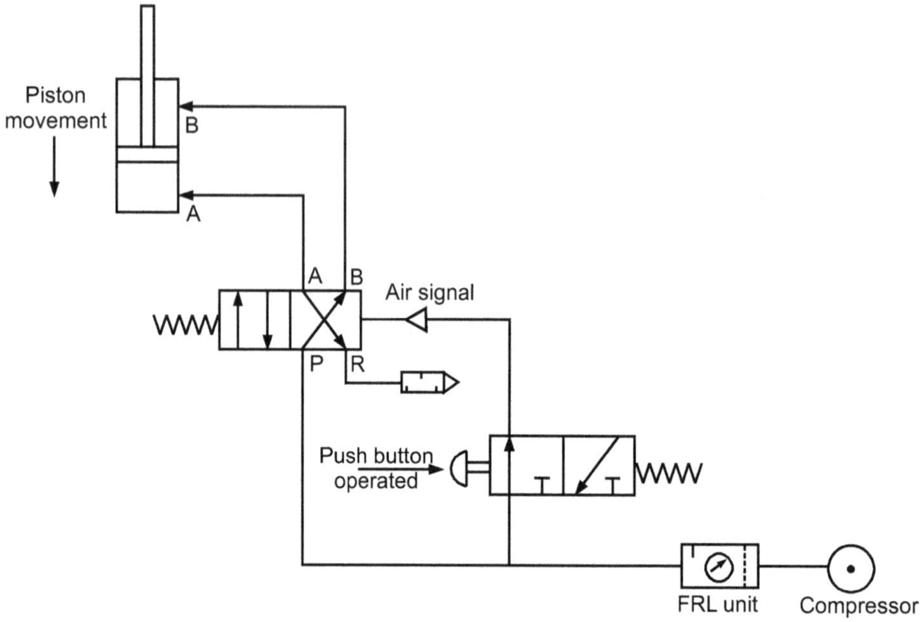

Fig. 6.6

6.1.5 Circuit No. 5 : Actuation of Pneumatic Motor (Air Motor)

[S-09, W-14]

- Now we will see Rotation (or Actuation) of unidirectional pneumatic motor using push button operated 3 × 2 DC valve.

Unidrectional motor:

- This motor rotates in one direction only i.e. clockwise or Anticlockwise.
- The push button is not operated. Pressure port P (compressed air) is not connected to port 'A' of motor. Hence motor is not rotating.

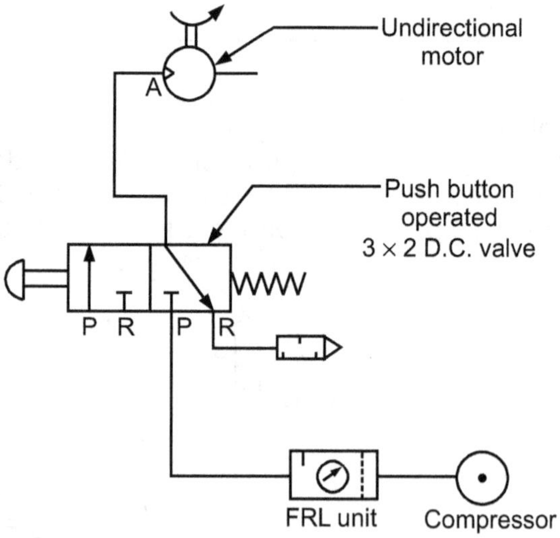

Fig. 6.7

Push button of DC valve is operated:

- Due to operation of push button of DC valve, the pressure port P will be connected to port 'A' of motor and motor starts rotating.

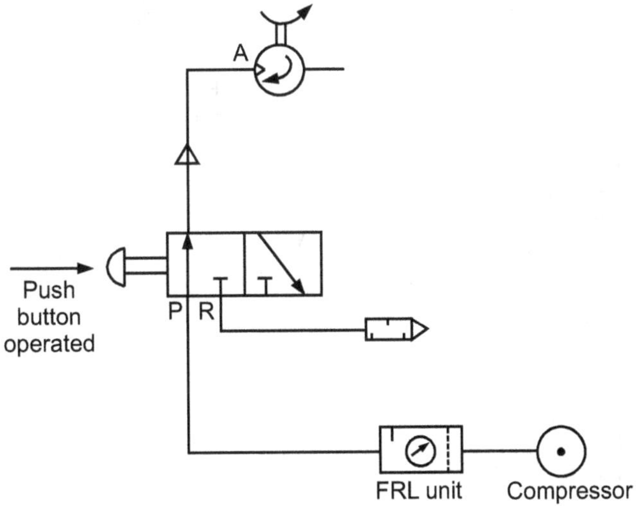

Fig. 6.8

Remember:

- See the symbol of unidirectional motor. The triangle apex is inside.

Fig. 6.9

6.1.6 Circuit No. 6 : Speed Control of DA Cylinder [W-10, 11]

- Speed control circuits are adopted to control speed of actuators. This speed control is achieved by adopting flow control valve in circuit.

Remember:

- **SA cylinder or DA cylinder:** The movement of piston of these two actuators is reciprocating. i.e. piston slides in a cylinder. This clearly means that the piston is having linear motion and hence naturally having linear speed. The linear speed can be expressed in meter/sec. (m/s).
- **Air motor:** These are rotary actuators. When air is admitted in the inlet port of unidirectional or bi-directional motor, then the shaft of motor starts rotating. Hence these actuators are having circular motion i.e. rotary speed. This speed can be expressed in RPM.

Following Circuit (Fig. 6.10) shows, speed control of a double acting cylinder (DA Cylinder)

- In this circuit we are using:
  (a) 4 × 2 DC valve – 1.
  (b) Variable flow control valve with built in check valve – 2 Nos.
- The two flow control valves are placed in two supply lines of DA cylinder. As position shown in the circuit (Fig. 6.10), Pressure port 'P' is connected to inlet port 'A' of DA cylinder. Now we can control the flow of air admitting in DA cylinder, through port A, by using variable flow control valve 'F_1'. If more flow will be allowed, then more air will enter in cylinder and piston will move towards right with more linear speed. If less air is allowed to flow, the linear speed of piston will be low.

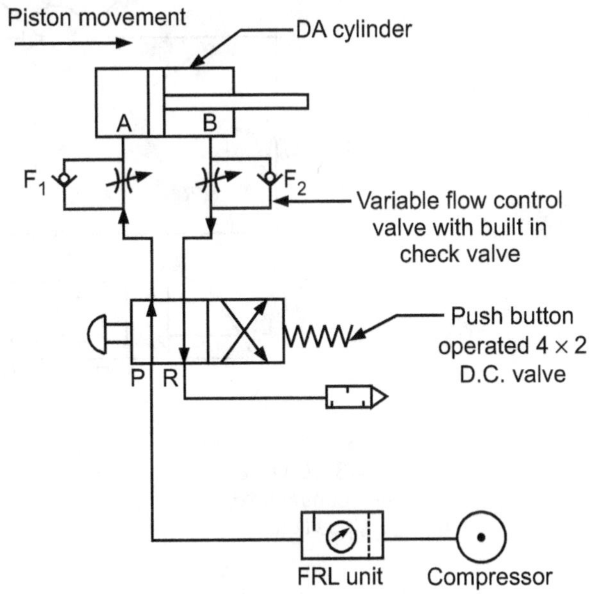

Fig. 6.10

- When push button is pressed (Fig. 6.11), then pressure port 'P' will be connected to port 'B' of DA cylinder and exhaust port R will be connected to port A of DA cylinder. Here also by controlling flow control valve F_2 we can control the speed of piston. During this stroke, the air from earlier stroke in the other side of piston will be exhausted through flow control valve F_1 via DC valve.

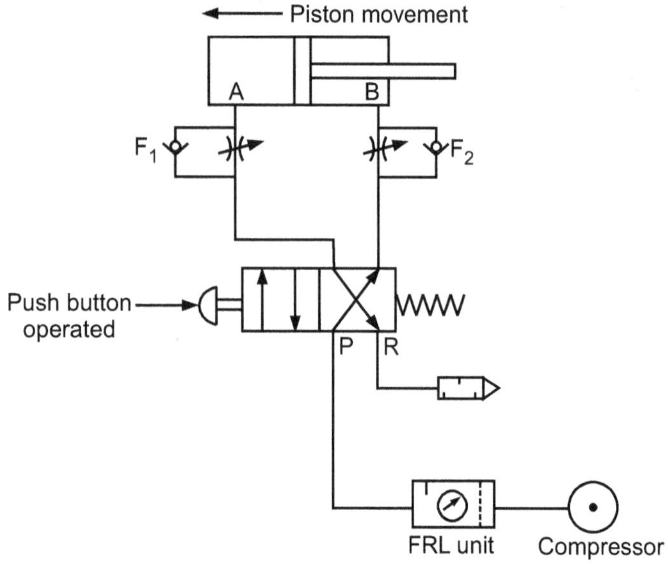

Fig. 6.11

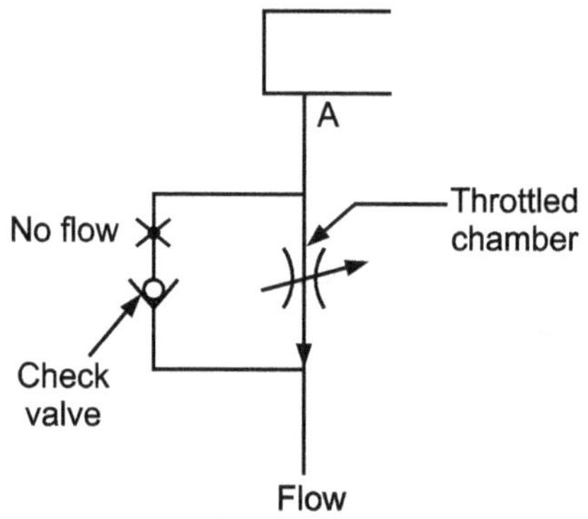

Fig. 6.12

- During this return, the air flows to DC valve through throttled chamber of flow control valve. Air will not flow through check valve because, it is undirectional valve.
- This arrangement of circuit design with non-return flow control valve is called 'exhaust throttling' or 'Throttle out'.

6.1.7 Circuit No. 7 : Speed Control of Bi-directional Air Motor

[S-10, 15, 16; W-10, 13]

- Bi-directional air motor rotates in clockwise as well as anti-clockwise direction. It is rotary actuator, the speed of which is measured in r.p.m. The speed of bi-directional motor is controlled as under:
- The following circuit in Fig. 6.13 shows speed control of motor by using variable flow control valves (2 Nos) having built-in check valve; and 4 × 3 DC valve having zero position or central hold position.

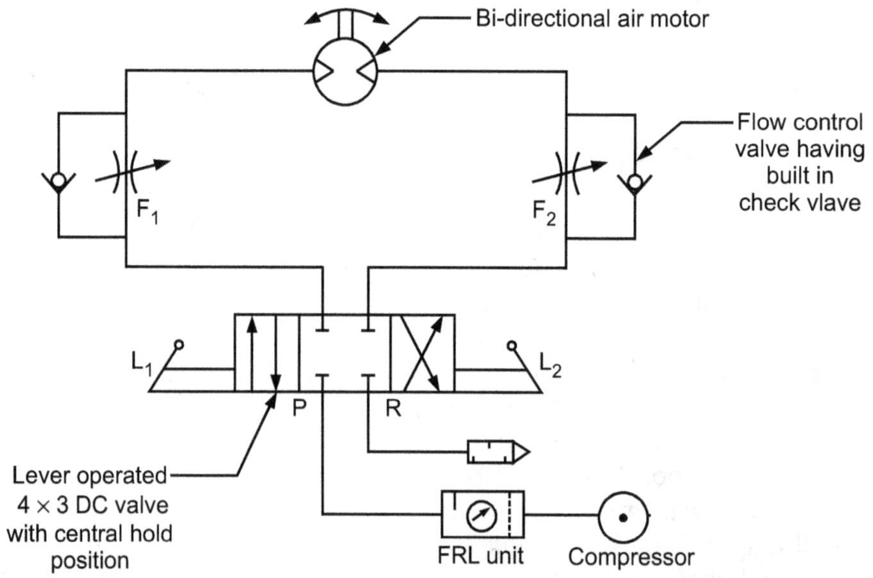

Fig. 6.13

- In this circuit 4 × 3 DC valve is used. The circuit shown is in central hold position. This clearly means that motor is not rotating. It is stand still.
- Now lever L_1 is operated.
- When lever L_1 is operated (Fig. 6.14), port P will be connected to port A of Air motor and motor will start rotating in say clockwise direction. Its speed can be controlled by using variable flow control valve F_1. Port B of motor will be connected to exhaust R. And the air in motor will be exhausted through port R via DC valve.

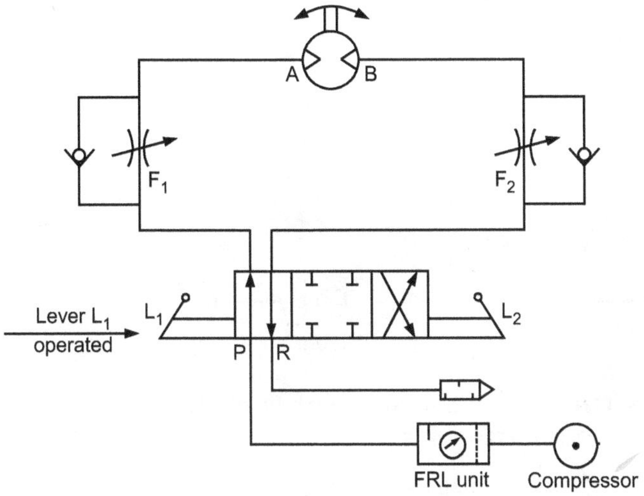

Fig. 6.14

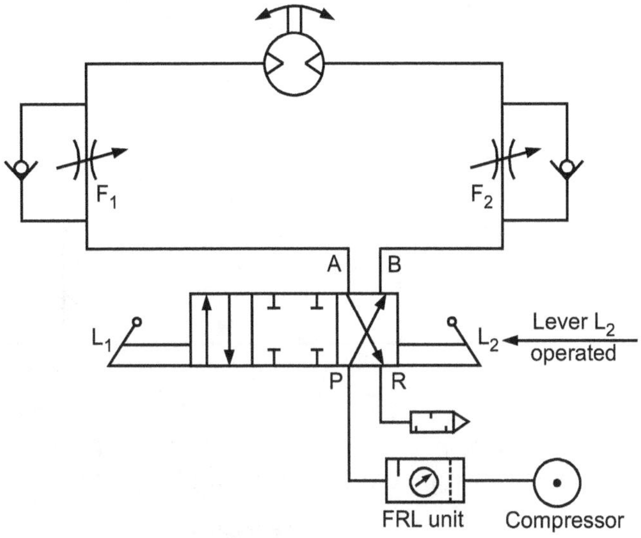

Fig. 6.15

- When Lever L_2 is operated. (Fig. 6.15)
- When lever L_2 is operated, pressure port 'P' will be connected to port B of motor and naturally motor will start rotating in anticlockwise direction. Port A will be connected to port R and air in the motor will be exhausted through port R via DC valve.

6.1.8 Circuit No. 8 : Sequencing Circuit of Two DA Cylinders [S-09, W-13]

- The word 'sequence' indicates that, the operations one after the another. In bottling plants (Pepsi/Coca Cola), the liquid with CO_2 Gas is filled in bottles and bottles are packed with crown cap. These plants are fully automatic and automization is done obviously with pneumatic power. In these plants, the movement of bottles needs 'sequence'. Many sequencing circuits are used in these plants. The actuators especially linear actuators are used to push the bottles for next operation in automation line.
- Now let us study a sequencing circuit using 2 DA cylinders. This circuit is having following units:

  (a) DA cylinders with cam – 1 No.

  (b) DA cylinder without cam – 1 No.

  (c) 4 × 2, pilot (Air impulse) operated DC valves – 2 Nos.

  (d) Push button operated 2 × 2 start valves – 2 Nos.

  (e) FRL unit.

  (f) Compressor.

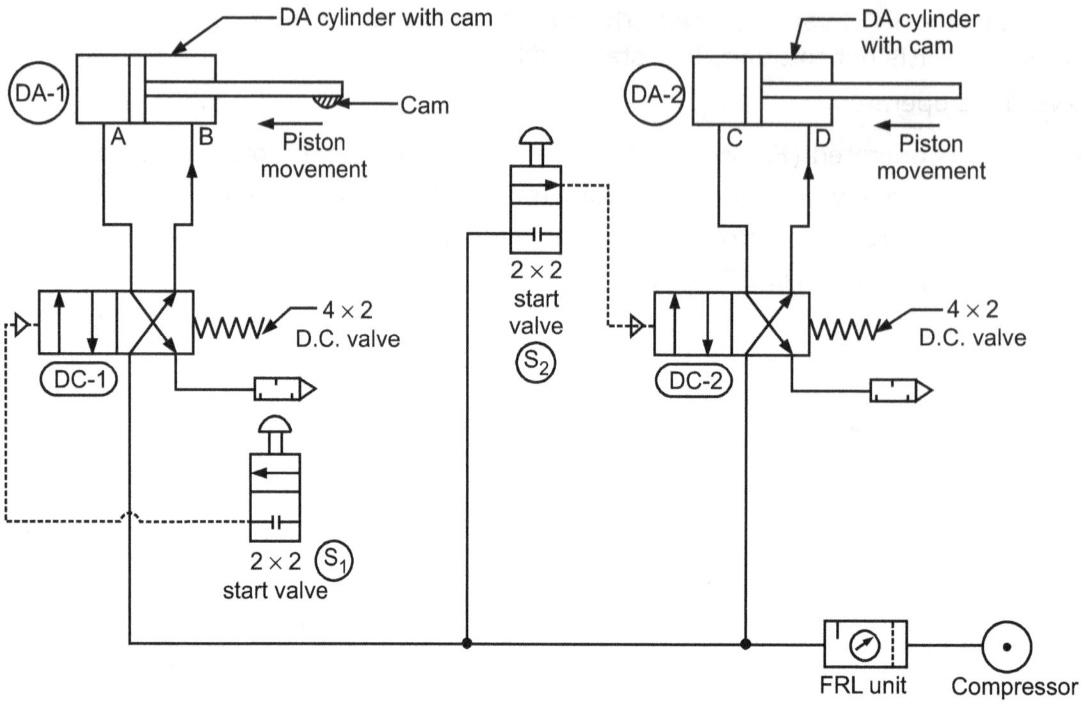

Fig. 6.16

- As shown in circuit, air is admitted at port B of **'DA-1'** cylinder and port D of **'DA-2'** cylinder. Hence both pistons are moving from right to left in both actuators.

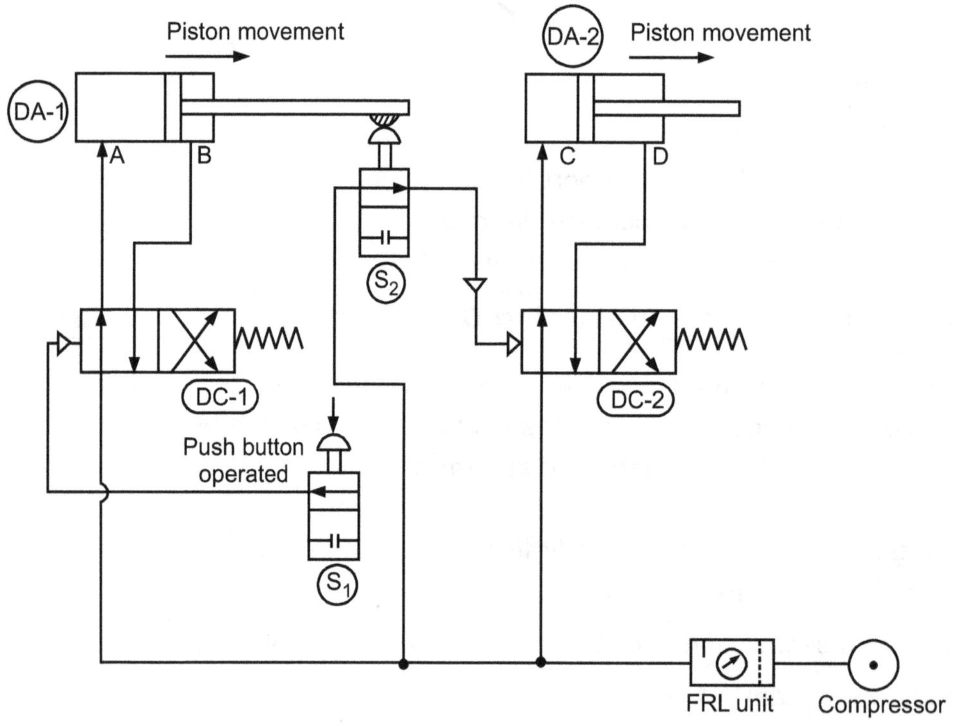

Fig. 6.17

- When push button of start valve **(S_1)** is operated, the air signal (Impulse) will be supplied to DC valve **(DC–1).** The air will be admitted through port 'A' of **DA-1** and piston will move towards Right. The cam is attached to end of piston rod. This cam will push the push button of start valve **(S_2).** Due to this air signal (Impulse) will be supplied to direction control valve **(DC-2)** and it will operate. Now air will be admitted through port 'C' of **(DA-2)** and the piston will now move from left to right.
- The sequence is achieved by cam of **(DA-1).** Unless and until **(S_2)** will not be operated with the help of cam, the piston of **(DA-2)** cannot move from left to right.

6.1.9 Circuit No. 9: Sequencing of SA cylinder and DA cylinder using Roller Operated DC Valve [S-15]

- Now let us study the circuit in which SA cylinder and DA cylinder are in sequence. The circuit uses roller operated 3 × 2 DC valve.

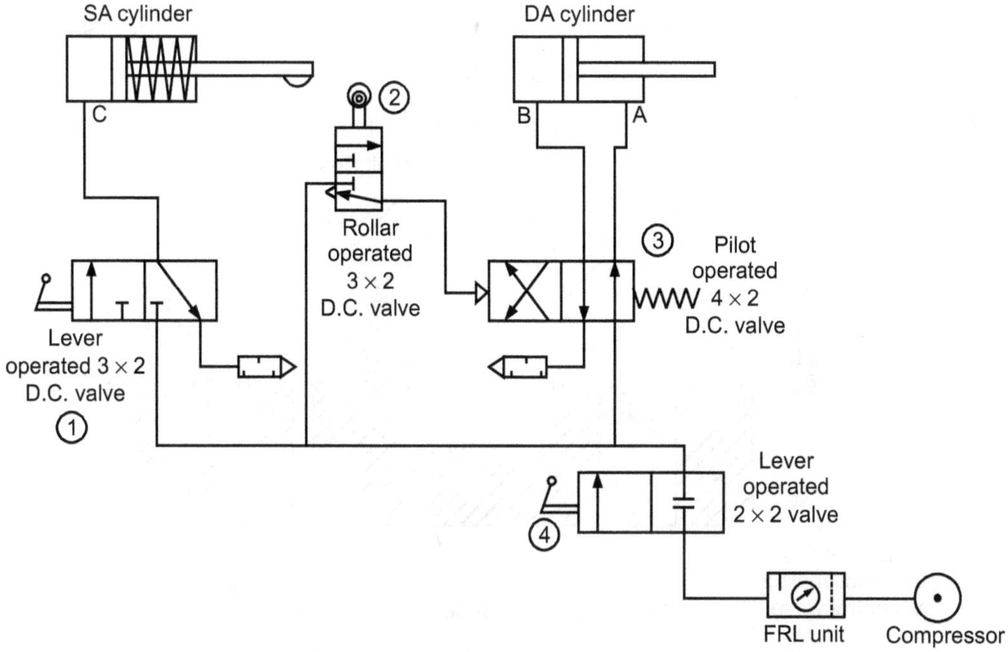

Fig. 6.18

- As shown in circuit (Fig. 6.18) there is no supply of air in any line because 2 × 2 valve No. (4) is not actuated i.e. it is in normal position (2 × 2 DC valve is normally closed).
- Now let us actuate 2 × 2 DC valve; by pushing lever. Due to this, the compressed air will be supplied to all lines in circuit and the actuated position will be as under.

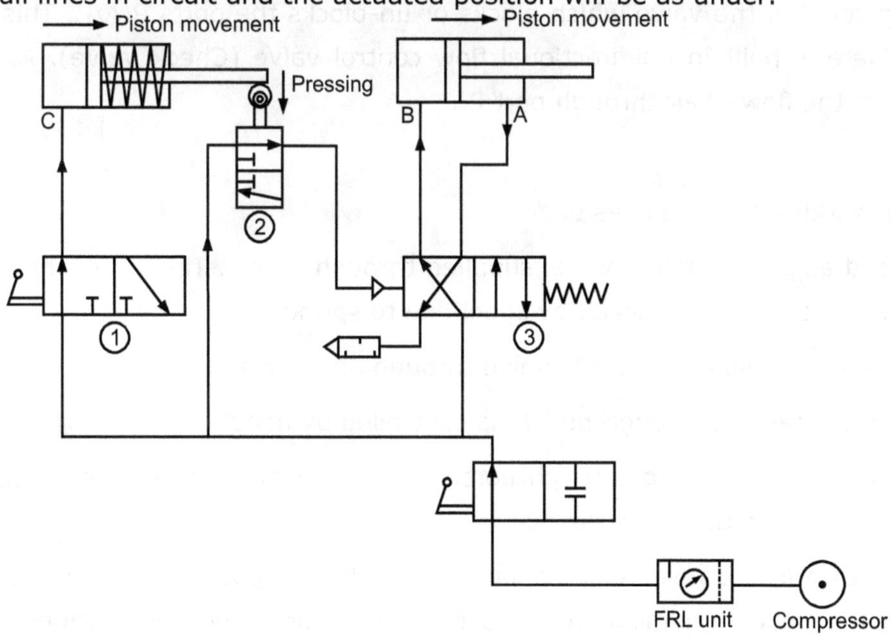

Fig. 6.19

- Due to air supply, the compressed air will be admitted through port 'C' in SA cylinder and piston of it will move from left to right. During this movement the piston rod of SA cylinder will foul with roller of valve (2) and the rod will press the roller lever down. This will operate valve (2) and air signal (impulse) will be supplied to valve (3). This valve will then actuate and compressed air will be supplied through port B. Due to this the piston of DA cylinder will also move to right.

6.1.10 Circuit No. 10 : Time Delay Operation [W-11, 13, 14, S-15, 16]

- A time delay in pneumatic circuit provides a delay between the operation of valve and the movement of actuator (piston or motor).
- Time delay valve is a main component of Time delay circuit. So to understand Time Delay Circuit we must first know what is Time Delay Valve.
- Following figure shows the internal construction of Time Delay Valve.

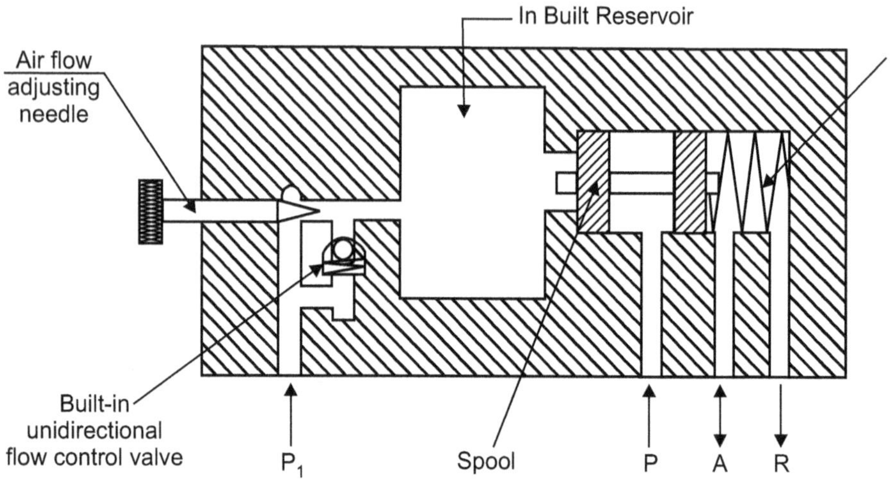

Fig. 6.20

Construction:

- Time delay valve is a combination of air reservoir and unidirectional flow control valve. There is a free sliding spool in the valve which blocks or un-blocks the ports P/A/R. This spool is spring controlled. There is built in unidirectional flow control valve (Check valve). Air flow adjusting needle, adjusts the flow of air through port P_1.

Working:

- The stepwise working of valve is as under:
  1. Compressed air (Pressurised Air) is supplied through port P. This air cannot passed through port A, because port A is blocked by spool due to spring.
  2. Pressurised air is also passing into valve through another port P_1.
  3. The flow of air passing through port P_1 is controlled by needle.
  4. The pressurized air passing through port P_1 will be accumulated in reservoir. The pressure starts building-up in the reservoir.
  5. When pressure in the reservoir is enough to push the spool (towards right) against spring pressure, the spool will move towards right and port P will get connected with port A, closing port R.
  6. The time required to build-up the pressure in the reservoir, is the amount of delay time offered by the time delay valve.
  7. With further increase in pressure, the inbuilt check valve opens, the air from reservoir gets exhausted and the spool regions its original position.

Symbol of Time Delay Valve:

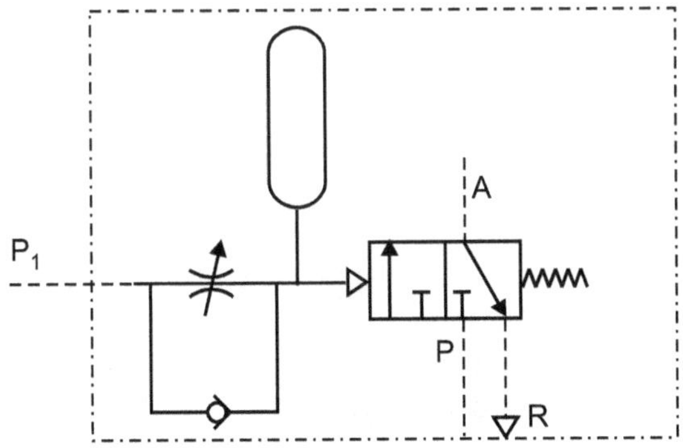

Fig. 6.21

- You can observe that the time delay valve symbol is a combination of check valve + Reservoir and 3 × 2 DC valve which is pilot operated (Air operated).

Time Delay Circuit

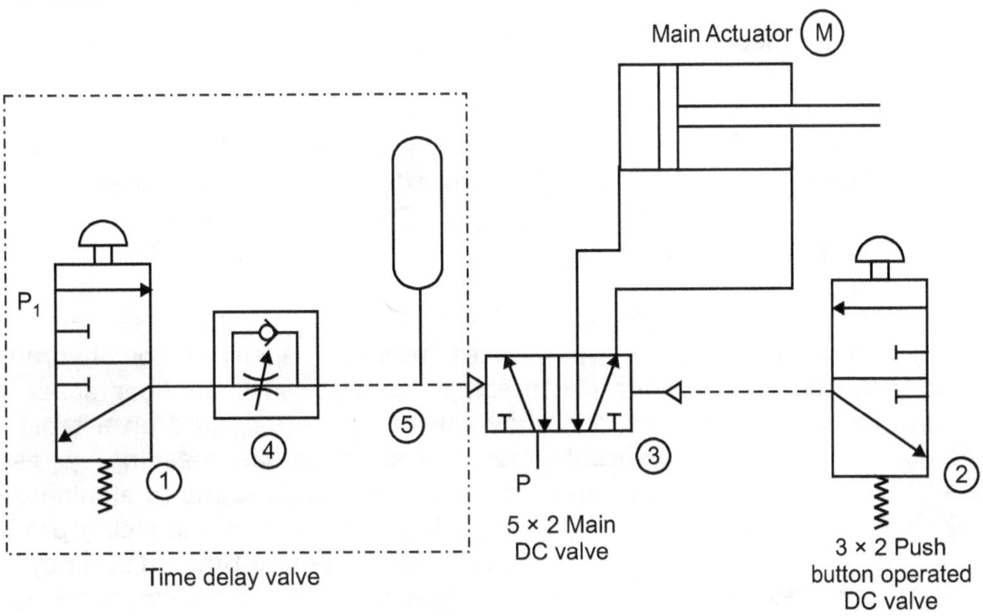

Fig. 6.22

Working:

(a) Push button of 3 × 2 DC valve (1) is operated i.e. pressed. Then P_1 will be connected to check valve (4) and air will start accumulating in reservoir (5).

(b) Pressure builds in the reservoir slowly, causing the time delay.

(c) Only when pressure in the reservoir is high enough, the air signal will operated 5 × 2 main DC valve and will connected port P to A thereby moving the piston of actuator (M) towards Right.

(d) When push button of valve (2) will be pressed, the 5 × 2 DC valve will switch, triggering the release of air and causing the piston to move towards left.

Remember:

- The length of time delay can be altered by using an adjustable unidirectional flow control valve.
- Do not confuse time delay with speed control. Time delay prevents the piston operating for a particular time period, whereas speed control reduces the linear speed of piston.

6.2 HYDRO PNEUMATIC APPLICATIONS

Hydro Pneumatic Press:

- When we discuss the most effective type of press, pneumatic presses are typically at the top of the list. But sometimes, you can make a great thing even better. In today's post, we'll be talking about the **pneumatic hydraulic press** and explaining how these two seemingly different systems actually work extremely well together.

Hydraulic and pneumatic systems may seem like opposites...

- Hydraulic presses and other hydraulic systems operate using liquids like water, oil, ethylene glycol, or even fire-resistant fluids. These fluids are pressurized and are moved by the pumps inside. Hydraulic systems like these can be found in your car's braking system, on the wing flaps of aircraft, in the arms of excavating equipment, and even in your own body's circulatory system!
- Pneumatics, on the other hand, use compressible air or gasses to operate. The liquids used in hydraulic systems are non-compressible, which means there's no delay in movement. In hydraulic systems, there is a slight delay due to the compression of these gasses. However, pneumatic press machines can actually move up to 10 times faster than hydraulic presses can! You have witnesses pneumatic systems at work if you have ever been in a dentist's chair and had your teeth drilled, used a nail gun in a construction project, or inhaled and exhaled air into your own lungs.

But they actually complement each other Beautifully:

- You might not understand how these two seemingly different systems could be combined into one. But hydro pneumatic presses are actually an extremely popular and effective option. They are sometimes referred to as "air over oil" presses. They utilize both of these concepts in order to operate - and they typically do so in a quieter, cleaner, and more affordable way than when each system is used alone.
- Because the primary source of power in these systems is compressed air, the use of a pneumatic hydraulic press drastically reduces required spending on hydraulic oils and components. These pneumatic hydraulic presses can also save on floor space due to their lightweight and compact design; in fact, they can easily be mounted on a table. In addition, these types of presses save a significant amount of energy while speeding up response times. That means the use of these presses make it easier for organizations to eliminate unnecessary waste and cement their commitment to energy efficiency without sacrificing production rates. They are also very easy to maintain, which equates to even more time and money savings.
- Hydro pneumatic presses can be used for numerous applications throughout a wide variety of industries. They can be used to make electrical switches and kitchen appliances, mold sheet metal, create rubber parts, and so much more. They are commonly used in the automobile and manufacturing sectors, but they cap, easily be utilized in other fields.
- While each of these separate systems has its benefits and specific uses, combining the two can make for an even more effective machine. To learn additional information about our pneumatic-hydraulic systems, please contact us today.
- Many Industries are using manual presses like toggle presses, fly presses, arbor presses for blanking, piercing, riveting, bending, embossing, etc. It is also used for forming two or more components in Manufacturing Industry. Manual operation of these presses gives fatigue to the operator, since he has to exert physical energy. Due to this the efficiency also gets lowered down. Considering these facts, the Hydro pneumatic press has been designed, which will reduce the wastage of Low Cost Automation in a Power Press human energy and increase the production rate. We know that hydraulic system is applied to attain precise movement. But in pneumatics, due to the compressibility characteristic of the air, the air cylinder tends to slow down on meeting an increase in load and to accelerate or jump forward when working against a load which suddenly decreases. So, pneumatic actuators are not suitable for the finer movement of the machine tool movements.

- These problems can be solved by combining the two fluids - air and oil. By the use of these two media, the quick action of air and the smooth high pressure action of oil blend ideally to provide a concept for the design of new Hydro-Pneumatic press. It should be made clear that although the two fluids are used in the circuit they are not to be mixed. In fact every precaution is taken to keep them separated.
- The basic laws of Hydraulic and Pneumatics are Pascal's law, Boyles law, Charles' law and Gay lussac's law. These laws are obeyed by our Hydro-Pneumatic press.

Construction of Hydro-Pneumatic Press

- Basically the hydro-pneumatic press consists of the following parts:
  - Hydro-pneumatic cylinder
  - Cylinder mounting plate
  - Supporting pillars
  - Base plate
  - Pneumatic control systems
  - Electrical control panel
  - Pneumatic pipelines

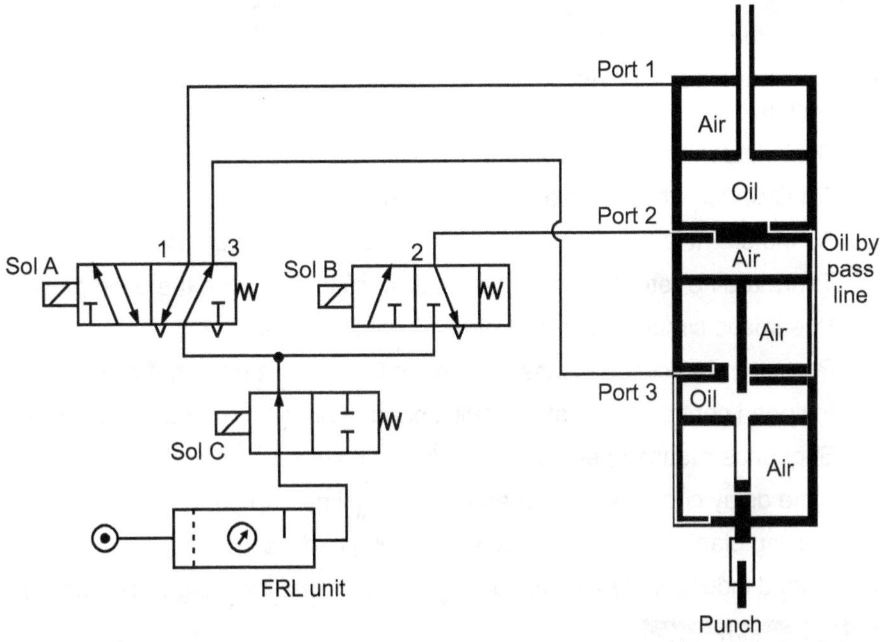

Fig. 6.23

Circuit Operation:

- When DCV A is energized, hence it shifts to position of the valve. Thus the pressurized air supply is given to opening of the cylinder and the air exhaust from the opening 3 of the cylinder. When DCV 8 is energized and hence supply is given to opening 2 of the cylinder. When both the DCV's are in de-energized condition, the supply is given only to opening 3 and hence it helps in retraction of the piston. When the emergency valve is energized the supply to the entire unit is stopped immediately.
- All the basic expectations of today's industry such as automation, increased production, low cost and appreciable quality are met by this hydro-pneumatic press. Low cost is proved by using pneumatic source and getting the same tonnage capacity of a hydraulic press.
- A single compressor is enough to run two or more hydro-pneumatic presses. The safety of the worker is also ensured by employing the Two Hand Safety system in the fluid circuit.
- This press is also portable and reliable as it has very small machine and structures can give the large tonnage. So, this press is expected to meet today's Industrial Manufacturing scenario.

Hydro Pneumatic system for a building:

- Hydro Pneumatic System is designed to deliver water continuously at all outlets in all Floors of High Rise Apartments aid Commercial Buildings with uniform pressure .The system is most ideal for High Rise Apartments and Buildings where the demand of water is variable and fitted with modern bathroom fittings like Jacuzzis, shower etc which require pressurized water for better functioning of the fittings the Hydropneumatic Systems are based on the principle of compressibility or elasticity of the air when it is subjected to pressure. Among the different water supply and distribution systems in buildings and installations, Hydropneumatic Equipment has proven to be an efficient and versatile option, with great advantages over other systems.

This system avoids building elevated tanks, placing a tank system partially filled with pressurized air. This makes the hydraulic network maintain an excellent pressure, improving the operation of washing machines, filters, showers, fast filling of toilet deposits, among others. It also prevents the accumulation of particles in pipes by flow at low speeds. This system does not require tanks or hydraulic distribution network on the roofs of buildings (avoiding moisture problems due to leaks in the network) that make the facades look so bad and leaving this space free for different uses.

Important Points

- Any circuit diagram is symbolic representation of system which operates on push button or pilot signals and the actuator operates so that we can obtain use full work.
- Air compressor/Air Receiver/FRL Unit/DC Valves/Flow Control Valves/Actuators are elements of Pneumatic circuits.
- SA Cylinder – It is Single Acting Cylinder.
- DA Cylinder – It is Double Acting Cylinder.
- Air Motor – It is rotary devices rotated by means of air.
- There are no return lines in pneumatic circuits. Because the used air is exhausted to atmosphere.
- Pneumatic circuits operates below 10 bar pressure.
- Pilot operated circuit means – the main DC valve is operated by using compressed air (instead of push button)
- In speed control pneumatic circuit, speed of piston or motor controlled by flow control valve.
- Sequence means operations one after the other.
- Time delay circuit is used when we need a time gap between two consecutive operations.
- Bottling plants uses pneumatic sequencing circuits.
- In food industry, the automation is generally done by using pneumatic circuits.

Practice Questions

1. List out elements of pneumatic circuit.
2. Draw the circuit showing control of DA cylinder using 4×2 DC valve. Explain the working.
3. What is impulse circuit? Explain.
4. Draw pilot operated DA cylinder circuit using 4×2 DC valve and 3×2 Pilot valve.
5. What is the meaning of unidirectional air motor and Bi-directional air motor? Explain with sketch and draw symbol of both.
6. How we can achieve speed control of any actuator? Explain the speed control circuit of bi-directional air motor.
7. What is sequencing operation? How you will sequence one DA cylinder and one SA cylinder using roller operated DC valve? Explain with circuit.
8. Explain time delay circuit.
9. Explain with neat sketch working of impulse operation pneumatic circuit.
10. Explain pneumatic circuit for controlling speed of double acting cylinder with circuit diagram.
11. Explain with circuit diagram of a pneumatic circuit for speed control of bidirectional motor.
12. Draw symbol of 'Time delay valve'.
13. Define F.A.D. of an air compressor. If an air compressor is delivering 0.15 m^3/min of compressed air at 12 bar pressure, calculate F.A.D.
14. How can be speed control of any actuator be achieved? Explain the speed control circuit of bi-directional air motor with sketch.

■■■

www.ingramcontent.com/pod-product-compliance
Lightning Source LLC
Chambersburg PA
CBHW061850130426
43041CB00043B/956
* 9 7 8 9 3 8 9 6 8 6 8 2 1 *